17.880

STRIP MINING OF COAL – ENVIRONMENTAL SOLUTIONS

STRIP MINING OF COAL

Environmental Solutions

William S. Doyle

NOYES DATA CORPORATION
Park Ridge, New Jersey   London, England
1976

Copyright © 1976 by Noyes Data Corporation
  No part of this book may be reproduced in any form
  without permission in writing from the Publisher.
Library of Congress Catalog Card Number- 76-2190
ISBN: 0-8155-0611-2
Printed in the United States

Published in the United States of America by
Noyes Data Corporation
Noyes Building, Park Ridge, New Jersey 07656

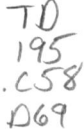

FOREWORD

This review contains carefully excerpted and collated data from diverse and difficult-to-locate sources. The information to be found in this volume—technological, geographic and environmental—is based on studies carried out by research teams, very often under the auspices of various governmental agencies. Private industrial research studies by companies interested in extracting the earth's surface or subsurface mineral wealth have also been a source of the information included in the book.

A complete list of the reports cited throughout the book is to be found in the references on pages 346-347. A valuable addendum is the list of State Agency Contacts on pages 348-350 where anyone interested may secure competent and responsible advice about strip mining in his state or area.

Advanced composition and production methods developed by Noyes Data are employed to bring new durably bound books to the reader in a minimum of time. Special techniques are used to close the gap between "manuscript" and "completed book." Industrial technology is progressing so rapidly that time-honored, conventional typesetting, binding and shipping methods are no longer suitable. Delays in the conventional book publishing cycle have been bypassed to provide the user with an effective and convenient means of reviewing up-to-date information in depth.

The Table of Contents is organized in such a way as to serve as a subject index and provides easy access to the information contained in this book.

CONTENTS AND SUBJECT INDEX

INTRODUCTION 1
SURFACE MINING LAND USE AND METHODS. 2
    Land Use and Reclamation Statistics. 2
    Surface Mining Methods and Their Impact on the Environment—Overview. . . . 3
    Strip Mining Techniques and Equipment. 7
        Area Mining. 7
        Contour Mining. 18

LAND RECLAMATION METHODS. 43
    Blasting Control 44
    Backfilling, Grading and Revegetation. 46
        Revegetation Problems 47
        Revegetation Preplanning 49
        Spoil Amendments 59
        Species Selection. 66
        Methods of Seeding and Planting. 68
        Revegetation of Arid and Semiarid Regions of the West 68
    References. 73

SEDIMENT AND EROSION CONTROL 74
    Control Measures. 74
        Sediment Control Basins. 76
        Primary Basin Design Criteria 77
    Pit Drainage. 79
    Bench Drainage 80
    Revegetation 80
    Coal-Haul Roads 81

REVEGETATION STUDIES 84
    Revegetation Studies at Three Pennsylvania Sites 84
        Preparation of Study Areas. 84
        Effect of Lime and Fertilizer on Grass-Legume Mixtures 85

Effect of Lime and Fertilizer on Trees 87
Relative Hardiness of 14 Species of Trees and Shrubs 88
Planting of 14 Tree Species on Pennsylvania Anthracite Strip Mine Spoils ... 89
   Materials and Methods. 89
   Results 91
   Discussion and Conclusions. 93
Vegetative Analysis of 81 Strip Mine Areas of Pennsylvania 94
   Description of Study Area. 94
   Vegetative Analysis. 95

SPOIL AMENDMENT STUDIES 98
Use of Sewage Effluent and Liquid Sludge 98
   Methods 98
   Results and Discussion 101
   Conclusions. 119
Use of Sludge and Dehydrated Sludge Product 120
   Materials and Methods. 120
   Experimental Results 122
   Discussion. 126
Use of Fly Ash on Kansas Strip Mine Soils 126
   Experimental Procedure 126
   Results and Discussion 127
Reclamation of West Virginia Acid Spoil with Fly Ash 131
   Materials and Methods. 131
   Results and Discussion 134

ACID MINE DRAINAGE 141
Coal Mine Drainage Pollution 141
   Sources 141
   Formation of Pollutants 142
   Water Quality Evaluation 143
   Damages 144
Prevention of Acid Formation. 144
Corrective Measures. 146
   Acid Mine Drainage Control 146
   Acid Treatment. 146
Infiltration of Water on Strip Mine Spoil Banks 149
   Materials and Methods. 150
   Characterization of Spoils. 151
   Field Procedures. 152
   Statistical Methods 152
   Results 152
   Discussion. 154
   Conclusions. 155

MINE SPOIL POTENTIALS FOR WATER QUALITY AND EROSION 156
Sulfur and Potential Acidity 157
   Sulfur Distribution in Overburden Material. 157
   Petrographic Evidence of Pyrite 158
   Modes of Occurrence of Pyrite and Associated Minerals 159
   Acid-Producing Potential of Pyritic Materials 160
Chemistry, Mineralogy and Weathering of Profiles 160

Chemical Analysis of Rock Chip Samples 161
Chemical Characteristics of Several Soil Profiles 164
Mechanical Analysis 164
Mineralogical Characteristics. 164
Mineralogy and Mottling of Soils. 165
Rock Weathering. 167
Simulated Weathering Experiment 167
General Observations. 168
Evidences from Old Mine Spoils 168
Microbiological Interactions 169
Interactions with Plant Covers. 169

STUDIES OF EFFECTS OF MINE DRAINAGE 172
Effect of Strip Mining on Water Quality 172
Lysimeter Study 172
Beaver Creek Study. 173
The Mining Operation 173
Materials and Methods. 174
Results and Discussion 176
Summary and Conclusions 179
Acid Mine Pollution Effects on Lake Biology 179
Description of Study Area. 179
Methods 181
Results and Discussion 182
Some Hypotheses of Ecological Succession. 195
Alternative Patterns of Stripland Utilization. 198

RECOVERY OF ACID STRIP MINE LAKES. 200
Methods and Materials. 201
Results and Discussion 202
Investigation I. 204
Investigation II 207
Investigation III. 212
Recovery of Acid Strip Mine Lakes—A Synopsis. 221
References. 222

AREA RECLAMATION PROJECTS 224
Grundy County, Illinois Demonstration Project 224
The Development Plan 224
The Project Document 228
Development of Cost. 228
Other Items of Interest 229
The Reclamation Project. 230
Spoil Relocation 232
General Work Procedures 235
Water Feature Alteration. 236
Topsoil Placement. 240
Acid Water Treatment. 243
Tree Transplanting 247
Conclusions. 247
Surface Mine Reclamation, Moraine State Park, Pennsylvania. 250
Purpose of Project. 250
Location 250

Hydrology...250
Condition of Sites Prior to Reclamation............................251
Project Specifications...252
Project Work...253
Vegetation of Project Sites.....................................256
Discussion..257
Evaluation of Pollution Abatement Procedures, Moraine State Park.......258
Project Background..258
Pollution Abatement Measures..................................258
Method of Investigation.......................................261
Discussion of Results..262

SURFACE MINED LAND RECLAMATION IN GERMANY..............268
The Importance of Lignite in West Germany..........................268
Brown Coal Mining Technology...................................269
Open-Pit Mines...269
Mechanization and Automation of the Lignite Mines..................270
Ground and Surface Water Control...............................271
Land Rehabilitation...272
Extent and Costs of Land Reclamation............................272
Forests and Lakes..273
Winning New Farmland.......................................274
Social and Economic Improvements...............................278
Agriculture..278
Villages..279
Recreation..280
Government Regulation and Supervision............................281
Historical Development.......................................281
The Brown Coal Committee....................................282
Implementing the Mining and Reclamation Plan.....................283
State Mining Office...283
Relevance to U.S. Surface Mining.................................284

RECLAMATION COSTS...287
Reclamation Costs Summary.....................................287
Eastern Surface Mining Reclamation..............................288
Western Surface Mining Reclamation.............................296
Strip Mining Economics.......................................299
Illinois Reclamation Cost-Effectiveness Analysis.......................302
Land Condition...304
Land-Use Alternatives..304
Reclamation Alternatives—Cost Analysis...........................305
Cost-Function Analysis.......................................305
Ownership of Disturbed Lands..................................312
Elkins Demonstration Project....................................313
Cost Analysis Procedures.....................................313
Discussion and Results.......................................315

AMD CONTROL FOR A SMALL COMPANY—COSTS AND EFFECTS.....320
Land and Water in Target Mine Area...............................321
Geography and Geology.......................................321
Water Quality..322
Target Mine's Water Quality...................................322

Ohio Water Quality Standards.................................323
Water Quantity..323
AMD Prevention and Abatement Programs Usable by Target Mine........325
Water Treatment Facilities................................326
Reclamation Costs...329
Industry and Target Mine Supply and Demand Relationships............329
Market Area of Target Mine................................330
Demand for Coal in the Market Area........................331
Elasticity of Demand for Coal from Southeast Ohio Producers........334
Target Mine's Demand Curve................................335
Target Mine's Supply Function.............................336
Impact of Mine Drainage Abatement.............................340
Short Run Output Adjustment...............................340
Long Run Adjustments......................................344

SOURCES..346

STATE AGENCY CONTACTS..348

INTRODUCTION

In light of the energy shortage, extraction of coal from the earth by strip mining techniques can be expected to continue and expand. According to a recent government report, it has been estimated that over 50% of U.S. coal production for the rest of this century will be mined by surface methods.

In the past strip mining was generally conducted with little regard for its effect on the environment. This has resulted in erosion and water pollution and left areas esthetically scarred and unusable for agriculture, grazing or recreational use.

Techniques are available to reclaim abandoned strip mined lands and to conduct mining in a way that prevents detrimental environmental effects. Authorities emphasize the importance of premining planning and point out that reclamation is less costly and more effective when integrated with the mining operation.

This book, based on 19 government reports issued from 1967 through 1974, describes coal surface mining land use and methods, land reclamation methods and sediment and erosion control. Acid mine drainage, its sources, prevention and correction as well as the mechanism of recovery of acid strip mine lakes are discussed. Specific studies on revegetation, use of spoil amendments and the effects of acid mine drainage are included. One chapter describes the Federal Republic of Germany's approach to the problem of reclamation of strip mined lands. The final two chapters discuss reclamation costs.

A complete list of the cited reports is given at the end of the book. This is followed by a list of state agencies to be contacted regarding strip mining and land reclamation.

SURFACE MINING LAND USE AND METHODS

The material in this chapter was excerpted from the following reports:

PB 219 259
PB 233 955
PB 238 538

LAND USE AND RECLAMATION STATISTICS

Mining, like agriculture, transportation, and urbanization, is ultimately a transitory use of land and is necessary for providing minerals essential to man's existence and welfare. The very nature of mining, and of surface mining in particular, requires disturbing the land surface.

Prior to the 20th century land areas disturbed by mining were small, land was plentiful, and little attention was given to reclamation. However, starting in the early 1900s, development of steam shovels and draglines facilitated removal of thicker overburden and utilization of lower grade ores. With corresponding technologic improvements and economic advantages in blasting, materials handling and increase in size of equipment, stimulated by constantly increasing demand for minerals, surface mining became more prominent and expanded to areas of more valuable land.

Various reclamation methods were employed to minimize the effects of surface mining and to restore the land to productivity. For example, the Indiana Coal Producers Association was founded in 1918 to revegetate part of the spoil banks from coal operations. Because of economic and technical limitations, not all surface-mined land was reclaimed. Some of the worked-out areas were abandoned without any attempt at reclamation. Most of the abandoned areas, known in the mining industry as orphan land or banks, were those of small operators.

Public demand for action to prevent unnecessary damage to mined lands and adjacent areas resulted in enactment of strip mine legislation, first by West

Virginia in 1939, followed by Indiana (1941), Illinois (1943), Pennsylvania (1945), Ohio (1947), Kentucky (1954), Maryland (1955), Virginia (1966), and Tennessee (1967). Other states as well as counties and municipalities have also enacted statutes controlling mining operations. Thirty-three states have surface mining and mined land reclamation laws.

A major portion of the coal resources of the U.S. lies in federal ownership and federal statutes exist with respect to mining on federal lands. The Department of the Interior has proposed new regulations for environmental control in the mining of federal coal. These regulations relate to coal exploration and mining operations, and reclamation of affected lands. The proposed regulations were published in the *Federal Register*, Vol. 40, No. 173, September 5, 1975, pages 41122-41138.

The total land mass in the United States comprises about 2.27 billion acres. Of this area, the domestic minerals industry, in a 42 year period extending from 1930 through 1971, utilized 3.65 million acres, or 0.16%. Also during this period 40% of the land utilized, or 1.46 million acres, was reclaimed. In 1971, 206,000 acres were utilized and 163,000 acres reclaimed. Thus the ratio between land used and land reclaimed doubled in 1971, compared with the ratio for the 42 years.

Fossil fuels accounted for 43% of surface land used during 1930-71. Mining of bituminous coal accounted for about 40% of land utilized in 1930-71 and 35% in 1971. Nearly 70% of the acreage reclaimed during 1930-71 was accounted for in the bituminous coal category and amounted to one million acres of the total 1.46 million acres reclaimed in the United States. In 1971, the coal industry, augmented by other organizations, reclaimed more acreage than was utilized for production of coal. In 1971, 94,600 acres were reclaimed by the bituminous coal industry.

SURFACE MINING METHODS AND THEIR IMPACT ON THE ENVIRONMENT—OVERVIEW

Stated in the simplest terms, surface mining consists of nothing more than removing the topsoil, rock, and other strata that lie above mineral or fuel deposits to recover them. In practice, however, the process is considerably more complex.

When compared with underground methods, surface mining offers distinct advantages. It makes possible the recovery of deposits which, for physical reasons, cannot be mined underground; provides safer working conditions; usually results in a more complete recovery of the deposit; and, most significantly it is generally cheaper in terms of cost-per-unit of production.

Surface mining is not applicable to all situations, however, because the ratio between the thickness of the overburden that must be moved in order to recover a given amount of product places a definite economic limitation upon the operator. While this ratio may vary widely among operations and commodities owing to differences in the characteristics of the overburden, types and capacities of the equipment used, and in value of the material being mined, it is nonetheless the factor that primarily determines whether a particular mining venture can survive in a competitive market.

The procedure for surface mining usually consists of two steps: prospecting, or exploration, to discover, delineate, and prove the ore body; and the actual mining or recovery phase. Topography and the configuration of the deposit itself strongly influence both. Exploration techniques generally employed consist of either drilling to intersect deeper-lying ore bodies, or excavating shallow trenches or pits to expose the ore. Although drill sites or excavations associated with exploration are usually small, their large number constitutes a serious source of surface disturbance in some of the Western States. Surface methods employed to recover minerals and fuels are generally classified as open pit mining (quarry, open cast); strip mining (area, contour); auger mining; dredging; and hydraulic mining. Of these strip mining and auger mining are used in the mining of coal.

Area strip mining usually is practiced on relatively flat terrain. A trench, or box cut, is made through the overburden to expose a portion of the deposit, which is then removed. The first cut may be extended to the limits of the property or the deposit. As each succeeding parallel cut is made, the spoil (overburden) is deposited in the cut just previously excavated. The final cut leaves an open trench as deep as the thickness of the overburden plus the ore recovered, bounded on one side by the last spoil bank and on the other by the undisturbed highwall. Frequently this final cut may be a mile or more from the starting point of the operation. Thus, area stripping, unless graded or leveled, usually resembles the ridges of a gigantic washboard.

Contour strip mining is most commonly practiced where deposits occur in rolling or mountainous country. Basically, this method consists of removing the overburden above the bed by starting at the outcrop and proceeding along the hillside. After the deposit is exposed and removed by this first cut, additional cuts are made until the ratio of overburden to product brings the operation to a halt. This type of mining creates a shelf, or bench, on the hillside. On the inside it is bordered by the highwall, which may range from a few to perhaps more than 100 feet in height, and on the opposite, or outer, side by a rim below which there is frequently a precipitous downslope that has been covered by spoil material cast down the hillside. Unless controlled or stabilized, this spoil material can cause severe erosion and landslides. Contour mining is practiced widely in the coal fields of Appalachia because of the generally rugged topography. Rim-cutting and benching are terms that are sometimes used locally to identify workbenches, or ledges, prepared for contour or auger mining operations.

Anthracite strip mining in Pennsylvania is conducted on hillsides where the coal beds outcrop parallel with the mountain crests. Although most of the operations are conducted on natural slopes of less than ten degrees, the beds themselves vary in pitch up to ninety degrees. Beds that are stripped are thicker than in the bituminous fields, most varying from 6 to 20 feet, and can be mined economically to much greater depths.

Because of the angles at which the beds lie, the methods employed may not be correctly identified either as contour or area mining, but rather as a combination of both. In a few instances, the operations may resemble open pits and quarries, while others are long, deep narrow canyons.

About one-half of the coal acreage mined has been by the contour method. In this type of operation, the waste material (overburden) cast down the outslope is on a steeper angle of repose than the natural slope upon which it rests, and extends downhill varying distances, depending on the steepness of the terrain. Spoil stacked at the outer edge of a bench, unless properly drained, causes water to accumulate on the bench between the spoil and the highwall. This accumulated water often becomes polluted and may overflow at the lowest point along the shoulder of the spoil bank during heavy storms, resulting in washouts, erosion, and stream pollution.

On the steeper slopes (in excess of 20 degrees), stabilizing the spoil material is difficult. It has been calculated that approximately 1,700 miles of outslope were affected by massive slides, some of which occurred after the areas had been reclaimed. Loose spoil absorbs large quantities of water, which not only act as a lubricant between the spoil and the original surface, but add an increased load on unstable slopes. The resulting slide not only covers trees and other vegetative cover, but may block highways and stream channels, and destroy valuable farmland and surface improvements.

About 20,000 miles of highwalls were created by coal mining in Appalachia alone and, in some cases, the walls completely isolated entire mountaintops. However, this condition is not confined to Appalachia, as precipitous highwalls are found throughout the Nation.

Area strippings create spoil piles or ridges with crests 50 or more feet high, 50 to 100 feet apart, and with side slopes that vary between 17 and 39 degrees. While the rate of erosion on these spoil banks is comparable to that of contour mining, a large percentage of the sediment is retained in depressions on the site. Thus streams and adjoining lands are not affected as severely as in contour stripping areas.

Large areas of woodland that helped to retard runoff in the anthracite region of northeastern Pennsylvania have been destroyed. Material washed from the spoil banks covered vegetation in adjoining downhill areas and choked many stream channels. Much of the surface water pooled in low places behind spoil banks, or trapped in strip pits, become acid. In many instances, water from strip pits and broken stream beds seeps into underlying mines, where it not only contributes to the volume discharged but also increases pollution from that source.

Auger mining is usually associated with contour strip mining. In coal fields, it is most commonly practiced to recover additional tonnages after the coal-overburden ratio has become such as to render further contour mining uneconomical. Augers are also used to extract coal near the outcrop that could not be recovered safely by earlier underground mining efforts. As the name implies, augering is a method of producing coal by boring horizontally into the seam, much like the carpenter bores a hole in wood. The coal is extracted in the same manner that shavings are produced by the carpenter's bit. Cutting heads of some coal augers are as large as seven feet in diameter. By adding sections behind the cutting head, holes may be drilled in excess of 200 feet.

As augering generally is conducted after the strip-mining phase has been completed, little land disturbance can be directly attributed to it. However it may, to some extent, induce surface subsidence and disrupt water channels when underground workings are intersected.

In the past, mining practices were all too often conducted with the purpose of removing minerals by the simplest and cheapest method possible, without plans for the preservation of land, water, and air, and with too little consideration for the rights of others. A mining company is in business to make a profit, but every company, regardless of the nature of its business, has moral obligations to reduce its undesirable effects on the environment and to safeguard the rights of others.

The problem of environmental degradation caused by surface mining is widespread and serious. As mentioned previously, thirty-three states have laws regulating surface mining. These laws vary considerably from state to state owing largely to the mining conditions within that state.

Even the states that are often cited as models of comprehensive regulations still have problems minimizing environmental damages. Uncontrolled surface mining presents a situation as critical to the well-being of society as any it has ever faced.

Many mining activities have imposed huge social costs on the public at large. These costs are long-range and are in the form of stream pollution, floods, landslides, loss of fish and wildlife habitats, unreclaimed land, erosion, and the impairment of natural beauty.

It has been estimated that of the 25,000 miles of contour bench in Appalachia, approximately 1,700 miles are affected by massive landslides. Additionally, 4,800 miles of streams and 29,000 surface acres of impoundments and reservoirs have been seriously affected by coal strip mining operations in the United States.

Coal is the Nation's most abundant and widely distributed fuel resource. Total reserves are estimated at 1,560 billion tons or over 2,500 years' supply at present consumption rates. The data denotes availability on broad, long term basis regardless of availability for mining or whether they are economically minable. Most of the coal reserves must be deep mined, as only 45 billion tons (or less than 3%) are economically minable by strip mining methods.

Although coal is abundant and widespread in the United States, resources of coal also have limits. In the extensively mined eastern coal fields, new areas containing thick beds of high rank coal are becoming scarce. Low-volatile bituminous coal used in the manufacture of coke constitutes only about 1% of the total resources. A large part of the total resources consists of lignite and subbituminous ranks, which yield less heat than bituminous coal. Another large part is contained in thin beds and in deeply buried beds that can be mined only with great difficulty and expense.

Since World War II, surface mining has emerged as a dominant force in the production of coal, bringing with it new and perplexing problems in land use and water control. The current expansion is due mainly to the greater production per man day, lower production costs; these result from new technology, which has produced equipment that has made strip mining highly productive and efficient.

In 1940, 9.4% of the total coal production was from surface mining, the 1972 figures show an increase to 49%. It is projected that for the remainder of this century surface mined coal will account for over 50% of the Nation's production. Strip mining can be done responsibly without permanent damage to the land and water. Technology exists for effecting the reclamation of mined lands and such reclamation is being performed in some areas.

STRIP MINING TECHNIQUES AND EQUIPMENT

To better appreciate the problems associated with surface mining, it is imperative that the stripping operation be understood. Surface mining is a very broad term and refers to any process of removing the earth, rock, and other strata in order to uncover the underlying mineral or fuel deposit. Strip mining is a type of surface mining in which the overburden is removed in narrow bands, one cut at a time. Strip mining methods employed to recover coal can be divided into two general types: area and contour.

Area Mining

In the United States, area stripping is characterized by giant earth moving equipment capable of handling several thousand cubic yards of material per hour. The load haul dump methods which are becoming increasingly popular for contour stripping may not be economically transferable to area stripping. Additionally, large strip equipment require that they be dedicated with enough coal reserves for several years operation to justify the capital expenditure for mining and preparation.

Simple overcasting, explained in Figure 1.1, is still the commonest form of stripping. However, the long-range plans for effective land use before, during and after mining, the mining and reclamation methods to practice, and the selection of stripping and reclamation equipment, etc., must be analyzed with respect to geological, social, technical and economic constraints.

Pit engineering to avoid unnecessary inventory and quenching delays becomes important as all other equipment (equipment for drilling, coal loading and hauling, clean up, etc.) must be carefully matched to the primary equipment and their capacities. Primary equipment selection is also difficult because of the availability of a wide range of equipment capable of working in all types of conditions.

In area stripping, shovels and draglines continue to be more popular, with draglines increasingly favored over shovels. They are available in a wide range of designs and capacities. The largest shovel and dragline in operation have bucket capacities of 180 and 220 cubic yards respectively. However, there is no trend toward selecting the largest equipment available and operators have continued to depend on equipment in the intermediate and high ranges which have proven performances.

Draglines provide greater flexibility, work on higher bank heights and move more cover per hour. Wheel excavators hold considerable promise where conditions are favorable. Ideally, this machine has the capability for continuous burden removal and selective placement of the topsoil. Designed capacities are up to 15,000 cubic yards per hour, though in practice figures of only 3,000 to 4,000 cubic yards per hour have been realized. The use of Kolbe or American wheels is more common.

In the West, where coal seams are unusually thick, open pit extraction techniques find application. At many operations, large conventional road excavating and grading equipment find wide use. Tractor scrapers, and bulldozers, while generally used for auxiliary stripping, have recently been used for primary stripping.

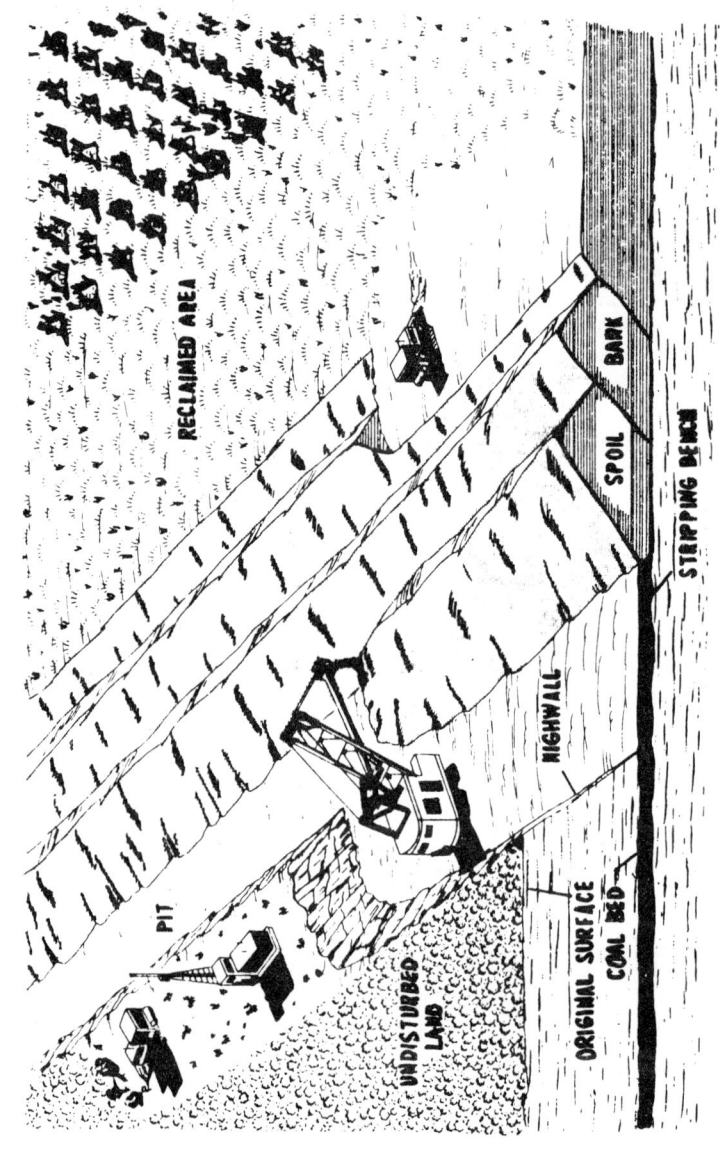

FIGURE 1.1: AREA STRIP MINING WITH CONCURRENT RECLAMATION

Source: PB 238 538

Surface Mining Land Use and Methods

Multiple seam mining is practiced where two or more seams occur close together. Such occurrences enable coal recovery in places where one seam by itself may not be economically mineable. Overburden removal practices include deployment of a shovel or a dragline operating alone, or in tandem to uncover the seams.

Drilling for fragmentation is commonly done with rotary type units capable of hole diameters from 5½ to 15½ inches with vertical drilling more common. Ammonium nitrate fuel oil (ANFO) mixes continue to be the leading explosive.

Stripping with a Shovel: A coal seam, about 4 feet thick, and overlaid by 120 feet of shales, sandstones, clays and limestone, is exposed by a 105 cubic yard bucket, 200 foot boom, shovel. Vertical drill holes 15½ inches in diameter spaced approximately on a 50 x 60 foot grid pattern, reach within 5 feet of the coal. Thirty to thirty-three 80 pound bags of ANFO are loaded into each hole. Usually, three rows of holes are shot with delays between each row. As can be seen in Figure 1.2, at any one time, a pit width of 180 feet is maintained. Coal is loaded by a 9 cubic yard shovel from a 54 foot cut into four 100 ton trucks.

FIGURE 1.2: STRIPPING WITH A SHOVEL

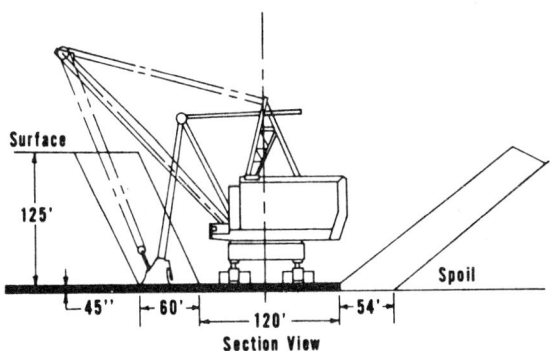

Source: PB 238 538

Stripping with a Dragline: Figure 1.3 shows a 220 cubic yard dragline removing 120 to 130 feet of overburden over a four foot coal seam. It is capable of working a pit 250 feet wide, and 185 feet deep. The overburden is prepared by bulk loading some 1½ to 5½ tons of ANFO into each hole, drilled on a 30 by 30 foot grid. A 14 cubic yard loading shovel with four to six 120 ton trucks is used for coal removal. The annual coal production expected from this mine is about 2.5 million tons.

FIGURE 1.3: STRIPPING WITH A DRAGLINE

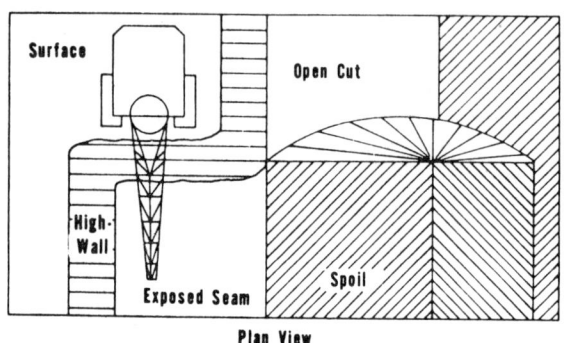

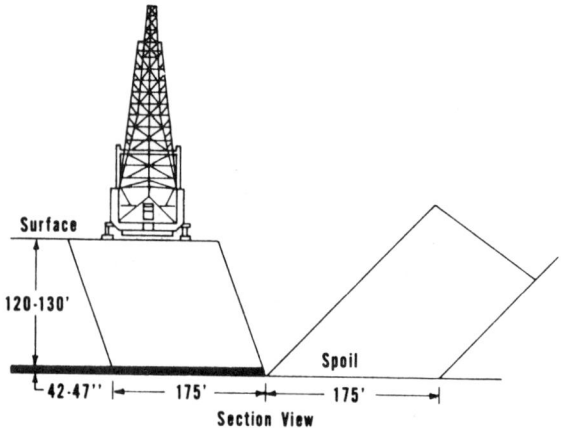

Source: PB 238 538

Shovel and Bucket Wheel Excavator Tandem Operation: The Kolbe wheel excavator can operate most efficiently by the frontal block digging method on benches of limited width. This must be so because the cutting boom and the discharge boom are in a straight line, and have no independent movement. Therefore, they swing in opposite directions, about the vertical axis of the machine. The wheel excavator is used to remove the loose top soil and soft beds whereas the harder beds are handled by a large stick shovel.

Surface Mining Land Use and Methods 11

As shown in Figure 1.4, the wheel excavator removes the top 54 feet whereas the shovel with a bucket capacity of 70 cubic yards removes the remaining 46 feet both equipment operating from the coal seam. The pit is about 1¼ miles long.

FIGURE 1.4: SHOVEL AND BUCKET WHEEL EXCAVATOR TANDEM OPERATION

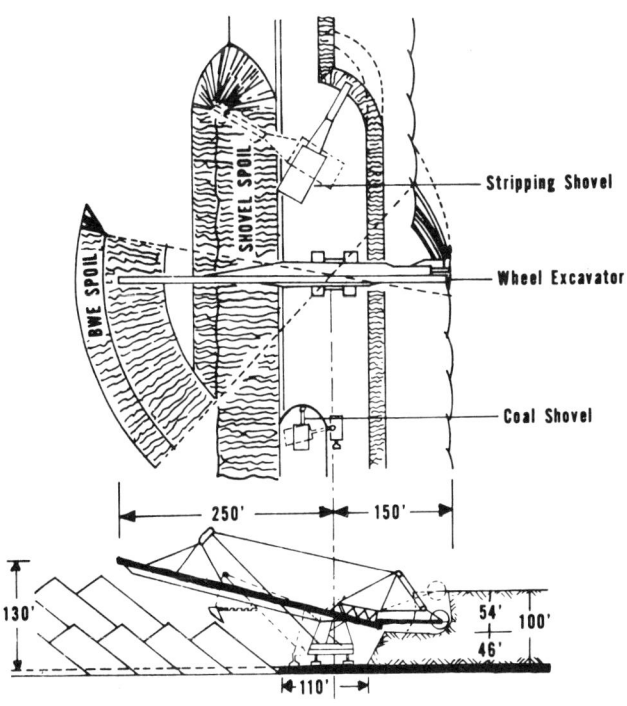

Source: PB 238 538

Bucket Wheel Excavator and Dragline Tandem Operation: In Figure 1.5 is shown a wheel excavator-dragline tandem operation in an Illinois mine. Both the equipment work from a bench 0 to 65 feet below the surface. The wheel removes the unconsolidated sand, clay and gravel beds above the bench.

Drilling (10½ inch diameter hole) is done on a thirty foot square grid to fragment the bench with ANFO explosive for removal by dragline. A six cubic yard loading shovel loads the coal onto four 100 ton trucks for hauling to the preparation plant, 3½ to 5 miles away.

FIGURE 1.5: DRAGLINE AND BUCKET WHEEL EXCAVATOR IN TANDEM OPERATION

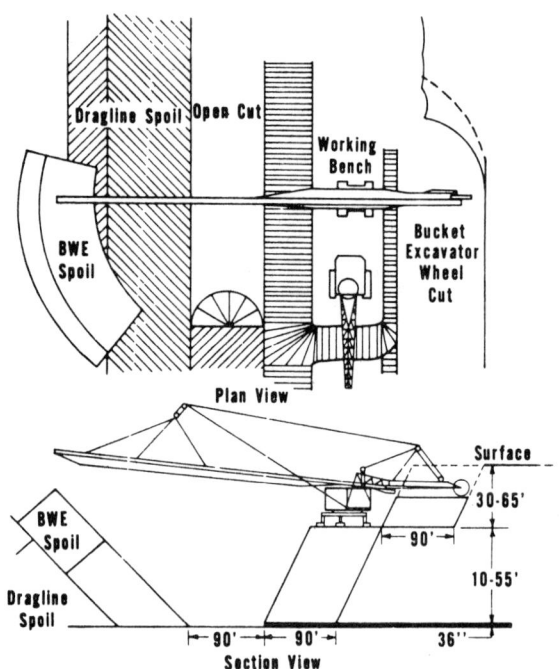

Source: PB 238 538

Stripping in the West: Three operations which employ conventional road construction equipment will be described. In the first a 32 foot coal seam in Wyoming is overlaid by soft and unconsolidated material. Bulldozers and scrapers remove the overburden atop the coal and a shovel loads the blasted coal onto trucks.

Figure 1.6 shows an open-pit operation where a 30 to 90 foot coal seam is overlaid by about 35 feet of soft overburden which requires no preparation. The coal in this pit is mined in two benches, each 30 to 45 feet high. The coal is blasted with ANFO, loaded into a single row of 6 inch holes, 16 feet from the face, and 24 feet apart. Two pan scrapers remove the overburden with the assistance of a dozer.

The dozer pushes the scrapers downgrade during the loading operations at point A in the figure, and another dozer pushes the dumped overburden into the mined out area of the pit at point B. The scraper route is represented by the dotted line. The coal is loaded by ten cubic yard front-end loader onto six twenty-eight ton trucks for a short haul, ½ mile to the power plant.

FIGURE 1.6: PLAN VIEW OF OPEN PIT COAL MINE

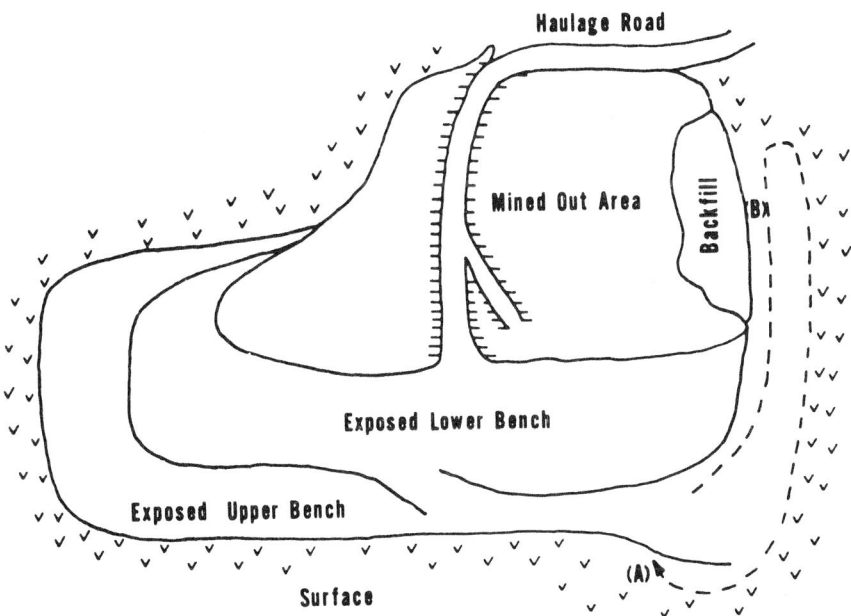

Source: PB 238 538

Figure 1.7 shows a multiple seam scraper operation. Several pan scrapers and bulldozers are used to remove the 70 to 130 feet of overburden over the ten foot thick Armstrong seam, and the 35 to 40 foot parting to the 50 foot thick Monarch seam complex.

The topography of the area is somewhat hilly, and the overburden requires blasting only when large rocks or boulders are encountered. When sufficient length of the upper seam is exposed (1,000 feet), some earth moving equipment is assigned to remove the parting and expose the lower seam.

The operation now resembles an open pit scheme with two benches. This is a massive earth moving operation. The scrapper haul is about 3,500 feet and this equipment combination has been used to depths of 250 feet.

Coal is loaded onto 9 to 50 ton trucks by two loading shovels of eight cubic yard and four cubic yard bucket capacities. Production is estimated to be two million tons (1.8 million metric tons) per year.

FIGURE 1.7: MULTIPLE SEAM SCRAPER OPERATION

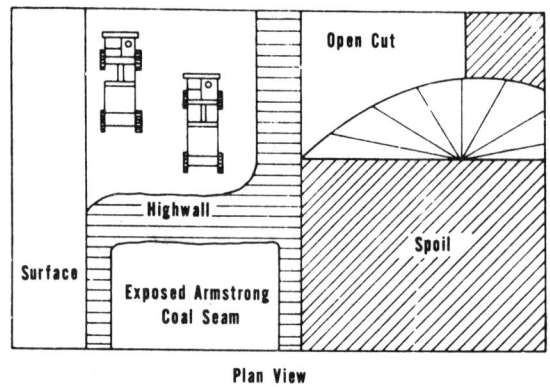

Plan View

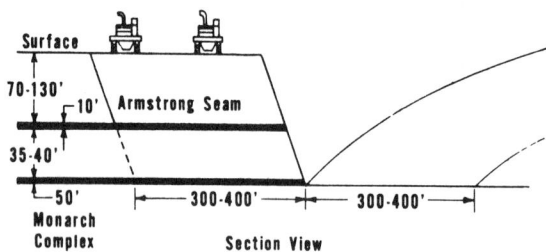

Section View

Source: PB 238 538

Multiple Seam Operation with a Shovel: In Figure 1.8 is shown a 65 cubic yard shovel uncovering two seams. The maximum overburden in the property is 120 feet with the parting between the seams varying from 3 to 18 feet. The shovel sits on the bottom seam, and uncovers both the seams during a single cut.

The overburden above the upper seam and the parting must be fragmented with explosives. In the overburden, 9 inch horizontal holes are drilled on 30 foot centers. In the parting, 9 inch vertical holes are drilled on a 12 x 14 foot grid.

The mine is operating with a 160 foot wide pit. First the shovel takes a 60 foot wide section of the parting exposing the bottom seam. The pit width now is approximately 80 feet. Then, the top seam is uncovered to the side, and a pit width of 70 feet is maintained on the upper level. Coal hauling is done by eight 12 ton trucks, loaded by two 10 cubic yard loading shovels.

FIGURE 1.8: MULTISEAM STRIPPING OPERATION WITH A SHOVEL

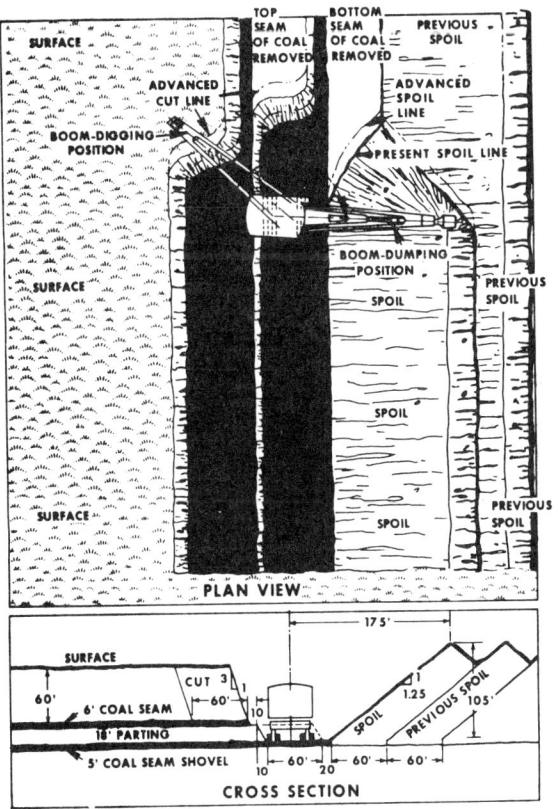

Source: PB 238 538

Multiple Seam Stripping with a Dragline: In the multiseam mining operation shown in Figures 1.9 and 1.10, a dragline with a 35 cubic yard bucket, and a 200 foot boom is used to uncover the 4 foot top seam. The clay parting between the top seam and the middle seam, which is about 0.7 feet thick, is removed with the help of two dozers and a front-end loader. For removing the parting to the bottom seam, the dragline operates from the leveled spoil on the low wall side.

In practice, the dragline exposes the top seam the entire pit length, and then moves to the spoil to uncover the bottom seam. This way coal recovery can be accomplished from both the seams at the same time and if there are quality requirements the two coal seams can be blended. Because the overburden is unconsolidated and soft, the dragline initially makes the keycut 100 to 150 feet long to the depth of the top coal seam, and establishes a safe slope for the future highwall. Because of this, and also because of the chopping operation (Figure 1.10), the dragline performance tends to be poor.

16 Strip Mining of Coal

FIGURE 1.9: MULTISEAM STRIPPING OPERATION WITH A DRAGLINE

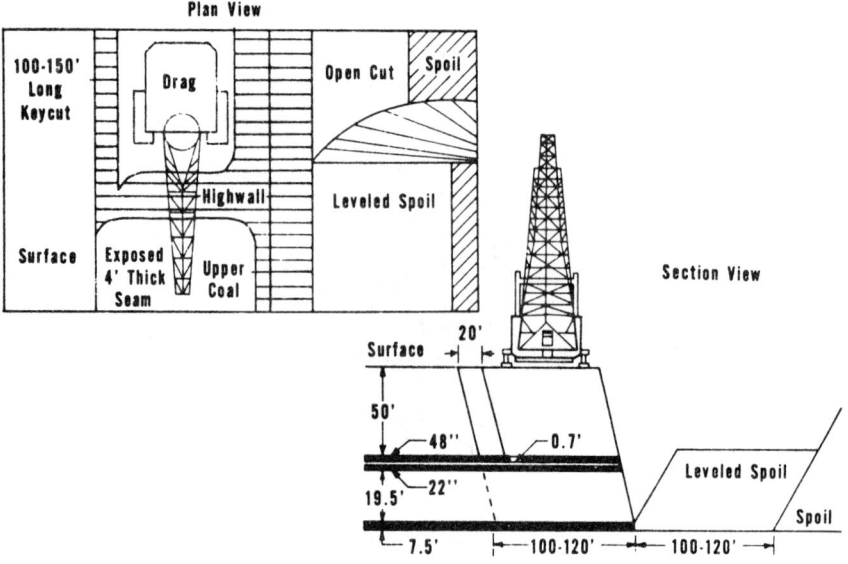

FIGURE 1.10: DRAGLINE EXPOSING LOWEST SEAM FROM LEVELED SPOIL

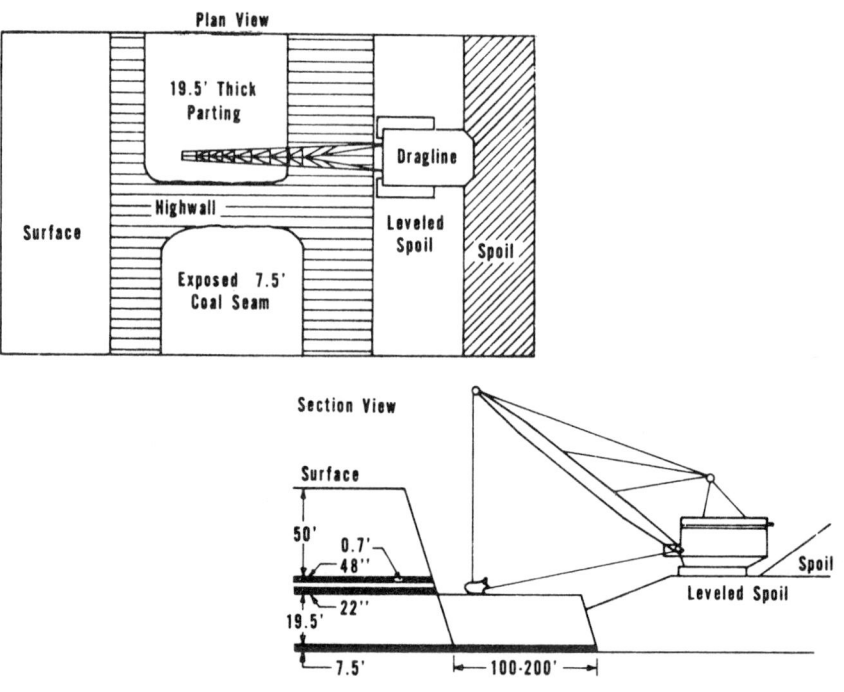

Source: PB 238 538

Surface Mining Land Use and Methods 17

Multiseam Mining with Shovel and Dragline: A shovel-dragline combination exposing three coal seams is shown in Figure 1.11. The dragline has a 100 cubic yard bucket, and a 275 foot boom. The 33 cubic yard shovel has a boom length of 113 feet and a dumping radius of 139 feet. The overburden and the partings are drilled on an approximately 30 x 30 foot grid, and shot with ANFO explosive.

The shovel removes overburden to expose the top seam, and the parting between the top and the middle seam, once the top seam coal is loaded out. The dragline operating from the spoil removes the parting between the middle and bottom seam.

Occasionally, the dragline may have to rehandle the shovel spoil to maintain the distance between the two machines for uninterrupted stripping. Two, ten cubic yard loading shovels and one front-end loader, alone with eleven trucks, are used for coal loading and hauling to produce 10,000 tons of coal per day.

FIGURE 1.11: SHOVEL-DRAGLINE TANDEM OPERATION FOR MULTISEAM MINING

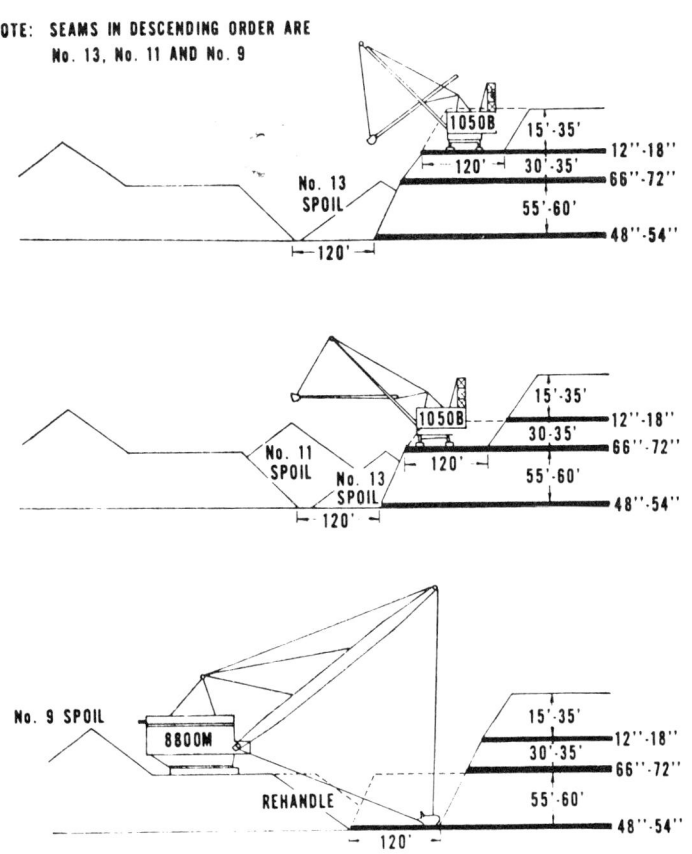

Source: PB 238 538

Area Stripping Reclamation Trends: Reclamation must be an inherent part of the method, and not an afterthought. Proper planning will permit burying toxic materials and soil modification at much reduced costs. This is also important because vegetation cannot be otherwise established, leading to air and water pollution in subsequent years.

In general, the pollution from area mines is not as severe as that from contour mines. Silt from erosion can often be confined to the mining area. The current legislative trend is to require restoration of the disturbed area to its approximate original contour with all spoil ridges and highwalls eliminated and no depressions left to accumulate water. Contour grading does not mean that all areas must be leveled, but rather the profile of the land must be put back to approximately the way it was before the strip mining began.

To accomplish contour grading, the spoil from the first cut is graded so as to blend into the contour of the adjoining land. Successive spoil piles are then graded with all materials pushed toward the last cut, where it is deposited in the final pit. Long slopes on the graded spoil must be interrupted by terraces and/or diversion ditches. All of the diversions and terraces must be constructed according to sound engineering principles and must end in suitable outlets.

Several states now require the operator to separate topsoil from the subsoil and to stockpile the two types separately so they will not be mixed during the excavation process. When mining is completed, the overburden can then be put back in its original sequence and revegetated to prevent erosion. Some operations remove the topsoil and immediately spread it on areas recently graded, thus handling the material only once. This provision insures that the best soil for plant growth is on top and not indiscriminately mixed with subsoils.

Some form of tillage of the site before planting is necessary. Any tillage measures must follow the contour of the slope and run parallel to the diversions or terraces. Chemical improvement of the soil in the form of liming and fertilizers is often needed for rapid establishment of vegetation.

Contour Mining

The conventional method of contour strip mining consists of removing the overburden from the mineral seam, starting at the outcrop and proceeding around the hillside (Figure 1.12). The cut appears as a contour line, thus, the name. Overburden is cast down the hillside and stacked along the outer edge of the bench (Figure 1.13). After the uncovered seam is removed, successive cuts are made until the depth of the overburden becomes too great for economical recovery of the coal. Physical limitations of equipment reach, capacity, etc., may also determine the strippable limit or cut-off point for mining.

Contour mining creates a shelf or bench on the side of the hill. On the inside it is bordered by the highwall, ranging in height from a few feet to more than 100 feet; and on the outer side the pit is bordered by a high ridge of spoil with a precipitous downslope that is subject to severe erosion and landslides. Because of the landslide problem, several states and the Tennessee Valley Authority have limited the bench width on steep slopes and forbid fill benches on slopes greater than 33 degrees.

Surface Mining Land Use and Methods 19

FIGURE 1.12: CONTOUR STRIP MINING

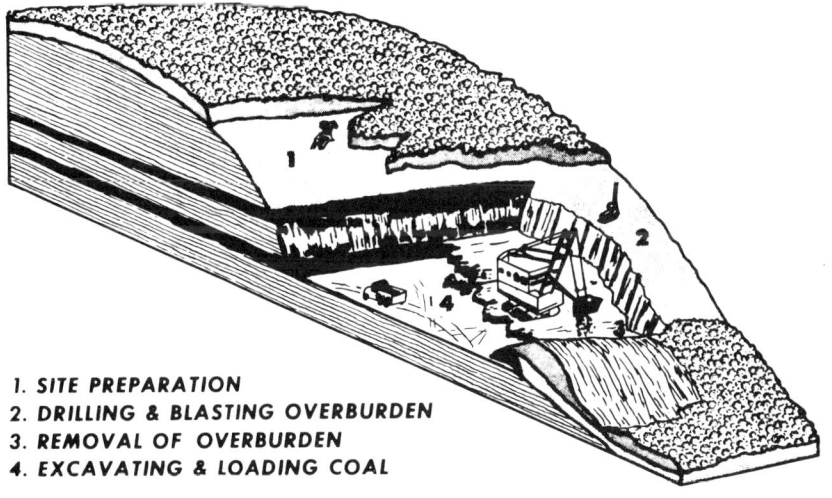

1. *SITE PREPARATION*
2. *DRILLING & BLASTING OVERBURDEN*
3. *REMOVAL OF OVERBURDEN*
4. *EXCAVATING & LOADING COAL*

FIGURE 1.13: CONVENTIONAL CONTOUR MINING

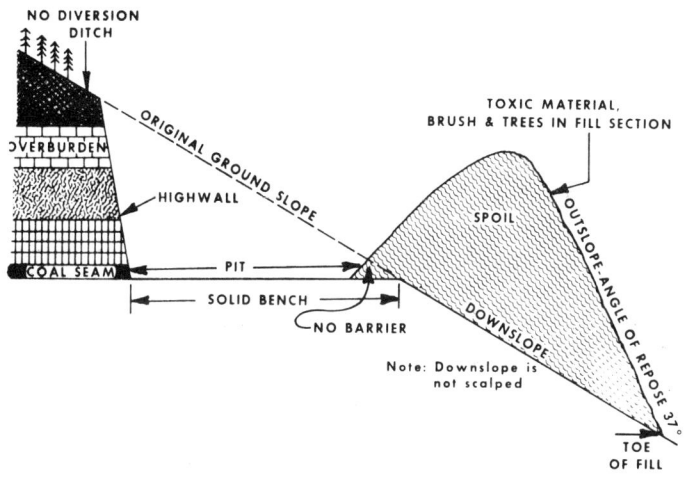

Source: PB 238 538

Even with these precautions, landslides still occur. Sediment slides coming off mining operations have uprooted trees, covered highways, destroyed farm land, filled up reservoirs and water courses, clogged stream channels, covered fish-spawning beds, caused flooding of adjacent lands, and destroyed farm buildings and homes.

Another problem inherent in contour strip mining is the toxic materials (i.e., pyrites, acid, soluble minerals, etc.), in the overburden. During the normal stripping operation, the high quality overburden near the surface is placed on the bottom of the spoil pile and then covered with low quality and often toxic overburden, leaving toxic material exposed to weathering and conversion to soluble acids and minerals that are carried away by water.

For a small extra cost, however, the high quality overburden can be set aside to cover the toxic material after grading and/or during excavation. By this means, the toxic material is not subject to weathering, and pollution can be reduced. Moreover, cover crops are difficult to establish on toxic overburdens, and therefore erosion damages occur. Erosion serves to prolong the mineral pollution problem by continuing to reveal new surfaces to weathering. However, when the toxic material is covered with a good material, cover crops can be grown to protect the surface.

The final cut in a contour strip mine can also be troublesome. Materials adjacent to the coal seam are often toxic. A final cut left uncovered is a potential pollution source; when it is covered, the danger from this source of pollution is reduced or eliminated.

Highwalls can also lead to pollution problems. An unstable highwall that sloughs off can ruin the natural drainage in a strip area. Material falling off the highwall can dam up channels and thereby prolong the contact between water and toxic material, or even force the water to seep through toxic spoil piles. Sloughing highwalls can open up new toxic materials to weathering. Highwall problems such as these can often be overcome by grading the spoil back against the highwall and knocking off the top of the highwall.

Often in the excavation of a strip area, a natural drainageway is cut across. Unless the water is diverted around the mine workings, the water enters the mine area, where it may become polluted. Problems such as these have been averted by not stripping the drainageway or by placing control structures such as drop boxes and concrete flumes to handle the water.

Diversion ditches with good, controlled outlets should be constructed along the top of the highwall to keep water out of the workings. Water that does enter the pit must be properly handled. Strategically located sumps and pumps of capacity sufficient to discharge the water rapidly through plastic pipe across the disturbed areas, to natural drainways or to treatment facilities can be used. This can reduce waterborne pollutant problems downstream.

Under some conditions, where a workable system can be developed, it might be better to catch the water on the bench and control the discharge to the treatment facilities. Drainage patterns should be established in the pit to facilitate water removal. Water discharge from the pit area should be through well-designed outlets and must not overload the natural drainageway. Proper management of

Surface Mining Land Use and Methods

water on the bench can markedly reduce the siltation and AMD problem.

It is critical that all efforts be made to locate underground mines adjacent to the surface mines. Cutting into abandoned or inactive underground mines can result in the discharge of large volumes of stored polluted water. The resultant, continued underground discharges into surface mining works during and following mining will aggravate the pollution problem. These conditions often make complete reclamation impossible, and in steep terrain, the underground mine can supply the water necessary for the development of slippage planes in the spoils. Where underground mines are adjacent to the proposed surface mines, barrier pillars should be left. When a deep mine is accidentally breached, the opening should be sealed as soon as possible by clay compaction, concrete, or any other method deemed necessary.

Removal and placement of the overburden are critical in environmental control. The nontoxic, nonacid, and fertile material should be stockpiled for later spreading or placed on top of the less desirable spoils already mined. The placement of the spoil should assure that long, steep slopes are avoided, that it is not on material subject to slippage, and that it does not produce high peaks difficult to regrade.

In very steep terrain, such as in eastern Kentucky and southern West Virginia, the spoil should not be placed on the outslope, but hauled to a fill area designed for that purpose or placed on the bench behind the operation. The existence of ground water seeps and natural springlines must be determined prior to spoil placement or slippages may occur.

Contour strip mines disturb an area of the earth's surface much greater than the area covered by the seam of coal extracted, and have environmental problems not experienced in area mining. Because of this, concerned federal and state agencies along with the coal industry have been working together to develop mining methods which minimize the adverse effects on the environment while allowing the maximum recovery of coal. These new methods (slope reduction, box-cut, head-of-hollow fill, mountain-top removal and block cut) are now accepted methods of mining on steep slopes. These new methods are not the final answer for all mining conditions and are being refined as more experience with varying conditions is gained.

Slope Reduction Method: The slope reduction method was developed on the theory that by reducing the weight on the fill bench and spreading the spoil over a large area, it would be less likely to slide.

7° Storage Angle — The overburden is purposely pushed down and distributed over the downslope with resultant slope of 7° less than the original slope. The storage area size is based on the original slope of the mountain. Overburden can be removed by either a one or two cut mining sequence.

Procedures for using the slope reduction method are as follows (see Figures 1.14 and 1.15):

(1) Scalp all organic material from the top of the highwall to the predetermined toe of the storage area. This procedure will insure a solid earth-to-earth bond between the pushed down spoil and the original surface.

FIGURE 1.14: SLOPE REDUCTION METHOD—TWO CUTS

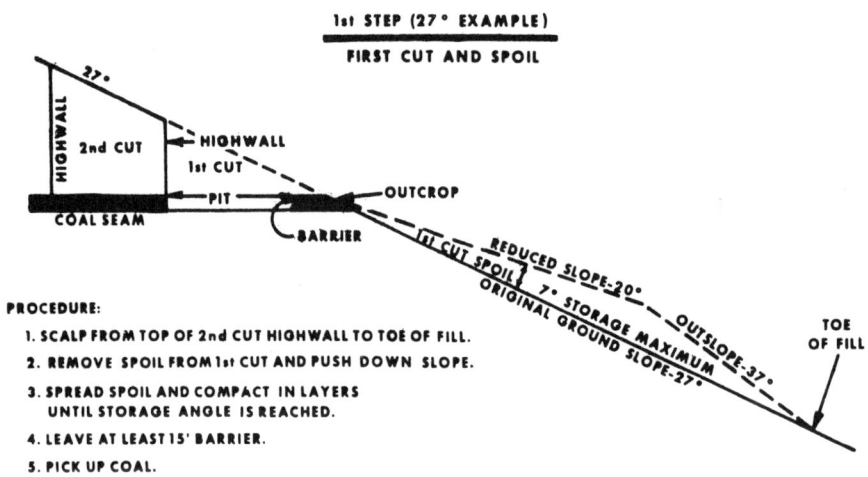

1st STEP (27° EXAMPLE)
FIRST CUT AND SPOIL

PROCEDURE:
1. SCALP FROM TOP OF 2nd CUT HIGHWALL TO TOE OF FILL.
2. REMOVE SPOIL FROM 1st CUT AND PUSH DOWN SLOPE.
3. SPREAD SPOIL AND COMPACT IN LAYERS UNTIL STORAGE ANGLE IS REACHED.
4. LEAVE AT LEAST 15' BARRIER.
5. PICK UP COAL.

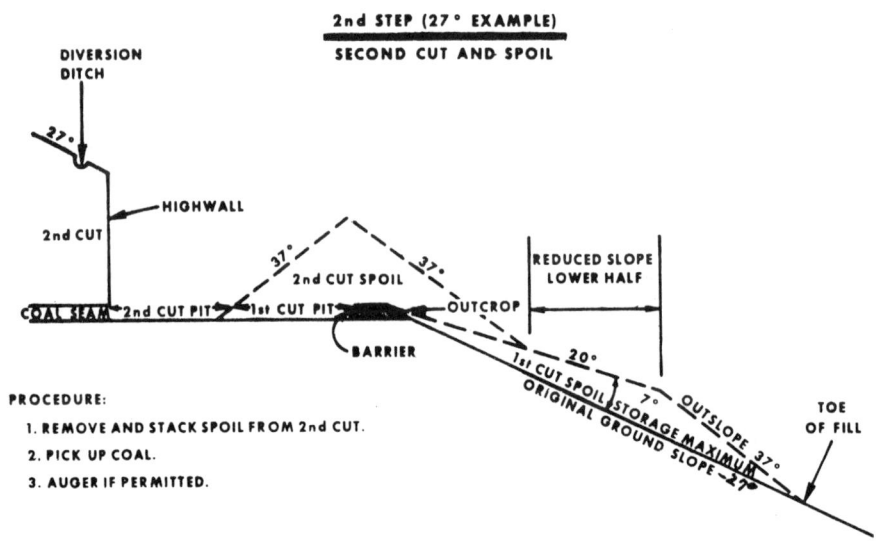

2nd STEP (27° EXAMPLE)
SECOND CUT AND SPOIL

PROCEDURE:
1. REMOVE AND STACK SPOIL FROM 2nd CUT.
2. PICK UP COAL.
3. AUGER IF PERMITTED.

(continued)

Surface Mining Land Use and Methods

FIGURE 1.14: (continued)

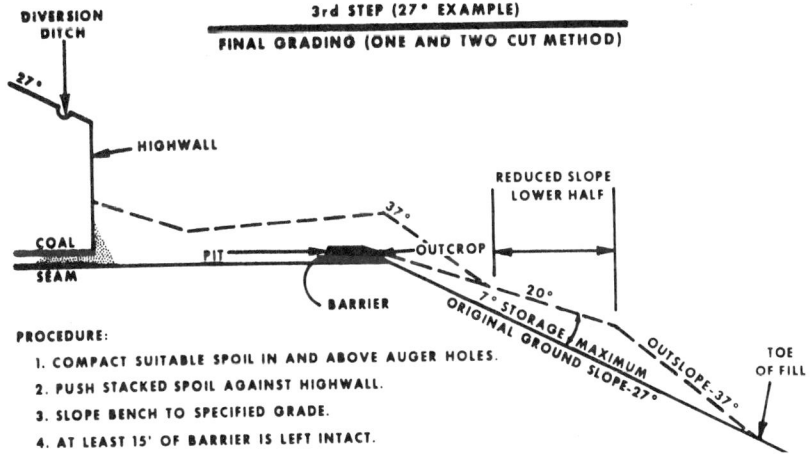

FIGURE 1.15: SLOPE REDUCTION METHOD—ONE CUT

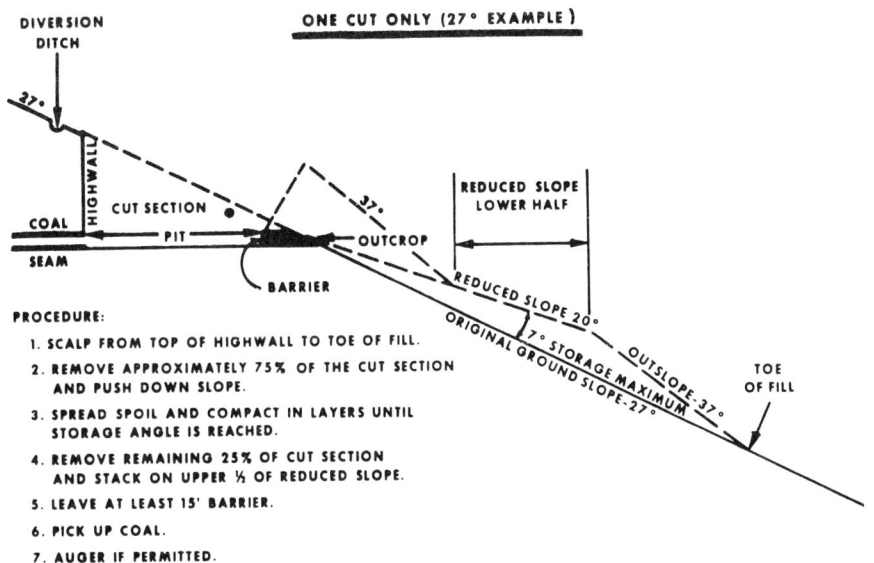

Source: PB 238 538

(2) Windrow all organic material at the toe of the spoil that will trap sediment eroding from the outslope.
(3) Push overburden from the first cut beyond the edge of the solid bench to the toe of the scalped storage area. This material is placed in 3 foot compacted layers until the slope is approximately 7° less than the original slope, measured from the seam down the hillside. Tables are available indicating the length of the storage area (Table 1.1).
(4) Install terraces on the contour during final grading to break up long slopes and to reduce the velocity of runoff. This procedure reduces erosion and assists in establishing vegetation.
(5) Do not disturb the area again after final grading, immediately revegetate the area, utilizing soil amendments, grasses, and trees.

TABLE 1.1: SLOPE REDUCTION (EXAMPLE)*

Bench width (feet)		Length of reduced slope (feet)	Length of outslope (feet)	Length from top of highwall to toe of fill (feet)		Linear feet of bench per acre	
One cut only	1st cut of 2 cuts			One cut only	1st cut of 2 cuts	One cut only	1st cut of 2 cuts
64	50	98	71	213	199	205	219
76	60	118	82	252	237	173	184
88	70	138	98	296	278	147	157
100	80	157	112	337	317	129	137
112	90	177	128	381	359	114	121

*Original ground slope, 27°; reduced slope, 20°.

This mining technique has been accepted as one method of contour mining in mountainous terrain. By reducing the weight on the fill bench and spreading the spoil over a larger area, slides have been minimized. Slope reduction is often the only practical method of reclaiming abandoned contour strip mines in steep terrain. It can be used to reduce the slope of any oversteepened spoil pile. It may be particularly effective for use on steep spoil and tailings slopes occurring at many western mines.

Parallel Fill — This method is a modification of the slope reduction method and has no storage angle. Overburden is pushed down the slope and compacted in three foot layers at the same angle as the original slope (see Figure 1.16). The depth of fill is determined from tables according to the degree of original slope.

Although parallel fill is still in the experimental stage, it may prove to be more successful than the storage angle method. No slides have developed, primarily because of the better friction plane which is more slide resistant.

Legislation at both the state and federal level is becoming more stringent and making it illegal to push overburden beyond the solid edge of the bench and

Surface Mining Land Use and Methods

FIGURE 1.16: PARALLEL FILL METHOD

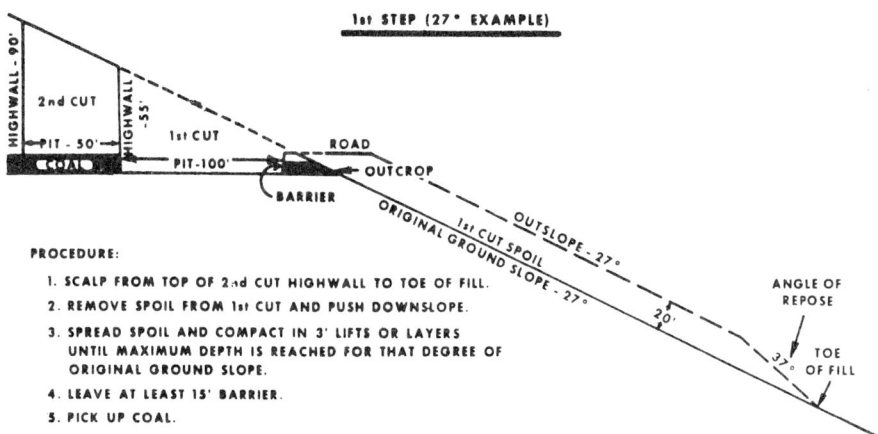

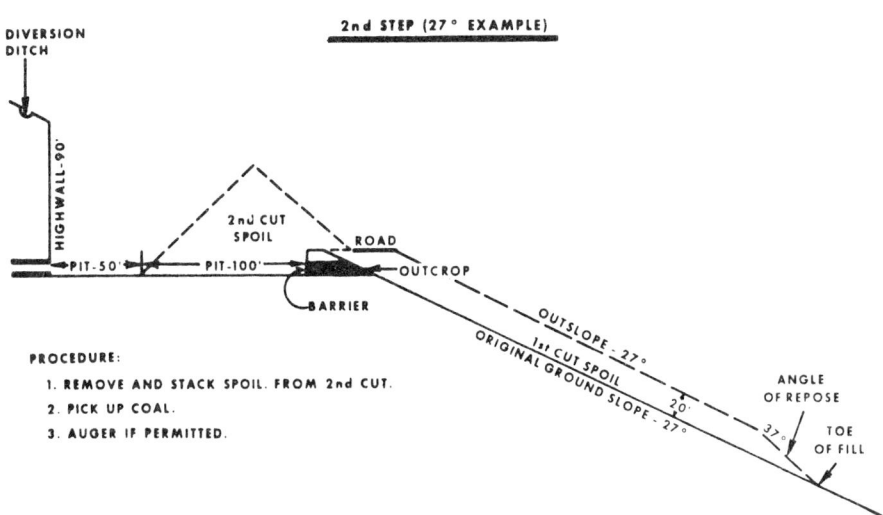

(continued)

FIGURE 1.16: (continued)

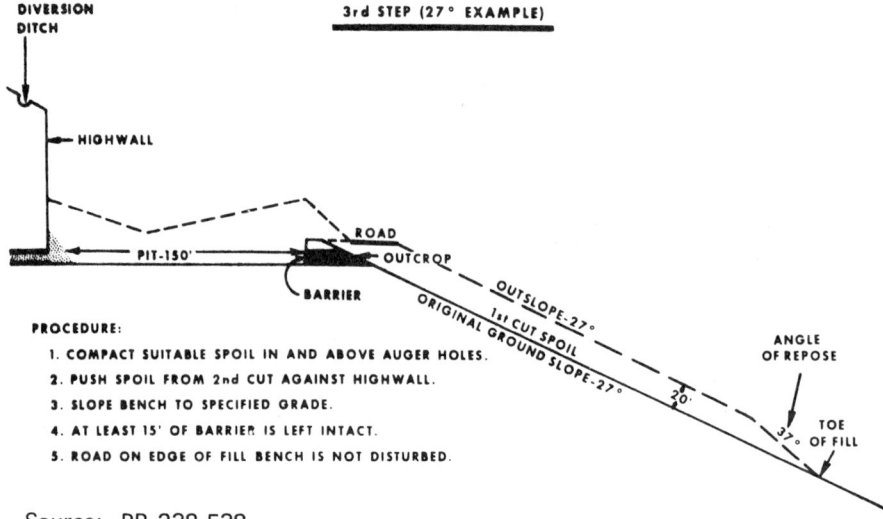

Source: PB 238 538

over the downslope. This type of restriction will ban the slope reduction method of mining. However, the theory of slope reduction has an interesting offshoot now being practiced by operators as an emergency measure when spoil begins to slide from the outslope. Bulldozers and/or pans are used to reduce the slide at its midsection. The resulting profile approximates that of the slope reduction method. Such an emergency measure is one practical and effective way to stop slides at an early stage when telltale tension cracks appear at the crest of the outslope.

Box-Cut Method, Two-Cut: The box-cut is one of the conventional contour strip mining methods. A box-cut is created by leaving an undisturbed section of the surface measured from the outer edge of the solid bench back toward the highwall. This barrier is at least 15 feet in width and provides a solid foundation on which to deposit spoil. It also helps to prevent water from running off the bench and percolating into the spoil on the downslope.

Basically, the two-cut box-cut method reverses the usual box-cut method by recovering the coal from the second cut first. This method was developed to prevent overloading the fill bench with second cut spoil and to make a more stable outslope. Procedures to follow: (see Figure 1.17).

  (1) Scalp entire area from top of highwall to toe of the outslope.
  (2) Drill and shoot the overburden.
  (3) Remove overburden from shot area to a point approximately 15 feet above the coal seam, making a flat bench from highwall to lip of spoil on the downslope.
  (4) Establish the permanent haul road on the outer edge of the bench. This road is located on the solid bench, if space permits, and will not be disturbed in future mining.

(5) Uncover the inside half of coal for the first cut. Stack the overburden on the flat bench between the road and the low-wall side of the first pit. After the coal is picked up and augered, push the stacked overburden into the pit.
(6) Uncover the outside half of the coal, stacking overburden against the highwall. Recover marketable coal, leaving the 15 foot barrier intact.
(7) Push spoil back into pit and slope bench to specified grade, leaving road on the outside undisturbed.

This method reduces the amount of overburden on the downslope, thereby reducing the incidence of slides and speeding up the final grading of the operation. However, the two-cut box-cut method places spoil on the downslope and would be illegal if legislation were passed that banned a fill section beyond the edge of the solid bench.

Head-of-Hollow Fill Method: The head-of-hollow fill method was developed to improve aesthetics, reduce landslides, allow for full recovery of one or more coal seams, and produce potentially valuable flat to rolling mountain top land that is suitable for many uses other than forestry.

The head-of-hollow fill method provides storage space for spoil from the removal of entire mountain tops and is also used as a waste area for overburden from contour benches. In the past, as the top coal seams were worked on the contour with a rim cut, islands of mountain land were left with no access. Many of these isolated areas of land left from previous mining operations are now being removed.

FIGURE 1.17: BOX-CUT METHOD—TWO-CUT

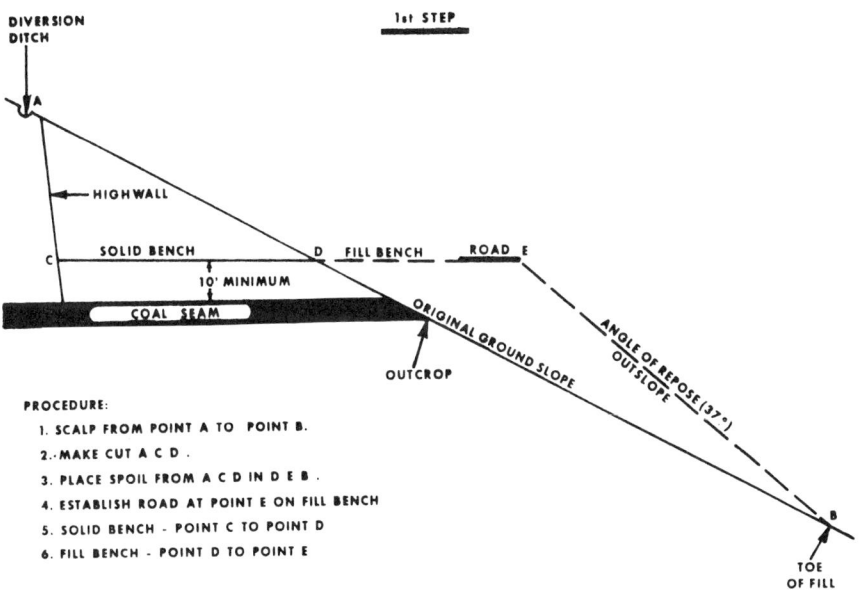

(continued)

FIGURE 1.17: (continued)

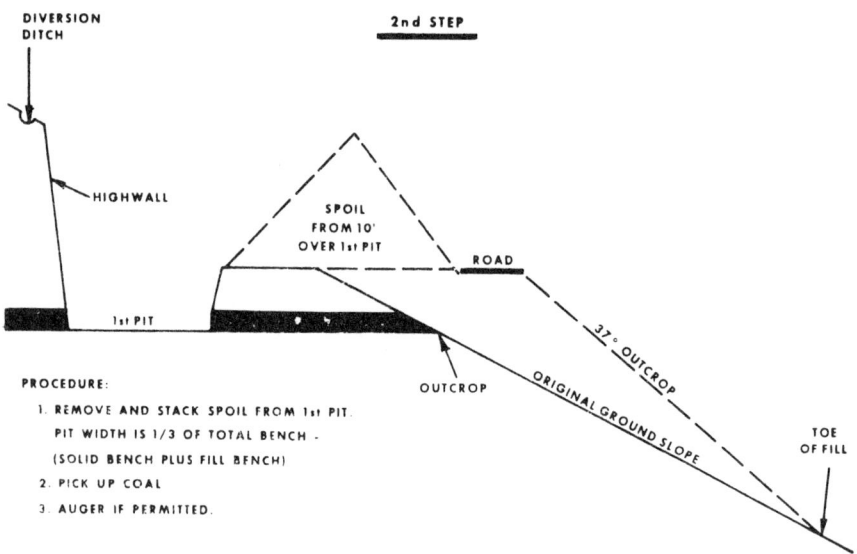

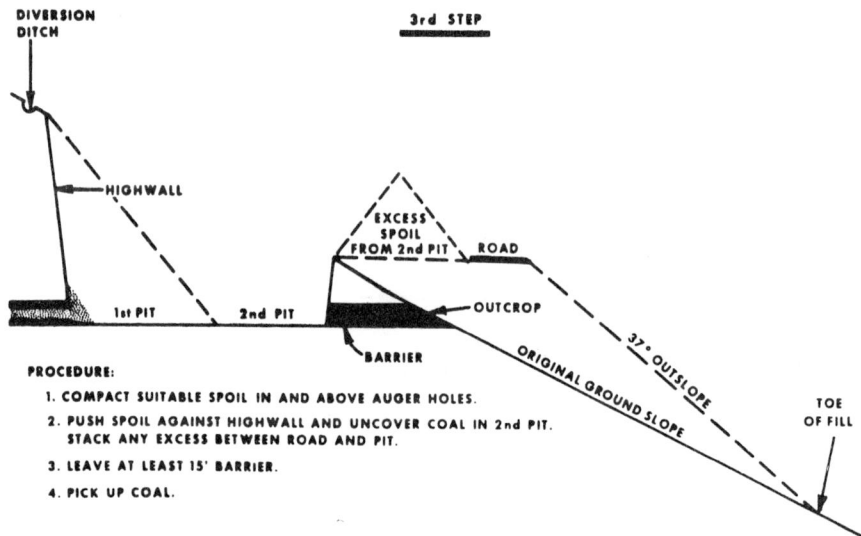

(continued)

FIGURE 1.17: (continued)

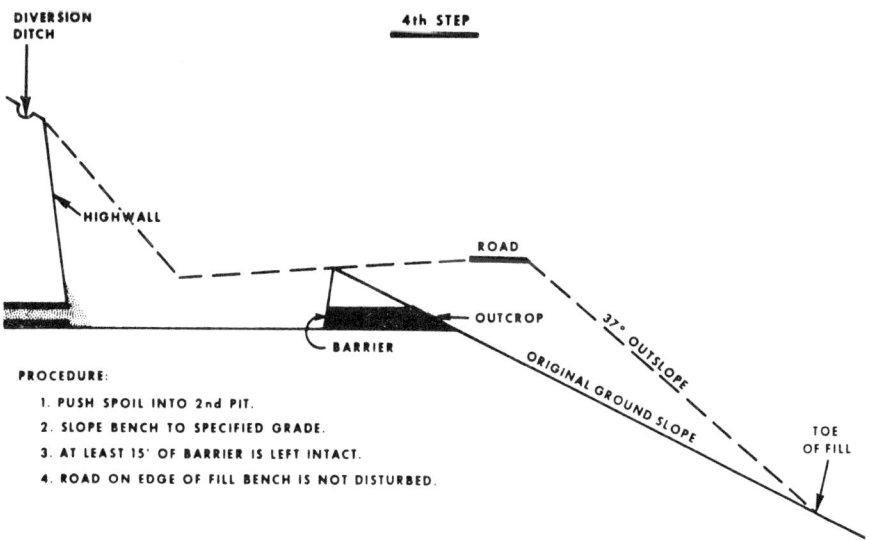

PROCEDURE:
1. PUSH SPOIL INTO 2nd PIT.
2. SLOPE BENCH TO SPECIFIED GRADE.
3. AT LEAST 15' OF BARRIER IS LEFT INTACT.
4. ROAD ON EDGE OF FILL BENCH IS NOT DISTURBED.

Source: PB 238 538

Narrow V-shaped, steep sided valleys that are near the ridge top, and are free of underground mine openings, seeps or wet weather springs are selected for filling. The size of the selected valley must be such that the overburden generated by the mining operation will completely fill the treated head-of-hollow (valley). Procedures to follow (see Figure 1.18).

(1) Scalp the vegetative cover from the area on which the spoil is to be deposited.
(2) Remove and store topsoil.
(3) Build French drains in all natural drainways that have been deepened by bulldozers, forming a continuous chain from the upper end of the valley at the mined bench, down to a point several feet below the toe of the base fill layer. These rock drains will provide for internal drainage of the fill and allow any water to percolate out instead of saturating the spoil and causing slides. The main drainway should be a minimum of 15 feet in width and composed of rock with a minimum dimension of 12 inches.
(4) After internal drainage is provided, the fill is placed in compacted lifts or layers beginning at the toe of the fill. All material is deposited in uniform horizontal layers parallel with the proposed final grade and is compacted with haulage equipment. The thickness of the layers should not exceed the maximum size of the rock used as fill material and in any case not be over four feet. Layering continues until the top of the fill is slightly higher than the established

bench level remaining after the coal has been removed. This slope should be no greater than 3%.

(5) The center of the completed fill is crowned so that drainage will be toward the highwall or bench level adjacent to it and then to a safe outlet away from the toe of the fill.

(6) The face of the fill resembles stair steps progressing from the base layer to the top of the fill. Each layer is a slightly crowned terrace that provides drainage to undisturbed land. The outer slope should be no steeper than 2 horizontal to 1 verticle.

(7) Check dams or silt control structures should be built downstream from the hollow fill.

(8) Revegetation of the hollow fill face should progress as the fill height increases; hydroseeding is a preferred method.

If constructed according to design, stability of the fill can be expected. The horizontal and vertical pressures should provide adequate friction to prevent a failure in the fill. Several head-of-hollow fills have passed through five winters with no slides and little or no erosion. Instead of miles of unstable outslope, with its potential for slides and erosion, or islands of isolated land with no access, a large, stable, fairly level area can be constructed with this method.

Some operators have graded the face of the fill to approximately 22° from the horizontal, eliminating the crowned terraces. By mulching and revegetating immediately after grading, erosion has been held to a minimum. However, it has been found that long slopes must be interrupted with diversion ditches to control surface runoff and excessive erosion. These diversions should be installed at a minimum of every 50 feet in vertical height of the fill. Flat ridges, depressions and old abandoned strip pits that commonly occur in Appalachia, are also used for storing spoil. These areas are particularly useful when starting a new operation.

FIGURE 1.18: HEAD-OF-HOLLOW FILL

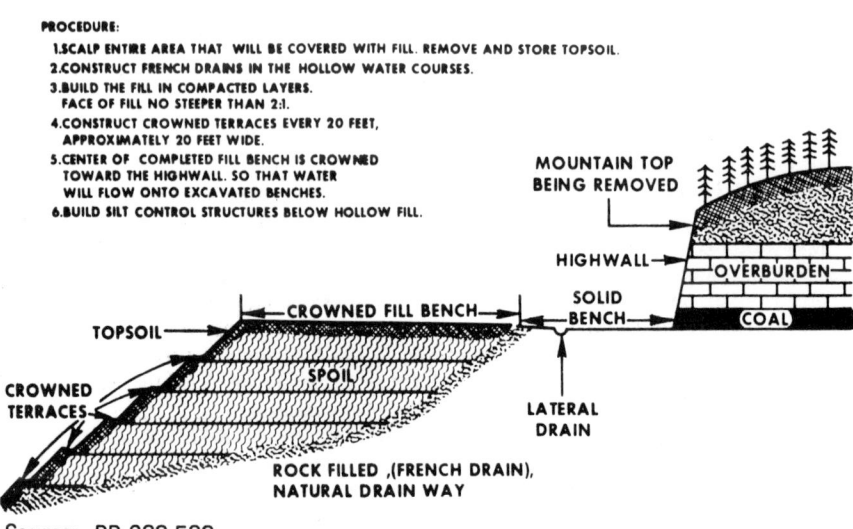

Source: PB 238 538

Multiple-Seam Mining: Recoverable coal seams often lie close together. Multiple-seam mining is the method in which more than one coal seam is strip mined at one time. This method is desirable, as all seams are mined in one systematic operation and it is not necessary to return at a later date and disturb the watershed again.

Method No. 1 — If the overburden from the upper seam will not reach the bench of the lower seam, treat each seam as a separate mining operation, mining the lower seam first. This bench may be used to store spoil produced during stripping of the upper seam.

Method No. 2 — If the overburden from the upper seam will reach the bench of the lower seam, mine the lower seam in advance of the seam above. Grading should be delayed on the lower bench in order to catch big rocks from the upper seam and bury them in the pit. In no instance can spoil from the upper seam extend more than one-half the distance from the highwall to the edge of the solid bench of the lower seam.

Method No. 3 — If both seams appear in the same highwall, separated by more than 25 feet, and two or more cuts are planned, the coal should be recovered from the bottom seam first. If the seams are separated by less than 25 feet, mine from the upper seam down, recovering both seams in one systematic operation (Figure 1.19). Lateral movement of the spoil is recommended.

Mountain-Top Removal Method: The mountain-top removal method of surface mining is an adaptation of area mining to contour mining for rolling to steep terrain. Where coal seams lie near tops of mountains, ridges, knobs, or knolls, they can usually be economically strip mined. The entire tops are removed down to the coal seam in a series of parallel cuts. Excess overburden that cannot be retained on the mined area is transported to head-of-hollow fills, stored on ridges, or placed in natural depressions. This mining method produces large plateaus of level, rolling land that may have great value in mountainous regions.

Cannelton Industries, Incorporated, is mining 2,010 acres, 25 miles from Charleston, West Virginia, using the mountain-top removal method. Overburden averages 110 feet over most of the property but ranges up to 294 feet on the highest point of the ridge. In filling up the voids and leveling off the top of the mountain, Cannelton is creating flat land that at one point contains a straight stretch of 7,000 feet. The potential new land-use area created by this plateau when mining is finished could accommodate a city of no less than twenty thousand people.

Many of the coal seams that lie high on the mountain cannot often be recovered by underground mining. Extreme surface subsidence, unsafe roof conditions, and the narrowness of the coal seams make these coal reserves recoverable only by mountain-top removal. Procedures for using mountain-top removal method:

(1) Select and prepare the hollows that will be used to store excess spoil (see Head-of-Hollow Fill). If ridges and natural depressions are to be used for spoil storage, they must be scalped of all organic matter and topsoil must be removed for later covering of graded areas.

FIGURE 1.19: MULTIPLE SEAM MINING METHOD

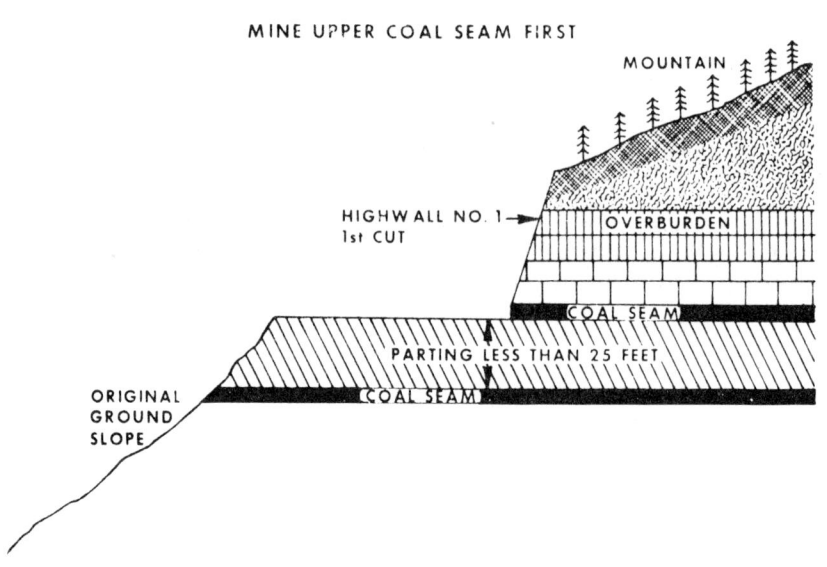

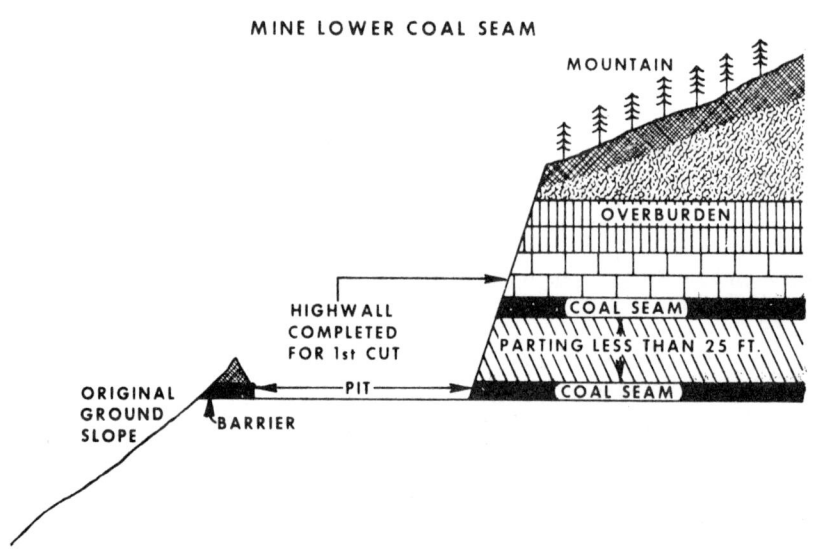

Source: PB 238 538

(2) The first cut is stripped as a box cut, leaving at least a 15 foot barrier of coal bloom undisturbed (Figure 1.20). This cut is made roughly parallel to the ridge. The barrier will serve as a notch to support the toe of the backfilled overburden from successive cuts. Overburden from the first cut is transported to the predetermined storage area.

(3) Once the first cut is completed, a second cut is made parallel to the first (Figure 1.20). However, the overburden from the succeeding cuts is deposited in the cut just previously excavated. The mountain top is thus reduced by a series of cuts parallel to the ridge line (Figure 1.20). Approximately 50% of the overburden would be transported to storage areas for disposal, and none would be pushed over the downslope. The mountain-top removal method can also be used by working around the mountain ridge from one side to the other.

(4) When mining is completed, the mountain top is completely covered with a 20 to 40 foot layer of spoil and is graded nearly flat (Figure 1.21).

(5) At least a 6 inch layer of topsoil is spread over the entire graded area.

FIGURE 1.20: MOUNTAIN-TOP REMOVAL METHOD

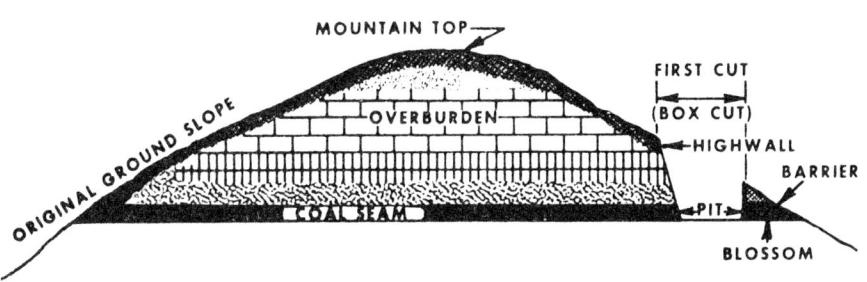

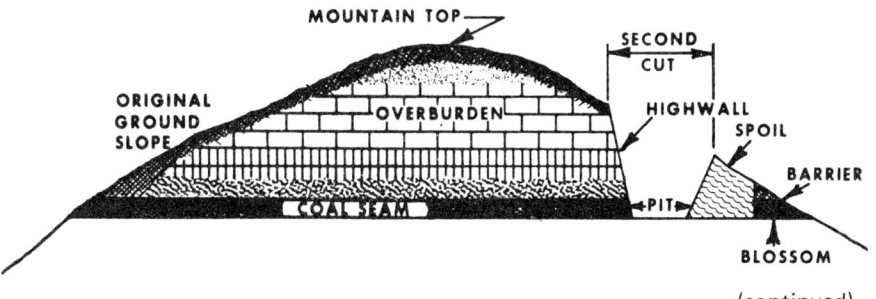

(continued)

FIGURE 1.20: (continued)

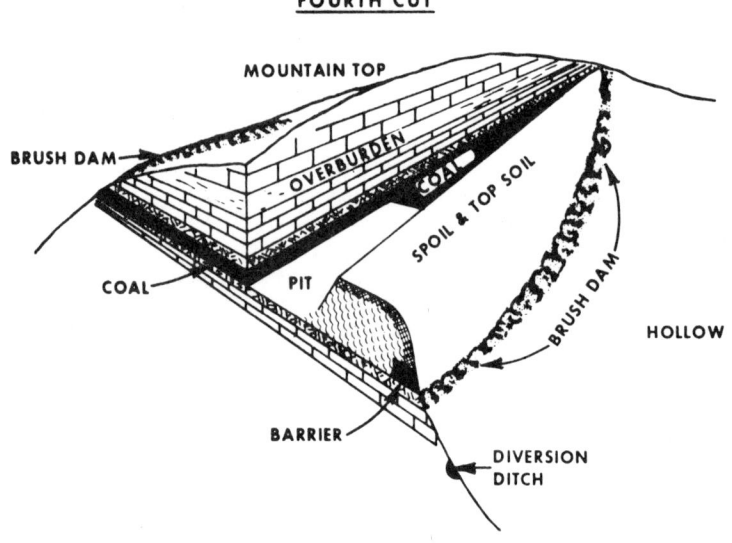

Source: PB 238 538

FIGURE 1.21: MOUNTAIN TOP AFTER FINAL GRADING AND TOPSOILING

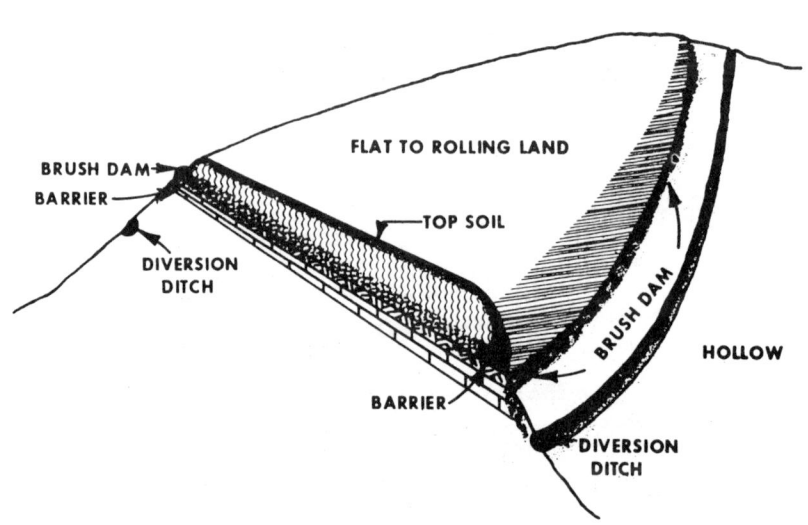

Source: PB 238 538

Surface Mining Land Use and Methods

Benefits and advantages of using the mountain-top removal method have been demonstrated at producing mines in various states and are as follows:

(1) Coal is recovered from areas that would not be mined because they are unsuitable for underground mining. Since all the coal is recovered, the reclaimed area will not be disturbed again by future mining.
(2) The method creates large, flat to rolling areas that are vitally needed in mountainous regions. The end result has an enormous post-mining land use potential when properly completed.
(3) Spoil has been totally eliminated on the downslope. Since no fill bench is produced, landslides are eliminated.
(4) Mined area is completely backfilled and is more acceptable aesthetically, as no highwall is left.
(5) Size of the drainage system is smaller and the number of sediment control structures have been reduced. Erosion is easily controlled because of the low velocity and quantity of surface water runoff.
(6) Overburden is easily segregated, topsoil can be saved, and toxic material can be deeply buried.

Disadvantages of mountain-top removal are:

(1) Detailed topographic maps must be available if proper preplanning is to be accomplished. Before mining begins the final spoil thickness above the bottom of the coal pit must be estimated. If the estimate is low, pits must be narrowed, and sometimes the operation will become spoil bound. The result of underestimating is unnecessary double handling of spoil material, which increases cost and ties up the earth-moving equipment.
(2) Investment costs for spoil haulage equipment are increased.
(3) Special precautions must be taken in scheduling the various phases of mining so as to realize maximum production and eliminate dead time.

Block-Cut Method: The block-cut method (haul back, pit storage, put and take, etc.) is a simple innovation of the conventional contour strip mining method for steep terrain (see Figure 1.22). Instead of casting the overburden from above the coal seam down the hillside, it is hauled back and placed in the pit of the previous cut. The method is not new and is known by various names, depending on the locality. Basically, the operational procedures are similar in that no spoil is deposited on the downslope below the coal seam, topsoil is saved, overburden is removed in blocks and deposited in prior cuts, the outcrop barrier is left intact, and reclamation is integrated with mining (Figures 1.23).

When beginning the mine, a block of overburden is removed down to the coal seam and disposed of (Figure 1.22). This first cut spoil can be placed above the highwall in some instances, or spread along the downslope as in conventional contour mining, or moved laterally and deposited in a head-of-hollow fill or ridge fill. The original cut is made into the hillside to the maximum depth that is to be mined. The width is generally three times that of the following cuts. After the coal is removed, the overburden from the second cut is placed in the first pit and the coal from the second cut is removed. This process is repeated

36 Strip Mining of Coal

FIGURE 1.22: CUT SEQUENCE IN BLOCK-CUT METHOD

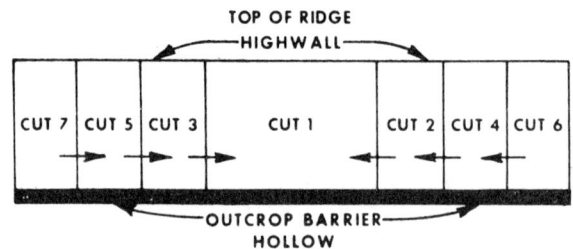

PROCEDURE:
1. SCALP FROM TOP OF HIGHWALL TO OUTCROP BARRIER, REMOVE AND STORE TOPSOIL.
2. REMOVE AND DISPOSE OF OVERBURDEN FROM CUT 1.
3. PICK UP COAL, LEAVING AT LEAST A 15 FOOT UNDISTURBED OUTCROP BARRIER.
4. MAKE SUCCESIVE CUTS AS NUMBERED.
5. OVERBURDEN IS MOVED IN THE DIRECTION, AS SHOWN BY ARROWS, AND PLACED IN THE ADJACENT PIT.
6. COMPLETE BACKFILL AND GRADING TO THE APPROXIMATE ORIGINAL CONTOUR.

Source: PB 238 538

FIGURE 1.23: BLOCK-CUT METHOD BACKFILLING PHASE

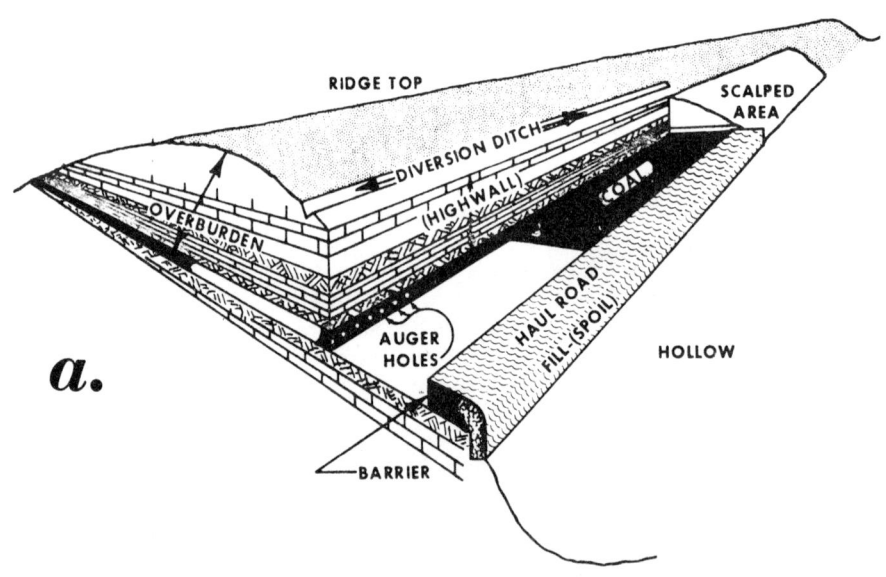

(continued)

FIGURE 1.23: (continued)

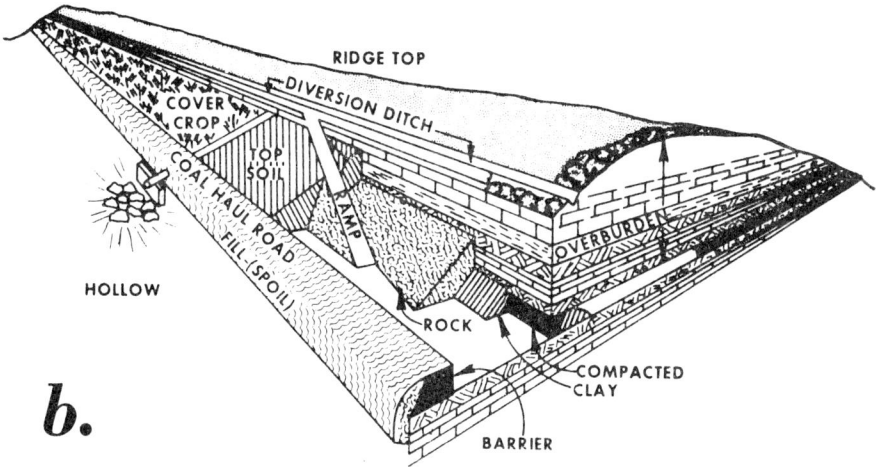

Source: PB 238 538

as mining progresses around the mountain. Once the original cut has been made, mining can be continuous, working in both directions around the hill or in only one direction.

The cuts are mined as units, thereby making it easier to retain the original slope and shape of the mountain after mining. In all cuts, an unmined outcrop barrier is left to serve as a notch to support the toe of the backfilled overburden. Block-cut mining makes it possible to mine on slopes steeper than those being mined at present without the danger of slides and with minimal disturbance. Approximately 60% less total acreage is disturbed than by other mining techniques now in use.

There is significant visual evidence that the block cut method is less damaging than the old practice of shoving overburden down the side of the mountain resulting in permanent scars on the landscape. The treeline below the mined area and above the highwall is preserved. Results of the mining operation generally are hidden and cannot be seen from the valley below. This cosmetic feature is only one of the advantages that contribute to making this an acceptable steep-slope mining method.

Existing or pending state and federal legislation makes it illegal to push overburden beyond the outcrop and over the mountainside and thus bans the conventional type of contour strip mining. However, the block cut or similar methods meet the criteria of this new legislation and allow for recovery of coal reserves in mountainous regions that would otherwise unmineable.

The block-cut method is no longer experimental and is now operational in several states. Enough information is available from active operations to show this method to be potentially feasible from an economic and environmental standpoint. Benefits and advantages of the block-cut method over conventional contour strip mining have been demonstrated at producing mines under varying conditions and are:

(1) Spoil on the downslope is totally eliminated. Since no fill bench is produced, landslides have been eliminated.
(2) Mined area is completely backfilled, and since no highwall is left, the area is aesthetically more pleasing.
(3) Acreage disturbed is approximately 60% less than that disturbed by conventional contour mining.
(4) Reclamation costs are lower, as the overburden is handled only once instead of two or three times.
(5) Slope is not a limiting factor.
(6) The block-cut method is applicable to multiseam mining.
(7) At present, this method does not require the development of new equipment. As new mining technology develops, however, modified or new types of equipment may be needed.
(8) Regular explosives are used, but blasting techniques had to be developed to keep shot material on the permit area.
(9) Bonding amounts and acreage fees have been reduced.
(10) Size of the disturbed area drainage system is smaller.
(11) Size and number of sediment control structures have been reduced. Total life of structure usefulness is increased.
(12) No new safety hazards have been introduced. However, the increased number of pieces of moving equipment in a more confined area may negate this point.
(13) Revegetation costs have been considerably reduced and it is easier to keep the seeding current with the mining. Bond releases are quicker.
(14) AMD siltation, and erosion is significantly reduced and more easily controlled because of concurrent reclamation with mining.
(15) Total amount of coal recovered is equal to that recovered by conventional methods.
(16) Overburden is easily segregated, topsoil can be saved, and toxic materials can be deeply buried.
(17) Equipment, materials, and manpower are concentrated, making for a more efficient operation.
(18) The method allows for early removal of equipment from the operation and placing it back in production at another site.

Disadvantages of the block-cut method are:

(1) Complicated and time-consuming methods of drilling and blasting are necessary to maintain control of the overburden and get proper fragmentation for the particular types of equipment being used in spoil removal.
(2) Economics may limit use of this method; i.e., thin seams of steam coal cannot be recovered profitably if the overburden must be shot.

Surface Mining Land Use and Methods

(3) Special precautions must be taken in scheduling the various phases of mining and reclamation so as to realize the maximum recovery of coal and at the same time eliminate any dead time for equipment.

(4) It is very important that the location of the initial box cut be properly selected. In some areas there will be no place to back haul the material taken at the beginning of the block cut or to dispose of the excess spoil at the end of the operation. Head-of-hollow fill is not always possible, as it can only be done in a restricted set of circumstances.

(5) Long-term environmental consequences are not known and will require a monitor program of a pilot block-cut operation to determine if stream siltation and mineralization can be eliminated.

(6) Investment costs for spoil haulage equipment are increased. Some small mines cannot afford this additional expense.

(7) The block-cut method develops no broad bench that has a high land use potential in mountainous terrain. No access is left for forest firefighting crews, timbering operations, or recreational purposes.

(8) Augering must be conducted concurrently with mining.

Perhaps the most salient feature of block cutting is that the removal of the overburden and the reforming of the original contour by backfilling are integral processes. As a result, the method tends to reduce many of the associated environmental impacts that occur by other methods. This new mining techinque has been accepted as one of the most significant breakthroughs made in contour mining in mountainous terrain.

Auger Mining: As stated previously auger mining is usually associated with contour strip mining to recover additional tonnage after the coal/overburden ratio has become too small to render further contour strip mining economical. When the slope is too steep for contour mining, augering is often performed directly into the hillside from a narrow bench. Augers are also used to recover coal near the outcrop that could not be extracted safely by underground mining.

Augering by itself disturbs less surface area than either contour or area mining, but it poses problems that are more critical, such as very poor resource recovery and providing access to underground mines for the entrance or exit of water. This water may be a prime source of AMD.

Theoretically, augering should recover considerably more of the coal seam than at present. It should be possible to drill longer holes with diameters nearly equal to the thickness of the coal seam, and the openings should be drilled so that little or no coal remains unmined between holes. This theory assumes coal seams of constant thickness and regularity, without undulations; but such is not usually the case.

Present augering equipment drills holes that sag gradually downward, eventually through the bottom of the coal seam into geologic formations beneath. In actual practice, each hole is drilled about 30% undersize to allow for the downward leaning of the hole, sagging caused by bending from the weight of the column of auger steel as it advances into the mountainside. Holes also begin near the top of the coal seam to allow for sagging. In practice, there are additional amounts of coal left unmined between drill holes because holes are often not parallel. For example, when the highwall is not a straight line, which is usually the case, there are pie-shaped blocks of the coal seam untouched by auger holes.

Wherever auger mining has been used to recover coal, the holes must be plugged. The objective is to prevent the flow of water in or out of the holes and to inhibit oxidation of the coal that was not recovered. If suitable material is compacted in each hole to a minimum depth of at least 6 feet, AMD and seepage problems similar to deep mine openings can be eliminated or minimized. In multiseam operations, the auger holes in each seam must be plugged. The exposed face of the coal seam, at the highwall, should be covered with selected backfill material and compacted at least 5 feet above the top of the holes.

Backfilling of all auger areas should be to the approximate original contour, or all highwalls should be reduced to a slope of 35° or less. If the operation is below drainage, a water impoundment may be granted for the final pit as an alternative plan for backfilling.

Minimal Overburden-Moving Mining Methods: With the exception of auger mining, all surface mining methods previously discussed depend on removing massive quantities of overburden to recover the coal. Some underground mining techniques and machinery may possibly be adapted for surface mining of coal at shallow depths. Coal companies are interested in new ideas for extracting coal from a highwall without moving overburden and without sending men underground.

Highwall Mining Method — Highwall mining is an automated variation of an underground mine cutting machine worked through the highwall following the stripping. It has been used only to a limited extent and needs further development to eliminate operational problems. The cutting machines are remotely controlled continuous miners designed to enter highwalls and remove coal up to 1,000 feet in depth at a rate of 3,000 tons per day. New entries are made at predetermined intervals along the outcrop until the end of the property is reached. At the present time, highwall mining using continuous miners is not considered feasible. However, technology has been developed that warrants further research, and chances for success are good.

Longwall Mining Method — Longwall mining is a method of coal recovery that allows the roof to be temporarily held up by jacks and then allowed to subside after the coal has been extracted. This method has been used successfully both in this country and abroad where deep competent cover exists.

The concept of applying underground longwall mining equipment to surface mining under relatively shallow cover was developed by the U.S. Environmental Protection Agency as a possible alternative to conventional strip mining. This shallow-covered coal could be mined without disturbing the overlying vegetation, all the coal could be recovered, and environmental problems such as uncontrolled subsidence and AMD would be greatly reduced. EPA feels that terrain is not a limiting factor, but unconsolidated roof conditions could preclude longwall mining.

Surface Mining Land Use and Methods

The idea is to work the coal cutting and removal equipment from a narrow bench. This equipment would operate back and forth along a wide coal face accompanied by self-moving jacks to prevent the overburden that subsides behind the operation from binding the cutting machine.

The theory of strata control for longwall stripping should be similar to that employed in conventional longwall mining underground. That is, the immediate roof strata above the coal must be supported and allowed to cave in a manner that allows controlled support and caving of the upper strata. The desired sequence of events that will take place as the longwall face advances would be:

(1) The immediate roof is relieved of the load of the upper overburden;
(2) The immediate roof sags away from the stronger, higher strata;
(3) The chocks advance and cause caving to occur with a breaker line formed at the rear of the chocks;
(4) The caved material expands to fill the void in the mined area and the upper roof, forming a span between the gob material and a line where the immediate roof has separated from the upper roof over and near the advancing wall face; and
(5) Most of the roof pressure taken by the solid coal ahead of the advancing face and the gob and the supports merely maintain the relatively light load of the immediate roof.

One requirement the EPA placed on the feasibility study was that standard off-the-shelf longwall equipment had to be used. It is possible that existing equipment will have to be modified or new equipment developed. Potential advantages of the longwall mining are:

(1) Abandoned surface mines can be reopened with little or no additional land disturbance.
(2) Coal that might not otherwise be mined will be recovered.
(3) Longwall mining will work well with other surface mining methods.
(4) Total resource recovery is possible.
(5) The need to overturn the entire earth surface to recover the coal is eliminated.
(6) Landslides are eliminated.
(7) Sediment and erosion problems are substantially reduced.
(8) Filling the voids left by removing the coal will reduce AMD, a major problem of underground mining.
(9) Subsidence can be controlled.

Potential disadvantages of longwall mining are:

(1) Mining method is not perfected.
(2) Expensive modification to existing equipment or development of new equipment may be necessary.
(3) Small operators will probably not be able to afford the cost of longwall mining equipment.
(4) Subsidence could disrupt numerous aquifers and alter underground water patterns.

(5) Subsidence could allow air to contact near surface coal seams creating spontaneous combustion problems. This is especially true in the lignite and subbituminous coal regions.
(6) A soft roof or bottom or a too-strong top that will not cave properly could preclude longwall mining.
(7) Outby control of the highwall is necessary to prevent slides.

LAND RECLAMATION METHODS

The material in this chapter was excerpted from:

PB 238 538

Mining as an extractive process alone is outdated and unacceptable to today's environmentally concerned public. Multiple land use must be considered as well. Only through effective preplanning can the full potential of reclamation result in a lasting asset for future generations.

The coal industry is now recognizing that reclamation of mined lands is part of the mining cycle that must be planned and carried out in a timely and orderly manner. It is also finding that planning reclamation in advance of mining is cheaper and more effective than waiting until mining is completed.

Preplanning involves coordinated efforts by the engineers, operational and management personnel, and reclamation specialists to develop mining and reclamation plans before actual disturbance. These plans are based on detailed studies regarding the physical, chemical, hydrologic, and biologic systems operative at the mining site.

Basic maps of the area are prepared and updated as prospecting information becomes available. The completed maps show the location of access roads, major waterways, sediment control structures, spoil storage areas, boreholes, coal outcrops, property lines, utilities, etc.

The preplan contains mining techniques for spoil segregation and placement, grading, erosion control, and water management practices along with plans for establishing vegetation on all disturbed areas as soon as possible.

It is essential that the geochemistry of the overburden be understood and considered in the preplan or reclamation will be a failure and result in environmental degradation.

BLASTING CONTROL

The goal of blasting is to get maximum fragmentation of the consolidated material in the overburden with optimum drilling and blasting cost. The amount of fragmentation required is determined by the stripping unit to be used in overburden removal. Many coal seams must also be broken by blasting; this is conducted before coal removal. There are environmental factors as well as due regard for public safety, health, and welfare that must be considered in choosing the blasting plan.

The blasting plan should be made during preplanning and is based on data from the overburden cores. Analysis of the data will help in determining the kind of drilling equipment and bit types that will be needed for overburden preparation.

A variety of complaints have always been received by industry pertaining to blasting. Since World War II, the population explosion and urban sprawl have acted to bring industry and the public into closer physical contact. In many cases, structures were built on property adjacent to surface mining operations. As a result, the number of complaints increased drastically and constitute a major problem.

Some complaints registered are legitimate claims of damage from blasting vibrations. The advances in blasting technology and a more knowledgeable blasting profession have minimized real structural damages. However, vibration levels that are completely safe for structures may be annoying and unpleasant for people. Though no actual damage is done, air blast pressures may cause windows to rattle and the loud noise can be intolerable. Repeated vibrations, such as those from a nearby quarry, may eventually cause damage.

Control of vibration damage to natural scenic formations is a very important environmental consideration in surface mining in the West. These wind-eroded formations are very fragile, and damage as far as one-fourth of a mile from the operation have been noted.

Where conventional detonating cord is used to link blastholes, most airborne noise results from these connecting trunk lines. A new, low-energy detonating cord has been developed that can be substituted for the conventional cord. A length of this cord makes about as much noise as one electric blasting cap or 2" of the conventional cord.

If detonating cord is used on the surface, noise can be reduced by covering the trunk lines with up to 10" of dirt. When detonating cord is used only in the holes to fire the primers, a shovelful of dirt at each hole will effectively cover the exposed cord and cap.

Millisecond delays can be used to decrease the vibration level from blasting, because it is the maximum charge weight per delay interval rather than the total charge that determines the resulting amplitude. Also many mines limit the number of holes per shot, using millisecond delays in series to minimize concussion and noise, especially near population centers, natural scenic formations, wells, water impoundments, and stream channels. Weather conditions can cause an increase in airborne noise. During temperature inversions blasting should be avoided.

Land Reclamation Methods

This condition exists frequently in early dawn and after sundown. Foggy, hazy, or smoky days are unfavorable for blasting. When the wind direction is toward residential areas, blasting should be postponed.

When blasting is performed in congested areas or close to a structure, stream, highway, or other installation, the blast should be covered with a mat to prevent fragments from being thrown by the blast. The possibility of dust problems from blasting is very remote.

The possibility does exist, however, and precautions must be taken to control dust pollution if the operation is close to high-use areas. During periods of dry weather, dust from explosions has been carried by air currents for many miles, and in certain isolated instances, it has been a public nuisance.

Several states, including West Virginia, Tennessee, Ohio, Montana, and Kentucky have established guidelines for preventing or holding vibration damages to a minimum. Most of the state laws concerning blasting pertain only to safety, storage, handling, and transportation of explosives.

When a blast is detonated, the bulk of energy is consumed by fragmentation and some permanent displacement of the rock close to the location of the drilled holes containing the explosive. This activity normally occurs within a few tens of feet of the blast hole. Leftover energy is dissipated in the form of waves travelling outward from the blast, either through the ground or through the atmosphere. The ground waves produce oscillations in the soil or rock through which they pass, with the intensity of these oscillations decreasing as distance from the blast increases.

One measurable quantity of interest that is caused by seismic waves or oscillations is particle velocity. This quantity defines how fast a particle (or structure) is moved by passing seismic waves, measured in inches per second. The results of a 10-year study program in blasting seismology by the U.S. Bureau of Mines concluded that particle velocity is more directly related to structural damage than particle displacement or particle acceleration.

It is not how much but how fast the ground under a structure is moved by the passing seismic waves that determines the likelihood of damage. Particle velocity, therefore, becomes the vibration quantity of greatest concern to those engaged in blasting activities.

They also concluded that a safe blasting limit of 2.0 inches per second peak particle velocity as measured from any of three mutually perpendicular directions in the ground adjacent to a structure should not be exceeded if the probability of damage to the structure is to be small (less than 5%). Kentucky now has a law based on seismographic measurements limiting vibrations adjacent to any structure to levels producing a particle velocity of 2.0 inches per second or less.

West Virginia uses the scaled distance formula, $W = (D/50)^2$, for control of vibration damages. W equals the weight in pounds of explosives detonated at any one instant and D equals the distance in feet from the nearest structure, provided that explosive charges are considered to be detonated at one time if their detonation occurs within 8 milliseconds or less of each other for maximum explosive charges. A blasting plan for each method for a typical blast must be submitted with the permit application. Citizen complaints concerning blasting on surface mining operations have been drastically reduced since the 1971 West Virginia law became effective.

Ammonium nitrate-fuel oil (AN/FO) blasting agents and slurries, used as breaking mediums for overburden, have greatly improved the efficiency of surface mine blasting operations and have reduced the cost of explosives considerably. It is an excellent heterogeneous fertilizer, since it contains readily available ammonia nitrogen and nitrate nitrogen and does not leave unfavorable residues in the soil.

A trend is developing for casting overburden with explosives. The goal is to cast as much overburden as possible into the parallel cut with blasting techniques. With proper loading, spacing, and detonation delays, a good portion of overburden can be moved, thus reducing backfilling costs.

This method also minimizes the need for recasting. Some mines report that 30 to 50% of their overburden is moved with explosives. This method works very well in deep narrow pits by casting overburden into the pit away from the highwall and up on the spoil pile on the low wall side.

BACKFILLING, GRADING AND REVEGETATION

Surface mining drastically alters the ecological characteristics of the area disturbed and in some cases has a decided effect on surrounding areas. Vegetation is removed, topographic features and characteristics are changed, and the original geologic overburden profiles are destroyed. Spoil banks generally are a heterogeneous mixture of rock fragments, rock particles, and soil-sized material derived from the overburden strata. With proper mining techniques, the various strata can be partially or completely segregated.

Segregation of overburden material offers the opportunity to bury the toxic, acid, or salt-producing strata under growth supporting material. In some situations, lower strata may have more desirable characteristics than surface material and can be placed near or on the surface. For example, limestone strata appear in some lower overburden profiles that have shale and/or sandstone near the surface. Also, some lower strata have higher nutrient levels that have been leached from the surface.

Experience has shown that natural revegetation is a very slow process on strip mined areas. Native vegetation may not be compatible with the environment on the mined areas for the following reasons:

> Low nutrients in the spoil;
> Toxic spoils (very acid or highly alkaline);
> Surrounding vegetation may be of the climax type and may not have pioneer- or primary-invader-type species present; and
> The seed source may be too far away from the adjacent mined areas.

Early attempts to revegetate strip mined lands with trees also proved to be unsatisfactory as they did not provide the initial ground cover required to stabilize the spoil.

Erosion control with trees only may take up to ten years before the canopies close and an effective cover is established. They are slow to form soil profiles and do not provide effective chemical pollution control until long after planting. A quick growing cover of herbaceous species is necessary to obtain quick stabilization and initial protection against erosion by reducing runoff and raindrop splash.

Vegetative cover will also build up a concentration of organic matter in the soil, which in turn will support high rates of aerobic bacterial activity. Such a layer will remove large amounts of oxygen from the soil atmosphere before it reaches the zone of pyrite oxidization.

A soil cover is important in developing alkalinity and plays an important role in preventing acid mine drainage (AMD). Vegetation also utilizes vast quantities of water in its life processes and transpires it back to the atmosphere, thus reducing the amount of water reaching underlying materials. Therefore, a suitable plant cover will not only control erosion, siltation, and dust, but it will reduce or eliminate acid formation.

Operations with proper preplanning, mining, backfilling and grading should present minimal vegetation problems. Orphan areas, on the other hand, generally have a hostile environment for establishing vegetation, and the problems are more complex.

Revegetation Problems

Spoil characteristics limit the use and treatment of surface mine areas. The main factors associated with the establishment and growth of vegetation on mine spoils have been identified as chemical properties, topographic factors and physical properties.

Chemical Properties: Acidity in mine spoil is due to the presence of sulfuritic material, particularly iron disulfide (FeS_2) in the coal and overburden strata. Either directly or indirectly it is one of the major factors limiting plant survival and growth. Below a pH of 5.0 the solubility of iron, aluminum, manganese, and other elements increases to the point that they may be toxic to plants. Low pH affects the ability of most plants to grow. Spoils below pH 5.0 usually require liming. The rise in pH will reduce the toxic levels of elements in solution and neutralize the acid-producing materials in the surface layer.

Saline and alkali spoils occur in the West when drainage is impeded and surface evaporation is excessive. Various soluble salts, especially calcium, sodium, and magnesium contribute to spoil salinity. The detrimental effects on plants is largely due to the toxicity of excessive sodium and hydroxyl ions in nonsaline or black alkali soils, whereas the concentration of neutral soluble salts (mostly chlorides and sulfates of sodium, calcium and magnesium) interfere with plant growth in saline soils. The latter group of soils usually has a pH below 8.5 because of the influence of the neutral soluble salts, whereas the alkali soils may have a pH as high as 10.

There are three general ways to handle saline and alkali soils to avoid plant injury: eradication, conversion, and control. Eradication is a method used to free soil of part of the excess salts.

Such methods as underdrainage, leaching or flushing and scraping are used. Conversion is the use of gypsum to change the caustic alkali carbonates into sulfates for leaching from the surface soil. Control is usually the retardation of evaporation. Soil mulches are one of the best methods. Frequent light irrigation is another. Salt-tolerant crops also are a useful control.

Topographic Factors: Slope — The length and percent of slope are important factors in erosion control and vegetative establishment. A general rule-of-thumb is that as the percent of slope doubles, soil loss increases 2.6 times, and as the length of slope is doubled, soil loss increases 3.0 times. Thus as the steepness and length of slope increase, the amount of erosion and soil loss increase, making it difficult to revegetate these areas (i.e., steep and/or long outslopes, highwalls, and ungraded spoil banks).

Steepness of slope affects all land uses and machinery operation. A 30% slope is maximum for farm use such as pasture, hay-land or row crops. Precautions must be taken to control erosion on sloping areas regardless of use. Diversions or terraces that break the slope length and remove runoff to a safe outlet also help in preventing stream siltation and damage to adjoining lands.

Aspect — The direction in which a slope faces is known as aspect. Slopes facing north and east are generally cool and moist and are not too difficult to vegetate. Survival and growth on the hotter, drier south- and west-facing sites is generally poor. Temperatures average about 10 to 12 degrees higher on bare slopes having southern or western exposures than on slopes with northern or eastern exposures.

Physical Properties: Physical properties for the most part present less of a revegetation problem than do chemical properties. With proper spoil segregation, placement, and topsoiling, the problem can be minimized. The major problem will occur on orphan lands where the spoil is a mixture of the entire overburden and is usually of coarse texture, stony and will not function to retain water at the surface, as required for a good vegetative cover.

Stoniness — Stoniness affects all land uses, particularly the operation of machinery for tree planting, tillage, and management activities. Mine spoil has been divided into four classes of stoniness (Table 2.1) in order to evaluate the land use potential and treatment needed to stabilize and vegetate them for future use.

TABLE 2.1: MINE SPOILS CLASSIFIED ACCORDING TO STONINESS

Stoniness Class	Criteria	Tillage Potential
1. Nonstony	<0.01% stone and boulders	Can be tilled
2. Stony	0.01 - 15% stones and boulders	Tillage limited, can be mowed for hay, pasture
3. Very stony	15 - 50% stones and boulders	Treat by hand
4. Extremely stony	>50% stones and boulders	Cannot use equipment

Source: PB 238 538

Texture — Texture refers to the particle-size distribution of sand, silt, and clay in spoil; it influences vegetation mainly through its effects on spoil moisture, aeration, and compaction. Sandy-type spoil has good aeration but poor moisture holding capacity. Clay-type spoil compacts easily and in some instances is impervious to water percolation and root penetration. Silty loams are the best spoil material for revegetation and provide very favorable moisture conditions.

Color — Spoil material may vary in color from very light to almost black. Because of these color differences, temperature and spoil moisture may constitute serious problems. Temperatures in excess of 150°F have been recorded on spoil surfaces composed of dark shales. High temperatures are especially intense on south and west exposures and can be deadly to recently germinated or young plants.

Nutrients — Most mine spoils are characterized by a low level or lack of nitrogen, phosphorus, and organic matter. Nutrients such as potassium or other trace elements may also be lacking, but not to the relative frequency of nitrogen and phosphorus.

Revegetation Preplanning

Before a successful revegetation program can be accomplished, there must be proper planning. Design criteria for reclaiming the mined land should become a part of the daily operational plan. The key to a successful reclamation program begins with the basic knowledge of the physical and chemical characteristics of the mineral seam and overburden, which is obtained by core drilling or prospecting with a bulldozer. The borehole data help to determine the proper handling, deposition, and segregation of the various strata in the spoil profile so that undesirable material is buried under clean fill and top soil is returned to the surface as a medium for vegetation.

Seeding Time: Until recently it was thought that the only time to seed and plant was in the spring. This meant that graded spoils would not be seeded for several months, losing the advantage of having a loose seedbed. These bare spoils, after a few rains, would become crusted and hard making it very difficult to establish an effective cover.

Erosion patterns such as rills and gullies will result unless mechanical measures were used to control runoff. However, graded spoil in some locations can be planted in the fall if certain precautions are taken. Perennial grasses often are better in a fall seeding, and legumes in the spring. Many failures are almost assured if species are planted out of season.

The U.S. Forest Service has found that cover can be established during mid-to-late summer by the use of annuals such as pearl millet, sudan X sorghum hybrids, Dorean and Kobe Lespedezas (legumes). These annuals provide only temporary cover, therefore, permanent cover perennial species must be sown either with the annuals or in a separate seeding the following fall or spring. Fall seeding is now commonplace and with the development of guidelines for the use of various summer annuals and perennials, quick cover can also be obtained in the summer.

Grading and Backfilling as it Affects Revegetation: Reclamation should follow closely behind the mining operation. The bare spoil and pit should be reclaimed as fast as possible, because the freshly moved material is easier to grade, handle and plant than older compacted material. In addition, bare spoils and pits are more susceptible to acid formation and erosion. Backfilling and grading should be kept current with the operation. Many state laws have current grading requirements. Revegetation should follow the grading as soon as possible in order to establish a quick protective plant cover on the barren spoil. Grasses and legumes should be planted on all areas.

Trees may be planted in combination with grasses, but not alone, as they require excessive periods of time to be effective for erosion control. Soil samples should be taken to determine the limestone and fertilizer requirements. Where possible, the ground must be loosened, the fertilizer and limestone worked in, and the grass seed planted. On steep slopes, blowers, hydraulic seeders, airplanes, and helicopters can be used to seed the area. A mulch may also have to be used. The type of grasses, legumes, and trees to be planted will depend on local conditions and long term use of the land.

Most enforcement agencies have the power to approve alternate plans of restoration, i.e., water impoundments in the final cut. High quality water impoundments are usually encouraged and such factors as the pH, temperature, dissolved oxygen, and mineral salts are considered in the determination of the water quality.

Major uses include water supply for domestic purposes, livestock, wildlife, fire protection, recreation, and irrigation. In some cases, a surface mining operation can provide a suitable site for developing a sanitary landfill. Under all circumstances, a water impoundment or sanitary landfill must be planned and constructed according to established standards.

The question of how much grading and backfilling should be performed on strip mined lands is very controversial. Although many of the problems are similar for contour and area mining, there are major differences. They are discussed separately below.

Contour Mining — Highwalls are the dominant physical feature of contour mining on steep slopes. They represent less than 15% of the total horizontal disturbed area. After many years of weathering, some highwalls will be reduced and covered with volunteer vegetation. However, in most cases, these scars will be there for many generations. They do not blend in with adjacent land and are considered by some people to be an aesthetic blight.

For many years backfilling was accomplished by simply pushing dirt into the pit, as shown in Figure 2.1. This technique proved to be unacceptable because of erosion and mine drainage problems, etc. Three types of backfills were then developed for contour strip mine benches: contour, pasture, and Georgia V ditch or swallow-tail. These basic types can be used singly, in combination, or modified to meet local conditions.

For contour backfill, the edge of the highwall is knocked off, and the spoil is graded back toward the highwall to approximate the original contour (Figure 2.2). This method is the closest approach to returning the area to its original topography and produces the most pleasing aesthetic effect. Contour backfills are preferred wherever possible. However, because it costs more than other types of backfilling, it is practiced only in states that have laws requiring its use.

Land Reclamation Methods

FIGURE 2.1: COMMON BACKFILLING PRACTICE

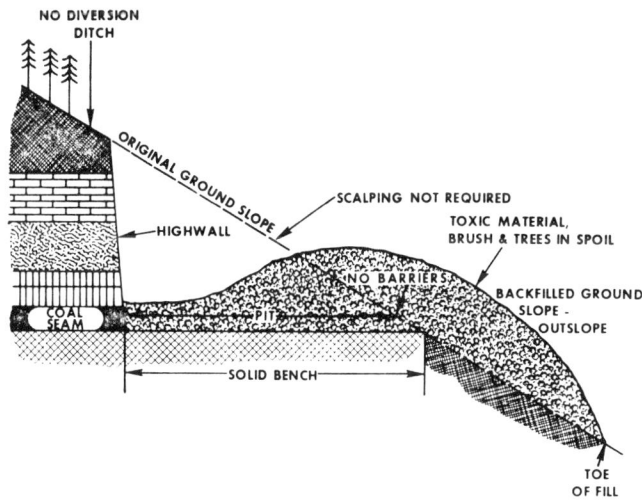

FIGURE 2.2: TYPICAL CONTOUR BACKFILL

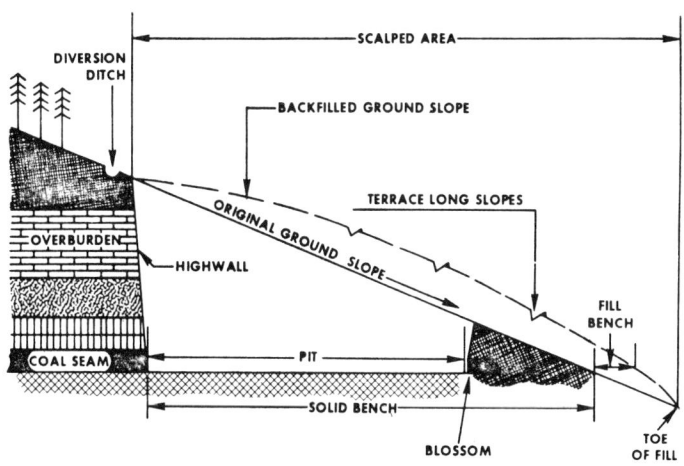

Source: PB 238 538

Long steep slopes that are subject to erosion can be formed if proper controls are not taken. The erosion problem can be solved by completing final grading across the slope and/or by the construction of a diversion ditch at the top of the slope and a series of terraces, diversions, or ridges across the slope. Each of these drainage control measures should have a constructed discharge outlet.

Rapid vegetation with grasses and legumes is critical for this type of backfill. By grading to the approximate original slope and reducing all highwalls, pollution attributable to exposed highwalls would be eliminated. Some of the major problems associated with exposed highwalls have been defined as follows:

> The area poses a safety hazard to people and animals.
> Areas of land above highwalls are inaccessible.
> Weathering causes sloughing that blocks drain ways.
> The area is amenable to fewer uses than before mining.
> Social and economic impacts are greater.
> Salts are dissolved by rainfall from the exposed highwall and
>   are then carried by runoff water, as pollutants, into tributaries.

When the highwall is reduced and covered, approximately 30% additional area above the highwall is disturbed in order to obtain the necessary fill material, unless the block cut method of mining is used. A modification of the contour backfill is the terrace backfill (Figure 2.3). The highwall is reduced and a terrace is formed. Precautions must be taken to control the velocity of surface runoff or excessive erosion may result.

Pasture backfilling calls for the grading of the spoil to cover the pit and any acid producing strata, but not the entire highwall (Figure 2.4).

FIGURE 2.3: TYPICAL TERRACE BACKFILL

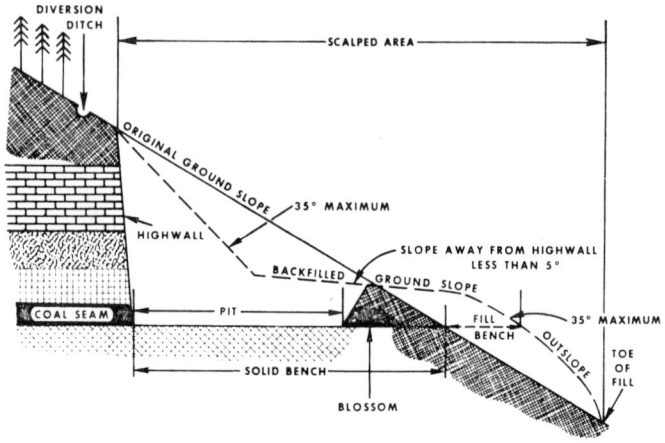

Source: PB 238 538

FIGURE 2.4: TYPICAL PASTURE BACKFILL

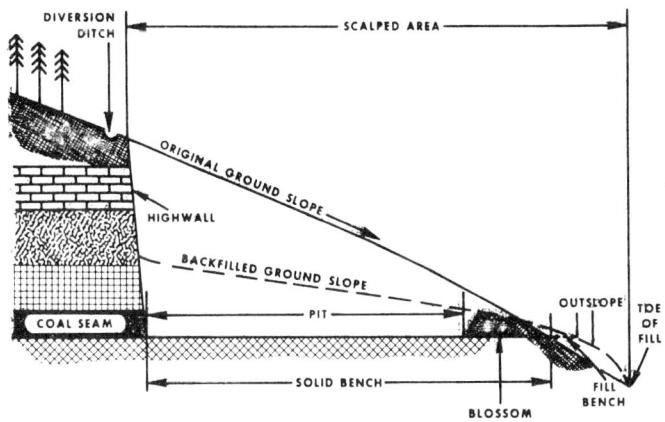

Source: PB 238 538

The slope of the graded spoil should be away from the highwall, and the slope of the outslope should be reduced to control the water running off the bench. Diversion ditches must be constructed along the top of the highwall to reduce the water entering the pit. Long slopes should be interrupted with terraces to control runoff and reduce erosion. This type of backfill is used to eliminate percolation of surface runoff water in areas that have been underground mined and/or augered.

Several states require that the slope of the graded spoil must be toward the highwall. A Georgia V ditch or swallow-tail backfill has proven to be the most satisfactory method (Figure 2.5). The drainage ditch is constructed on the solid bench parallel to the highwall and of sufficient distance from the highwall to assure that any material falling off it will not obstruct the drainage way. The ditch should be laid on a nonerosive grade, and in some cases it must be lined to prevent excessive erosion of the channel. The ditch should carry the runoff to a properly designed discharge structure. If highwalls are not reduced and the benches are properly reclaimed, they can provide land conducive for:

    Pasture development

    Access roads or trails that can be used as:
      Forest fire breaks
      Entrance to remote areas for forest fire control crews
      Logging activities
      Recreation such as horseback riding, hiking, camping, hunting and fishing

    Openings for wildlife (including food, cover and water)

    Housing and industrial sites.

FIGURE 2.5: TYPICAL GEORGIA V DITCH (SWALLOW-TAIL) BACKFILL

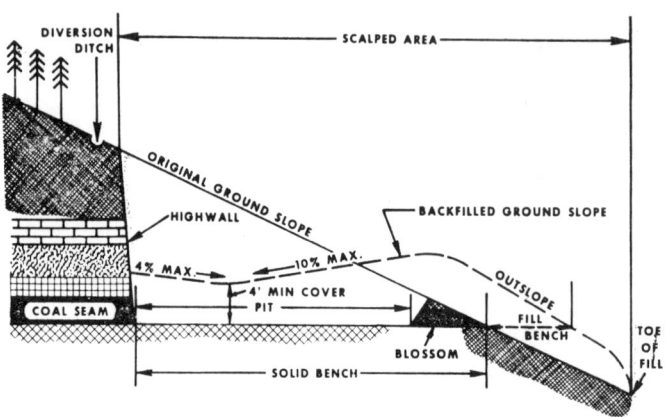

Source: PB 238 538

Area Mining — On lands where the method of operation does not produce a bench (area strip mining), complete backfilling and grading to the approximate original contour is generally required.

Backfilling and grading to the approximate original has been interpreted to mean that lands are considered to be completely backfilled and graded when the contour of the land conforms approximately to the contour of the original ground. However, the final surface of the restored area need not necessarily have the exact elevations of the original ground surface.

A flat surface or a surface with less slope than the original ground surface is also considered to comply with backfilling and grading to the approximate original contour requirement. Some of the points stressed for spoil segregation, placement, topsoiling, and complete grading are:

> Land can be made productive more easily and quickly by being able to use conventional farm implements and mechanical reclamation equipment for restoration purposes.
>
> Toxic and undesirable materials are buried and should not cause future pollution problems.
>
> Light-colored materials can be put on the surface to help decrease spoil temperatures and evaporation. However, the dark-colored A horizons are generally preferred when available.
>
> Topsoiling provides proper techniques to assure rapid establishment and growth of suitable vegetation that plays a significant role in erosion control.
>
> Compaction can be controlled by large discs, subsoilers or rippers that will break up the surface for seedbed preparation, provide a rooting zone, and permit the penetration of moisture.
>
> Management of the graded areas is easier, cheaper and more profitable.

Land Reclamation Methods

Grading is easier and cheaper because the rock present in the overburden is buried and covered by segregation of materials during mining.

Grading to the approximate original slope is more pleasing to the eye, and almost any vegetation, properly selected and planted, will take care of the aesthetic problem.

No one can predict, with any degree of certainty, the land use requirements of future generations. However, if the disturbed area is graded back to the approximate original slope, with adequate safeguards to control runoff and erosion, it will be available when needed without additional treatment.

Mechanical Spoil Manipulation: Basically, spoil manipulation is the grading and shaping of the mined area to produce as many flat surfaces with short slopes as possible, and at the same time to leave the spoil surface in a rough or furrowed condition on the contour. The resulting topography will increase rainfall retention, infiltration and percolation, increase leaching of spoil salts, and reduce runoff and erosion.

By increasing the infiltration rate, spoil moisture for vegetation is available during dry periods. Smooth, compacted surfaces with long slopes should be avoided, as they contribute to excessive runoff, severe erosion, and sedimentation; very often, the seed and fertilizer are washed off.

Terraces can effectively control runoff, erosion, and sedimentation and conserve moisture for vegetation. Curtis (1), using a Rome bedding harrow to mechanically form terraces on a strip mine bench in Breathitt County, Kentucky, found that the peak flows on a terraced plot averaged 65% less than on the control plot, sediment yield averaged 52% less, and total runoff averaged 42% less.

A visual comparison indicated much better vegetative cover on the terrace plots. This was attributed primarily to effective seedbed preparation during terrace formation and increased moisture retention. Used in conjunction with the Rome bedding harrow and with ridges 8 feet apart, a medium-sized dozer can cover about 4 acres per hour.

Water-retarding basins have been used with varying degrees of success on strip mine benches. These basins are shallow depressions made by a bulldozer or highlift to trap or slow down runoff water on the bench. Silt is deposited in the basin and is thusly prevented from reaching receiving streams.

Trapped water evaporates and infiltrates into the spoil providing moisture for plant growth. Location of these basins is very important, as they could cause AMD problems if they are constructed on the high side of the coal seam and water percolates into underground mines.

Furrow grading is the result of successive parallel passes by a bulldozer, with spoil spilling from the end of the angled blade. The furrows are generally on the contour and range from 2 to 3 feet in height and are 3 to 4 feet between peaks. Furrow-graded and conventional smooth-graded spoils were comparatively studied for a ten year period [1962 to 1972 by Riley (2)]. Objectives of the research were to determine: the nature of chemical changes occurring to furrow-graded spoil materials; the effect of increased rainfall retention on leaching of of soluble salts, sulfates, and certain metals from the spoil surface; and the survival and growth of selected plant species as an indicator of site improvement.

The data clearly reflect the beneficial effect of an increased rate of leaching of chemical components of spoil materials as the result of increased rainfall retention and infiltration on the furrowed spoil surface. As a result of increased leaching of soluble salts, sulfates, and other chemicals inimical to plants, site improvement was reflected by better plant survival, growth, and reproduction.

Three mechanical spoil manipulation treatments were implemented by Sindelar et al (3) in a semiarid region of Montana. Each of the mechanical treatments was designed to retard surface flow and simultaneously increase infiltration into the spoils. In addition, the treatments should result in an improved seedbed for broadcast seeding. The three treatments are as follows:

(1) Gouging — Gouging is a surface configuration composed of many depressions and is accomplished with a specially constructed machine that has hydraulically operated, 25" diameter disc scoops that alternately raise and lower while being drawn by a tractor. The three disc scoops create elongated basins on the contour approximately 14" to 16" wide, 3' to 4' long and 6" to 8" deep. This pattern is amenable to gradual slopes and flat areas. It creates a cloddy seedbed ideal for broadcast seeding.

(2) Dozer Basins — Dozer basins are large depressions designed to accomplish what terracing is intended to do, but without the characteristic precision, hazards, and expense. Dozer basins are 15' to 20' long and are formed by dropping the bulldozer blade at an angle at intervals and bulldozing on the contour. The resulting basins are approximately 20' to 25' from center to center and are about 3' to 4' in depth. Basins are constructed in parallel rows with about 20' between rows. Precipitation intercepted within each mine drainage accumulates in the basin limits. The increased soil moisture availability assures the establishment of a nucleus stand of vegetation the first growing season; from this nucleus, it can spread between basins to make a complete cover.

(3) Deep Chiseling — Deep chiseling is a surface treatment that loosens compacted spoils for a depth of 6" to 8". The process creates a series of parallel slots on the contour to effectively impede water flow and increase the infiltration rate. Deep chiseling uses a modified Graham-Hoeme plow with 12 chisels to form a rough cloddy seedbed. This treatment is effective on relatively flat slopes and is very beneficial in loosening spoil before gouging or following dozer basin construction.

Performance of each treatment was evaluated and results indicated that gouging stored more water in the upper 4' of spoil, greatly reduced moisture stress days, and had better plant survival than chiseling or dozer basins. Erosion damage appeared greater on chiseled plots.

Topsoiling: Topsoil is the unconsolidated mineral matter naturally present on the surface of the earth that has been subjected to and influenced by genetic and environmental factors of parent material, climate, macro- and microorganisms, and topography, all acting over a period of time and is beneficial for the growth and regeneration of vegetation on the earth's surface. Several states require topsoiling as part of their strip mine reclamation program. Topsoiling is the process of removing a separate, developed layer of desirable soil material from areas to be mined and keeping it in such a condition that it will not deteriorate until it is returned as the top layer, after the operation has been backfilled and graded.

Of all the natural resources that support life, soil is possibly the most important. A mature soil system takes thousands of years to develop and is as complex and integrated as the plant community that develops on it. Life in the soil depends on having a good supply of organic matter readily available. Organic matter is the main supporter of soil microorganisms, which are necessary for soil development.

Despite the rejuvenating ability of soil and its potential as a growth media, operators have shown a remarkable reluctance to practicing overburden segregation during mining and returning the topsoil as the final act of backfilling. Cost has been given as the reason. In most cases, the additional costs are insignificant when one considers the cost of additional lime and fertilizer for making subsoil suitable to plant establishment, replanting bare areas, and continual maintenance of revegetated areas.

The United States Department of Agriculture, Agricultural Research Service, Mandan, North Dakota report that the topsoil is very important in the surface mining reclamation process. Their research at Mandan and elsewhere has shown that in most instances revegetation is doomed to failure if this productive part of the profile is not conserved and replaced in sufficient quantity on the surface as the medium for revegetating the land.

The underlying materials are sterile. They often contain excess salts and toxic elements (heavy metals, etc.). They are massive in structure and will not take water. The difficulties and costs of revegetation are increased several orders of magnitude if these underlying materials end up on the surface.

The directors of reclamation for the states of West Virginia and Pennsylvania attribute the success of their restoration program mainly to the topsoiling techniques that are required and being practiced on all operations.

During mining, when strata in the overburden are found to be toxic or limiting to plant growth, they must be effectively isolated from the root zone in the established spoil profile. If topsoil is placed directly on toxic overburden, the topsoil may become polluted by upward movement of harmful salts. There must be a layer of clean fill between the root zone and the underlying toxic or undesirable materials so that plant establishment and growth will not be impeded.

On all slopes that will be covered with topsoil, it is essential that the topsoil be firmly bonded to the existing spoil surface to prevent slippage downslope. This bonding can be increased by scarification of the slope before topsoiling.

In area mining, topsoiling has been accomplished with draglines, bucket wheel excavators and scrapers. The topsoil is usually placed on top of the adjacent cut spoil during the mining operation to prevent rehandling. In contour mining, the topsoil is often stockpiled and replaced after grading (except in block cut mining, where topsoil is removed and placed on graded areas in one operation). Topsoil can only be stockpiled for a limited time or it will lose its ability to enhance vegetative growth. Topsoil should be saved, even if there is only limited amounts available.

Mulches: Various mulches such as straw, hay, wood chips, and shredded bark have been used successfully as aides in establishing vegetation on graded, surface mined spoils. They provide insulation against intense solar energy thus lowering ground temperatures. Evaporation rates are reduced, thereby minimizing the accumulation of toxic salts on the surface.

A more favorable moisture supply is assured in the growth media. Mulches also break down after ground cover establishment, supplying valuable organic matter to the soil, which in turn promotes microorganism buildup. On steep slopes and highly erodable material, mulches will reduce raindrop impact, help control erosion, impede the flow of runoff water, and hold seed in place.

Straw and hay mulch can be applied by hand on small plots and by mulch-blowing equipment on larger areas. It is applied at rates of 1.0 to 2.0 tons per acre. Straw and hay mulch should be tacked to insure against excessive losses by wind and water. Liquid and emulsified asphalt are the most commonly used mulch tack. However, anchoring of straw and hay mulch by mechanical equipment is used quite extensively.

A mulch-anchoring tool is composed of a series of notched discs that punch the mulch into the spoil. This method not only anchors the mulch, but it also incorporates organic matter into the spoil and increases infiltration rates. The spoil should be loose to permit disc penetration to a depth of 2 to 3 inches and should be used on the contour for erosion control.

Other methods of anchoring that are simple and have been proven to be effective are: pushing the straw or hay into the ground with a shovel at approximately 12-inch intervals; and placing shovelsful of earth on top of the mulch at about 24-inch spacings.

Wood chips are produced by processing tree trunks, limbs, branches, etc., in wood-chipping machines. As a mulching product on newly seeded areas, wood chips may be applied by hand on small plots and by mulch blowing machines on larger areas. Application rates of 60 to 100 cubic yards per acre are recommended. Mulching with wood chips has proven successful when used with late fall seeding that requires protection over winter.

Wood chips use nitrogen in their decaying processes, and as a result, 20 pounds of nitrogen per ton of wood chips should be added to the spoil. This nitrogen is in addition to that required for spoil fertilization.

Normally, vegetation on areas to be mined is removed and burned, or covered with spoil. This is a waste of a natural resource that can be recycled back to the soil as a wood chip mulch for plant establishment and a source of organic matter when it decays. Shredded bark can be used in much the same way as wood chips, but bark does not require nitrogen in its decaying processes because of the absence of cellulose. Other than this difference, it is similar in its properties to wood chips.

Land Reclamation Methods

Small grains and annuals will provide quick, temporary stabilization until permanent cover is established, produce food and cover for wildlife, add organic matter to the soil in the form of roots, and leave considerable surface mulch. In the past, the use of mulches, small grains, and annuals as spoil amendments has been largely ignored. However, many states now recognize their importance and are requiring that they be used in the revegetation phase for reclaiming strip mined lands.

Spoil Amendments

If the surface mining operation has been properly preplanned and reclamation procedures incorporated into the mining method before disturbance, then acid conditions that will effect revegetation should not be a major problem. The goal is to prevent acid conditions from developing, rather than to correct the problem after it has been created. Acidity of the spoil material is one of the most important factors limiting establishment and growth of plants on many strip mine areas.

Limestone: Neutralization with agricultural grade limestone is the most common method of treating acid spoils. Liming, by increasing the pH to a minimum of 5.5, will also eliminate toxic concentrations of iron, aluminum, manganese, and other elements in solution. At the pH level of 5.5 and above, the biological reactions that form surface mine acid are inhibited. The agricultural grade limestone should contain sufficient calcium and magnesium to be equivalent to not less than 80% calcium carbonate.

Lime should be applied and worked into the spoil as far in advance of the seeding as possible. This will allow time for it to react with the spoil and to be deep enough to be available for plant use. The amount of lime required per acre is generally not excessive. However, over a period of time, additional lime may have to be applied so as to maintain a good plant growth. Lime requirements are based on the results of soil tests for acidity rather than on pH.

Grube et al (4) has developed techniques for determining lime requirements for strip mine spoils. Methods used by most soil testing laboratories are not suitable for mine spoils. The operator can receive guidance from the local soil and water conservation district, the county agent, or the university extension service.

Fertilizer: Soil analyses of spoil banks generally show an insufficient supply of nutrients for plant establishment and growth. Nitrogen and phosphorus are the nutrients most commonly deficient.

Davis (5) reports that plant establishment and subsequent growth on spoil banks in eastern Kentucky is enhanced by fertilizer applications. In no cases were fertilizer applications detrimental to plants. Nitrogen is nearly always low in spoil materials. Often, however, there is no response to nitrogen applications until phosphorus is also applied.

Supplementing phosphorus with nitrogen usually produces an additional growth response on most spoils. Though the rates of fertilizer to apply will vary from spoil to spoil, fertilizer applications of nitrogen and phosphorus are recommended for all spoil banks. Fifty-six kilograms of phosphate (P_2O_5) per hectare and 112 kilograms per hectare of nitrogen (ammonium nitrate) applied at the time of seeding are usually helpful in establishing initial plant cover. Sindelar et al (3) report on the necessity of fertilizing Montana mine spoils with both nitrogen and phosphorus. The absence of either nutrient limited production in spite of the concentration of the other nutrient. The optimum rate of fertilization for winter wheat at this site is a combination of 84 kg of available nitrogen per hectare and 112 kg of available phosphorus per hectare. Higher rates of fertilizer did not produce significantly greater plant response.

Bengtson et al (6) reports that results of studying one-year-old loblolly pine seedlings on strip mine spoil in northeastern Alabama show that fertilization is necessary to get maximum survival per acre with maximum height growth. They also found that both phosphorus and nitrogen were insufficient to support vigorous growth of the planted pine.

Applications of phosphorus alone stimulated tree growth somewhat, but maximum growth was attained with application of both nitrogen and phosphorus at 112 kilograms of each element per hectare. For fertilizing strip mines, Kentucky requires a minimum of 68 kilograms of available nitrogen per hectare and 112 kilograms of phosphate per hectare.

Despite the fact that fertilizer application of nitrogen and phosphorus has proven to be necessary for the establishment and growth of plants on spoil banks, this practice is still meeting resistance in some areas. Bennett (7) states that if the fertility and management needs of a particular species are met, almost any grass species can be grown on strip mined areas.

Normally, the fertilizer should be applied at the same time seeding is done, or within a few days following seeding. One exception would be during the dormant season (winter), where a seeding is made with the intention that the seed will not germinate until spring. In such a case it would then be better to wait and apply the fertilizer at about the time the seed is expected to germinate.

If the fertilizer is mixed with seed in a hydroseeder, the mixture should not be allowed to sit for more than a few hours, for it is possible that the salt solution formed from water and fertilizer could damage the seeds, especially grass seed.

Another problem that could result from mixing fertilizer and seed in a hydroseeder is a reduction in the effectiveness of the inoculating bacteria for legumes. Inoculation of legume seed increases their chance of success by insuring the presence of needed nitrogen-fixing bacteria. The spoil found on most contour mines lacks sufficient bacteria to naturally supply legumes, therefore, seeds must be coated with the bacteria cultures.

If the pH of the fertilizer slurry is below 5.0, most of the inoculation bacteria will be killed within 30 minutes. If the pH is above 5.0, about 25% of the bacteria will still be viable after about 2 hours. Therefore, the slurry should be kept at pH above 5.0 and spread as soon as possible after mixing.

Fertilizer can be applied dry with cyclone spreaders, aircraft, and lime-spreading trucks. It is also possible to dry mix seed with fertilizer and spread both together. However, some separation of seed from fertilizer could result if the mix were hauled for a long distance over rough roads. If the seed and fertilizer are dry mixed together before application, the fertilizer should not be allowed to get damp.

The high concentration of salts going into solution could quickly damage the seed.

Fly Ash: Fly ash is a powdery residue product when coal is pulverized and burned in boilers for electricity generation. Most fly ashes are mildly to moderately alkaline so that fly ash can substitute for limestone in strip mine spoils neutralization. Lignite ashes are characterized by a Ca-Mg content and are relatively high in neutralizing power, i.e., 5 tons of lignite fly ash are equivalent to 1 ton of $CaCO_3$, while the neutralizing power of bituminous coal fly ash ranges from 15 to 200 tons of ash equivalent to 1 ton of $CaCO_3$.

Thus, more fly ash is required to perform the same neutralization level as limestone and the surface mine operator, if he is to assume the full cost burden, will choose ash only if it is competitive or he requires it for its other properties.

Mixing large quantities of fly ash with spoil also effects physical changes that enhance plant survival and growth. The decreased bulk density of the mixture increases the pore volume, the moisture availability, and the air capacity, thus improving conditions for root penetration and growth.

Where vegetation may be difficult to establish on some surface mine spoils because of nutrient deficiencies, unfavorable moisture regimes, acidity, excessive salts, toxic substances, and poor physical condition, the application of fly ash as an ameliorating material to modify or correct these factors offers an attractive opportunity to improve the spoil and establish a good cover.

There is further potential for utilizing fly ash in reclamation by employing the backhaul concept wherein trucks that haul coal to the power plant deliver fly ash to the surface mine area on the return trip. In many cases fly ash can be obtained for a nominal loading charge and the cost of transportation from the plant to the site.

A mutually beneficial arrangement between the coal operator and the power company provides the operator on the one hand with a material that aids in reclamation and on the other hand gives the power company the opportunity to usefully dispose of a troublesome waste product. The application of the backhaul concept to other waste products, such as sewage sludge, cement kiln dusts, feedlot manure, composts, etc., as well as fly ash, presents additional opportunities for recycling wastes at reduced costs compared to the straight haul charge.

Results of greenhouse and field studies indicated that application of selected fly ash samples to soils either completely or partially corrected boron, molybdenum, phosphorus, potassium and zinc deficiencies in plants. However, detrimental effects on plant growth including boron toxicity, soluble salt damage, and nutrient deficiencies due to increases in soil pH, are possible when higher than optimum amounts of fly ash are applied.

Experiments by the Central Electricity Generating Board in England on pulverized fuel ash utilization and its effect on plant growth showed that boron was the major plant toxin in the ash. Bennett (7) also found that when quantities of certain fly ashes (90.7 to 181.4 metric tons) are used on strip mine spoils severe toxicity symptoms on plants appeared, possibly from soluble boron. This condition can be overcome by using boron tolerant plants or the vegetation can be cut and hauled off and the area plowed. This procedure removes a considerable amount of boron and will improve the fertility of the ash. Old ash that has been weathered does not have this problem. In any case, grazing animals should not be given the hay or forage until it is tested and declared safe to use.

All fly ash is deficient in nitrogen and this element must be supplied by the use of fertilizer. Nitrogen deficiency can be made up in part by the use of certain legumes, such as bird's-foot trefoil, sweet clover and crown vetch. Organic matter in the form of sewage sludge has been used with toxic fly ash and is of great value in establishing normal soil populations on the treated areas.

Research by the Morgantown Energy Research Center of the U.S. Bureau of Mines has proved the technical feasibility of reclaiming acid-surface mine spoil using fly ash. On a site absent of natural vegetation, fly ash was spread at rates of about 336 metric tons per hectare on bench and slope. Heavier application of fly ash was placed against the highwall. Fertilization and seeding was carried out according to the research plan. Although farm equipment was used for part of the study, the most efficient spreading and mixing of fly ash and spoil was by large earth moving machines.

Fly ash analyses show that the power-plant waste contains many of the trace elements essential for plant growth, hence, the material should be useful as a fertilizer to correct nutritional deficiencies. Plants require considerable quantities of P, K, Ca, Mg, and N, for example, and lesser or even trace amounts of Mn, Fe, Mo, Cu, Zn, and B.

Cost with fly ash depends on several variables, including the terrain, soil type and age, acreage, equipment used, legislative requirements, and degree of reclamation. Based on experience at 65 acre test site (Stewartstown), the cost of vegetating areas devoid of growth with grasses is estimated at about $300 per acre ($741 per hectare).

Fly ash application rates for use in the field were determined empirically in the laboratory by measuring the pH of equilibrated soil-water mixtures on a 1:1 by weight basis. A rule of thumb for fly ash application is that 1 inch cover of ash equals 100 tons per acre.

In U.S. Bureau of Mines studies the fly ash treatment has been effective in increasing pH, enhancing water-holding capacity, and improving soil texture. The grass and hay yields produced on fly ash treated spoils were comparable to average values for West Virginia. Rye grass, red top grass, orchard grass, Kentucky 31 fescue, and bird's-foot trefoil showed good promise.

Strip mine spoils have been shown to be not only suitable disposal areas for large quantities of fly ash sewage sludge, compost, etc., but in addition, applications of these and other wastes can create suitable sites for future agricultural, forestry and recreational enterprises.

Sewage Sludge: Sewage sludge is another waste product that is being recycled on strip mine areas to supply nutrients for establishment and growth of plant cover. Sewage sludge is a dark grey liquid containing 2 to 5% solids as finely divided and dispersed particles. Its physical and chemical properties vary according to the composition and treatment of the sewage and the processes used to treat the sludge. It includes all or part of the solids removed in primary, secondary, and tertiary treatment of sewage. There are several methods of stabilizing sludge. One of the most popular methods is to treat sludge in 15 day, heated anaerobic digesters to stabilize the solids, thus eliminating obnoxious odors and fly problems after application on land. The pathogenic contamination hazard of digested sludge can be reduced to nil by lagooning the material for 30 days before land application.

Investigations conducted by Dick et al (8), Department of Civil Engineering, University of Illinois, show the average percentage die-off of fecal coliforms in digested sludge after 30 days to be 99.9%. Since the flow properties of freshly digested sludge vary little from water, it is easily transferred by pipes, using ordinary pumping techniques and equipment. Several methods for applying sludge have been developed:

> Furrow irrigation on properly contoured and contained areas.
> Sprinkler irrigation on irregular, temporary, or nonengineered sites.
> Flooding an entire area that is self-contained or surrounded by dikes.
> Placing liquid sludge beneath the soil surface.

Liquid digested sludge contains nitrogen, phosphorus, and potassium. Two inches of liquid digested sludge applied intermittently throughout the year would satisfy the average corn crop requirements of 168 kg of nitrogen per hectare, 45 kg of phosphorus per hectare, and 90 kg of potassium per hectare.

Digested sludge contains additional growth-promoting ingredients. Being a natural organic material, it imparts the same favorable characteristics to soils that are normally attributed to its natural humus content. Sludge increases soil fertility, improves soil structure, increases water-holding capacity, and controls moisture supply. It contains vitamins and trace elements essential to growth: sodium, boron, calcium, magnesium, manganese, iron, aluminum, sulfur, copper, zinc, molybdenum, chloride and silicon.

The U.S. Forest Service has recently completed the field evaluation of test plots treated with liquid sludge and planted with grasses. These tests were conducted on the Shawnee National Forest in southern Illinois on orphan strip mine spoils (Palzo Project). The spoils were very acid (pH 2.45) and had virtually no vegetation since mining ceased in 1961.

Conclusions were that sludge produced a vigorous growth of grasses and improved the subsurface drainage water quality. The test plot results indicated that a minimum-maximum limit of 200 to 250 dry tons of sludge per acre should be applied. One inch of liquid sludge per week was the best rate to apply.

At regular intervals during application, accumulated solids should be incorporated by disc into the first 9" to 12" of soil. Discing on the contour will help increase infiltration rates, provide protection against erosion and minimize odor problems should they occur. Rest periods should be provided between applications to help dry the soil.

The Rand Development Corporation used liquid sludge in 1966 to reclaim diked plots of extremely acid spoils in the vicinity of Canton, Ohio. The plots represented spoils with different degrees of acidity from pH 2.3 to near neutrality.

Liquid sludge was applied to plots by flood irrigation and left on the surface to dry and form a seedbed for a mixture of grasses and legumes. It formed cracks as it dried where the seeds germinated and grew, extending roots into the spoil. The grass and legumes were growing vigorously 6 years later. Not only can sludge treatment reduce acidity problems of strip mine spoils, it can also reduce severe alkalinity problems. In 1969 sludge was used to treat an alkali sand filled lagoon (pH 10.5) near Ottawa, Ill.; 170 dry tons/acre of sludge incorporated into the sand surface and planted rye, orchard and brome grasses resulted in a dense vegetative cover. The sludge application reduced sodium concentration, the main deterrent to plant growth. This showed that sludge applications can reclaim sterile alkaline land in less than 1 year; it could have application possibilities in the alkaline spoils of the West.

By using the most up to date reclamation techniques available, complete restoration of the strip mined lands to levels equivalent to those characteristic of productive soils will take many years under normal agricultural practices. Efforts by the most conscientious operators cannot put the humus layer back to its original position in the soil profile. It will be mixed with other materials during the mining and reclamation phases and this storehouse of plant nutrients will not be available for plant use.

To replace this soil organic matter, stabilized sludge is outstanding in its ability to increase the humus content of soils quickly. The results of studies indicate that the organic material produced in a 15-day heated anaerobic digestion process has properties very close to that of natural soil organic matter of humus. Digested sludge is one of the few materials that can be used to effect a rapid increase in the humus content of soil. It is the only substance with these properties that is available in quantities.

Nitrogen contained in digested sludge is usually the first factor to limit rates of application. Adding excess nitrogen to spoil involves the risk of polluting groundwater with nitrates. Hinesly (9), indicates that about 2 inches of liquid digested sludge would satisfy the nitrogen needs of nonleguminous crops without producing excessive nitrate in percolated water. Higher loading rates can be made on strip mined lands because they have much greater assimilative capacities for plant nutrients and nonessential trace elements than most soil types.

Many of the trace elements in sludge are essential to plant growth, but nearly all can be toxic if the concentration is high enough. Hinesly (10), states that higher applications of sludge can be applied on strip mine lands without encountering trace element problems than might be the case when sludge is applied on soils.

In the University of Illinois study, 150 tons of sludge were applied per acre over a five year period to corn plots without causing toxicity. Where sludge is to be used for reclamation of surface mines, it will probably be a short term treatment and the volume used will not reach toxic limits. Where sludge is to be used in a continuing management program for crop production, toxic levels must be considered.

Hinesly reports the outlook is promising for mixing 200 dry tons per acre of lagooned sludge into the surface foot of cultivated strip mine spoils during a four year period without significantly affecting nitrogen content of water supplies. With such a program the top surface foot of reclaimed spoil bank will contain a humus content of 4 to 5% to a depth of one foot within a four year period.

Land Reclamation Methods

The cost of removing water from liquid sludge is high enough to cause liquid sludge handling and application to be preferable for most communities, but for some, mechanical dewatering may be feasible. Sludge cake or partially dried sludge can be hauled in rail cars, barges or trucks. Sludge can then be spread with a manure spreader or a bulldozer.

If spraying is a more feasible method of application and water is available at the reclamation site, reslurrying the sludge may be possible and used in hydroseeders. The most economic methods of transportation and applying sludge depend upon the amount and kind of sludge facilities available, and other local conditions.

Thus sewage sludge has several qualities that make it desirable for reclaiming spoils. It adds not only the nutrients needed to establish vegetation, but also a stable organic matter that will form a humus in the surface layer or serve as a mulch. Adding organic matter improves the spoil structure, water-holding capacity and ion exchange capacity, and creates a more favorable root zone for grasses and legumes. Sludge buffers extremes in the spoil pH and immobilizes ions that may be present in toxic concentrations.

Although constraints that limit the rate of sludge applications to agricultural crops and pasture lands may also limit the amount that can be applied for reclamation, the permissable rate is much higher on strip mines. Plants that are tolerant of relatively high concentrations of metals in spoils are good for revegetation purposes. The opportunity for controlling percolation and runoff water to prevent nitrate pollution of groundwater is greater than it is in a regular farming operation. Under drainage, terraces, dikes and catch basins can be constructed during the shaping of strip mined areas. Public exposure to pathogens, which could be present in treated mine spoils, should not be a problem.

Compost: Composting is the biochemical degradation of organic materials to a humus-like substance, a process constantly carried on in nature. It is a sanitary process for treating municipal, agricultural and industrial wastes.

Properly managed windrow or enclosed, high-rate digestion composting will produce a product safe for agriculture and gardening use. Compost has the remarkable ability to provide soils with better tilth, water-holding capacity, improved nutrient-holding capacity and due to its high organic content is a good soil conditioner. Present technology of composting will permit the recycling or organic waste materials back to the soil without significant pollution of water or land resources.

Compost plants are operating successfully in all parts of Europe, some for as long as forty years. One use has been as a soil builder for reclamation and recultivation of lands devastated by strip mining.

Composting of municipal refuse has received almost no attention in the United States in spite of the fact that it is being successfully practiced in other countries. The reason cited most frequently for the lack of interest is that no market exists for compost. However, the high organic content which makes it a good soil conditioner could find a market in strip mine reclamation, if it is competitive with other materials now in use. Compost mixed with sewage sludge would be an ideal material to use in producing an artificial soil for orphan spoil bank reclamation and could be most helpful in the West where organic matter in the original soil is very low. The major drawback could be the high cost of transportation if it had to be shipped a long distance.

Four years of tests by the Tennessee Valley Authority proved the effectiveness of composted municipal wastes in producing vegetative cover on coal strip mine sites in Virginia. Examination of organic layer development on the test sites indicated that to obtain a stabilized organic layer over mine spoil in two years, application rates between 26 and 71 tons per acre would be required. Fifty tons per acre left substantial residue and initiated a humus layer after two years and also resulted in good vegetative cover.

Manure: In a few incidences local farmers have reclaimed surface mined lands using manure instead of commercial fertilizers. Manure was applied either directly or by holding animals on the area. Manure has also been used to build up the organic material in tailings dams in order to help establish vegetation. Manure should be considered along with sewage sludge and compost as a material to build organic matter in spoils.

Species Selection

During the early years of strip mining the acreage disturbed was small and land was plentiful. No thought was given to reclamation or returning the land to some form of productive use. Strip mined areas were left in a rugged, irregular, and harsh condition consisting of hills, valleys and peaks. The very steep slopes limited future land use to forestry practices. It was not until after World War II that surface mining as is known today began to develop.

Reclamation began a fews years later and consisted mainly of tree planting. Trees were first selected because some of the species such as pine and locust grew at low pH and did not require soil amendments in order to survive. Reclamation costs were exceedingly low, less than $30 per acre. The main objective was to cover the ugly scars and provide for a potential economic return.

Trees alone do not provide quick stabilization in their early stages of development. Excessive runoff and erosion took place until the canopies closed and afforded protection from direct rainfall impact. Up to ten years are required to get crown closure and adequate ground protection to control erosion. Curtis (1) reports that watershed studies indicate high sediment yields during and immediately following mining. During this critical period, trees alone are of little value as ground cover. Experience has shown that a quick growing, herbaceous cover with tree planting is indispensible if maximum site protection is to be obtained immediately.

Some species of grasses and legumes are more competitive than others with trees. In any revegetation program, it should be borne in mind that the objective is to stabilize the area as quickly as possible after it has been disturbed. Plants that will give a quick, protective cover and enrich the soil should be given priority.

In some instances, these initial plantings should be considered only as a tool in the land management process of obtaining maximum land use, and not the end result. Many mixtures of seed and techniques for establishment are available that will not only not hinder tree growth but give good survival, and quick stabilization. The available knowledge on what species will or will not grow, where they will grow, what is required to make them grow and their effective use for reclamation purposes is broad. This is particularly true for the eastern United States; far less is known of the arid and semiarid western situations. It is fortunate that many of the plant species that have been successfully used in revegetating eastern strip mined areas are also desirable economic crops and accelerate the natural succession of plants.

Legumes are of special value in vegetating strip mine spoil because of the low nitrogen level in spoils; they should be included in all seed mixtures. When legumes are properly inoculated, they develop nodules on their roots and are then able to fix atmospheric nitrogen that can be used by plants. Grasses especially need nitrogen, which legumes are able to supply without annual treatment of fertilizer.

In addition, legumes are taprooted and can incorporate organic matter deeper than grasses. Grasses are fibrous rooted and bind the soil together better than legumes, especially in the critical years following germination. Both should be included in all seed mixtures in order to secure the benefits offered by each species. Over a long period of time additional lime and fertilizer may be required to maintain good growth of grasses and legumes. This is no different from what would be expected on any agricultural soil.

Establishment of vegetation on an area must be done as soon after grading as possible so as to provide a quick protective cover for erosion control. Such covers may include species of little or no commercial value. Fast growing site stabilization species are given priority when making up the planting plan. A mixture composed of several species is preferred over planting of a pure single species. By using mixtures, species can be selected that will yield better economic return than either alone and provide quick and more complete long-term cover.

Mixtures are also less susceptible to disease and insect epidemics and reduce danger of frost heaving of legumes. Species that have compatible growth rates and have proven successful in the particular area under similar site conditions should be used, and legumes should be included in all mixtures because of their ability to fix nitrogen. Unfortunately, no one mixture can be recommended that will establish successfully in all kinds of spoil, topographic, and climatic conditions.

In most coal mining states, the Soil Conservation Service has developed or assisted in development of handbooks to guide revegetation procedures. The handbooks are based on data obtained from State and U.S. Department of Agriculture research findings and are in the form of guidelines for use of plant materials according to a spoil classification system.

Soil Conservation Service technicians can predict plant performance for specific site conditions and recommend the cultural and management techniques needed for establishment. These technicians are available for assistance through local Soil and Water Conservation Districts.

The Soil Conservation Service Plan Materials Centers are continuously searching for new plant material that can be field tested to determine its site adaptability. As soon as its performance is confirmed, the plant is keyed to the vegetation guide according to use and site requirements.

Methods of Seeding and Planting

Many factors must be considered when selecting the methods for seeding and planting. These include: access to the area by vehicles; location—availability of water, distance to airport; slope, especially steep outslopes; seedbed conditions—age of spoil, rainfall, and time since final grading; topsoiling; size of area; and time of year. The conditions pertaining to each site where seeding is to be done will determine the method to be used.

Areas accessible by vehicles may be treated using conventional farming equipment, mechanical tree planters, and hydraulic seeders. Use of mechanical equipment is limited during wet, muddy, freezing, and thawing weather, which is the ideal time to seed. The freezing and thawing action loosens the soil and works the seed that has been broadcast on the surface into the soil.

Early germination and growth is thus obtained. Using conventional farming equipment, the area is treated with agricultural lime, which is worked in with discs, harrow, etc. Fertilizer may be applied in the same manner or applied with the seed. Seed is either broadcast or drilled.

Helicopters with motor-drive spreaders have been used successfully in rugged terrain and in places where wheeled vehicles cannot operate because of wet, muddy spoil. A heliport must be readily accessible to vehicles bringing in loads of blended seed and fertilizer. The helicopter can hover and reload from hopper trucks. A large area can be seeded in a short time when seedbed conditions are most favorable. West Virginia uses the helicopter in rugged mountain areas and has seeded several thousand acres (hectares) by this method.

Fixed wing aircraft have been used with varying degrees of success in several states. However, special detailed planning and preparations are required. A landing field must be nearby, and mixing equipment for blending seed and fertilizer is needed. Seeding areas must be flagged, and weather conditions, particularly wind, have to be favorable. Large areas can be seeded fast and with low per acre costs.

Spoils can be seeded during late winter periods of freezing and thawing, when seedbed conditions are excellent. This type of seeding lends itself very well to area mining and to orphan bank reclamation because accessibility is not a prerequisite. If the area is small, it can be seeded with a hand operated cyclone seeder.

Regardless of the planting and seeding methods used, if the spoil is not loose, then seedbed preparation is necessary. The crusted, hard surface must be scarified before seeding. This can be done with a disc, harrow or ripper. A heavy-duty seed drill has proven to be very outstanding on crusted spoils. In areas of low rainfall, drilling of seed is mandatory to get acceptable germination.

Revegetation of Arid and Semiarid Regions of the West

Reclamation in the West is a new field of endeavor. Research has been conducted on a very limited trial and error basis. The Bureau of Land Management, U.S. Department of the Interior, in their land use studies for the Bull Mountain and Buffalo Creek area of eastern Montana (1973) recommended that certain coal

Land Reclamation Methods

beds not be mined. This area is rimrock country and most of the mining would have to be a contour type with some augering. Damages would be extreme, especially to the many scenic and natural wind-eroded formations. Slopes are steep, making reclamation very difficult and costly. It would be impossible to restore the drainage patterns and slopes to their original forms. Extraction would create a relatively high degree of surface disturbance per ton of coal mined.

Problems of revegetating strip mine areas in the arid and semiarid West differ drastically from those in the humid areas of the East. From the standpoint of plant growth, climatic conditions are extreme. Seventy-five percent of the area receives less than 20 inches annual precipitation available for plant growth. Along with limited precipitation are seasonal temperature variations from $-60°$ to $120°F$, short frost-free periods, wide variations in overburden material and lack of adequate topsoil.

In some cases the saving and spreading of topsoil can do more harm than good, for instance, where the calcium carbonate layer underlying much arid land soil is mixed with the nitrogen-rich organic layer and the biologic carbon-nitrogen balance destroyed. Since evaporation is at the surface, minerals dissolved in the soil water are precipitated in the upper strata and may cause highly saline conditions that are toxic to plant establishment.

Water is a key factor to any successful reclamation program especially in the West. Ample moisture at planting and during establishment is critical for revegetation success. Irrigation should be used sparingly and in such quantities that the plants will not be conditioned to the extra moisture. However, irrigation may be necessary during peak plant demand and low rainfall the first year to ensure survival, particularly for shrubs and trees.

Other factors that must be considered in establishing vegetation on strip mined areas include exposure, aspect, slope, pH and salt content of the spoil material, texture, and climate. Any one or a combination of these factors could be limiting and critical to plant growth.

Lang (11) set up a study area in the Kemmerer coal fields in southwestern Wyoming. The study area is a part of the northern desert scrub region and receives an average annual precipitation of 9.42 inches. Much of the precipitation occurs as snow, with an average fall of 56.6 inches. In this area, snow has several peculiarities. It occurs after the ground is frozen, so any snow that melts is subject to runoff rather than percolation into the soil.

Snow is also blown about and distributed in uneven patterns. Many areas catch little snow, while others, such as gullies and leeward sides of wind obstructions receive large amounts. Snow also is vulnerable to sublimation, a process by which solids pass directly into a gaseous state without being transformed to liquid. It is not uncommon for 60 to 80% of a snowfall to be sublimated, leaving 20 to 40% of the moisture content to be transformed into water, which may or may not penetrate the soil surface. In short it is not unreasonable to estimate that of the 9.42 inches of annual precipitation, less than 5 inches are available for plants.

Three studies were conducted on the Kemmerer coal spoil banks: (1) use of various species of trees, fertilizer, and irrigation; (2) use of four species of grass seed and various means of holding moisture; (3) transplanting sod chunks and sprigs or two rhizomatous species with different means of holding moisture. Conclusions from the tree planting study showed that watering the trees during mid and later summer greatly increased survival percentage, growth, and vigor. Survival differences as high as 50% were observed between trees on irrigated plots and those that were not irrigated, those that were not fertilized, or those that received no treatment (control).

Conclusions from the grass-seeding study, which included treatment of test plots with jute net, barley straw, mulch, snow fence, irrigation, and combinations of these are as follows.

> Available moisture was a principal limiting factor in plant establishment, but this could be supplemented by snow fences and irrigation from nearby permanent ponds. Snow fences were effective means of acquiring additional moisture for plant growth only when placed on the leeward side of large, open, level areas.
>
> Mulch, necessary for good seedling establishment, required some means of holding it in place. Annual plants grown for a nurse crop served both as mulch and for erosion control, but they in turn depended on ample precipitation for optimum growth.
>
> Jute netting served as a means of erosion control and as a partial mulch. Stabilization of erosion was a prerequisite for successful revegetation on slopes.

Conclusions from the grass transplanting study using sprigging and sodding are as follows.

> The best time of year to plant was in the spring. Early fall planting proved least successful, and late fall planting was not tested.
>
> Sodding produced far better results than did sprigging. Roots within sod clumps stayed moist and were protected by the surrounding soil, whereas in sprigging, roots were damaged and moisture was lost from plants being prepared for planting.
>
> Planting behind snow fences resulted in slightly better survival because of early spring snow melt behind the fences and the wind break provided by the snow fence.
>
> The most limiting factors influencing vegetative establishment was the amount of precipitation received just before, during and immediately after planting time.

The overall conclusions of Lang were as follows.

> Ample moisture at planting and during establishment was critical for stand success with seeded grasses, planted trees, and grass sod or sprigs. Irrigation and/or the use of snow fences to accumulate extra moisture increased the percentage stand establishment of all types of vegetation.
>
> Russian olive and caragana were the best of the tree species tested, and the top part of east and northeast-facing slopes were the best sites for their establishment.

Intermediate and crested wheatgrass appeared to be the best adapted of all the cool season grass species seeded. The most satisfactory stands of all species were obtained where mulch with some type of netting to hold it in place was used with the seeding or where the seeding received additional moisture benefits from being on the lee-side of a snow fence.

Sodded grasses were most effectively established on the flat top of the spoil piles, whereas tree species and seeded grasses were more effectively established on northeast- and east-facing slopes.

Nitrogen fertilization did not significantly affect establishment of either grasses or trees.

Saulman (12), after a three year study of snowdrift management at an open range site in eastern Montana, indicated that a standard snow fence can effectively induce snowdrifts on water-harvesting catchment basins. Although water loss by evaporation from induced snowdrifts averaged 50%, runoff was increased by an average 4.4 inches during the winter season.

Water harvesting is defined as the practice of collecting and storing precipitation from an area that has been treated to increase runoff. Acceleration of snow melt by applying lamp black, pulverized lignite or other heat adsorptive dust could reduce evaporative exposure time and increase runoff yield. The use of snow fences has proven to be a feasible method of accumulating extra moisture that is helpful for plant establishment and growth during critical periods. In nearly all instances, snow fences have increased the percentage of stand survival of all types of vegetation.

Several systems of water retention on spoils have been tested and evaluated by Sindelar et al (3). The section entitled Mechanical Spoil Manipulation contains a detailed discussion of gouging, dozer basins, and deep chiseling, which have proven to be successful in trapping moisture for seed germination and survival, controlling erosion, relieving compaction, and improving the spoil moisture reserve.

Direct seeding of most trees and shrubs is unsatisfactory under arid and semi-arid conditions. They must be established by planting seedlings or transplants. Hodder (13) has developed three dryland planting techniques for trees and shrubs. They are condensation traps, supplemental root transplants, and tubelings.

Condensation traps are made by digging a small basin for each plant. The entire basin is covered with a plastic sheet and heeled in around the edge to contain a large amount of air. The foliage of the plant is guided through a hole in the plastic sheeting. Rocks are placed on the tarp around the plant to provide protection and to weight the plastic and keep it taut in a funnel form. Condensate collecting on the underside of the plastic sheet trickles down to the plant location and effectively irrigates it.

Supplemental root transplanting is accomplished by carefully removing a pair of interconnected seedlings of a rhizomatous shrub species. The top of one seedling is pruned off at the crown, leaving two root systems connected to the uncropped seedling. The horizontally connected root systems are then planted in a vertical attitude, one being placed down in deep soil moisture, and the other planted in a normal manner in the drier surface soil. Tubelings are plant seedlings planted and nursery developed in two-ply paper cores or tubes. The tubes are 2½" in diameter and 2' long. The paper core is reinforced with a ½" square mesh plastic sleeve. When the root system develops and extends from the bottom and sides of the tube, the tubeling is ready for transplanting. A powered soil auger is used to drill holes in the field. Tubelings are dropped in, sealed around the top, and abandoned without further care or maintenance. Seeding of grasses and legumes include the following methods:

> Drilling — Wherever possible, grasses and legumes should be drilled. The recommended planting depth is ½" to ¾" and is best accomplished by using a drill equipped with depth bands and packer wheels.
>
> Broadcast Seeding — Broadcast seeding is satisfactory for small or inaccessible areas. The surface should be rough enough that the seed will be covered. Roughening is best accomplished by dragging with a harrow, disking, or dragging with a heavy-spiked chain. Broadcast seeding may also be satisfactory when seeding immediately after construction, before the surface has become crusted, or before mulching.
>
> Hydroseeders — Hydroseeding has not generally given good results under climatic conditions similar to those where open-cut mining will be done in Montana. Accessibility and available water are also limitations to their use. Mulching is usually necessary.
>
> Aerial Seeding — Aerial seedings have not generally been satisfactory in the precipitation zones that will be encountered in most of Montana's surface mined areas. However, if the surface is rough enough that the seed will be covered by rain or wind action, satisfactory stands may possibly be obtained. Helicopters are superior to fixed wing aircraft.

Seeding rates and species should be obtained from local agriculture agencies who are familiar with local conditions.

It is assumed that no mining will commence until detailed mining and reclamation plans have been approved by the responsible permitting agency. These plans should, as standard procedures, require spoil segregation and placement according to the core drill analysis and soil analysis made during the preplanning phase.

They should also include topsoiling, slope restoration, and the regeneration of the native, self-sustaining plant community. Introduced species may permit quick stabilization, but ultimately, the dominant cover should be native species. This means that a seed source must be developed to furnish the large quantities that will be needed.

Through research and experience, unproven reclamation trends have developed that can be used as guidelines for preplanning the mining operation; if such guidelines are followed during mining, they should make restoration of the disturbed land possible. At this point in time it must be accepted, however, that certain areas cannot be mined because of the limited knowledge and techniques for restoring mined land in the arid and semiarid West. Assuming that reclamation is effective, there would still be a considerable delay before land could be returned to its original use.

Until the vegetative cover is firmly established, grazing would be discouraged and must be controlled.

REFERENCES

(1) Curtis, W.R., Vegetating Strip-Mine Spoils for Runoff and Erosion Control. Paper presented at the symposium on Revegetation and Economic Use of Surface-Mined Land and Mine Refuse, Pipestem State Park, West Virginia, December 1971.

(2) Riley, C.V., Furrow Grading—Key to Successful Reclamation. Paper presented at the Research and Applied Technology Symposium on Mined-Land Reclamation, Pittsburg, Pennsylvania, March 1973.

(3) Sindelar, B.W., Hodder, R.L., Majerus, M.E., *Surface Mined Land Reclamation Research in Montana, Progress Report 1972-1973*. Montana Agricultural Experiment Station, Montana State University, Bozeman, Montana.

(4) Grube, W.E. Jr., Smith, R.M., Singh, R. and Sobek, A.A. Characterization of Coal Overburden Materials and Mine Spoils in Advance of Surface Mining. Paper presented at the Research and Applied Technology Symposium on Mined-Land Reclamation, Pittsburg, Pennsylvania, March 1973.

(5) Davis, G., *Strip-Mine Reclamation in Appalachia* (Review Draft). U.S. Department of Agriculture, Forest Service, Northeastern Forest Experiment Station, July 1971.

(6) Bengston, G.W., Mays, D.A., Allen, J.C., Revegetation of Coal Spoil in Northeastern Alabama: Effects of Timing of Seeding and Fertilization on Establishment of Pine-Grass Mixtures. Paper presented at the Research and Applied Technology Symposium on Mined-Land Reclamation, Pittsburg, Pennsylvania, March 1973.

(7) Bennett, O.L., Grasses and Legumes for Revegetation of Strip-Mined Areas. Paper presented at the Symposium on Revegetation and Economic Use of Surface-Mined Land and Mine Refuse, Pipestem State Park, West Virginia, December 1971.

(8) Dick, R.I., and Associates. "Influence of Soil Moisture on Fecal Coliform Survival." *SW-30d,* U.S. EPA, Cincinnati, Ohio, 1971.

(9) Hinesly, T.D., Braids, O.C., Molina, J.E., "Agricultural Benefits and Environmental Changes Resulting from the Use of Digested Sewage Sludge on Field Crops." *SW-30d,* U.S. EPA, Cincinnati, Ohio, 1971.

(10) Hinesly, T.D., Jones, R.L., Sovewitz, B., "Use of Waste Treatment Plant Solids for Mined Land Reclamation." *Mining Congress Journal,* September 1972.

(11) Lang, R., "Reclamation of Strip Mine Spoil Banks in Wyoming." *Research Journal,* 51, Agricultural Experiment Station, University of Wyoming, Laramie, Wyoming, September 1971.

(12) Saulman, R.W., "Snowdrift Management Can Increase Water-Harvesting Yields." *Journal of Soil and Water Conservation,* 28 (3), May-June 1973.

(13) Hodder, L., Surface Mined Land Reclamation Research in Eastern Montana. Paper presented at the Research and Applied Technology Symposium on Mined-Land Reclamation, Pittsburg, Pennsylvania, March 1973.

SEDIMENT AND EROSION CONTROL

The material in this chapter is excerpted from:

PB 238 538

Sediment is one of America's greatest pollutants. More than a billion tons of sediment reach the major streams of the Unites States annually. Damages are reflected in the reduced carrying capacity of streams, clogged reservoirs, destroyed habitat for fish and other aquatic life, filled navigation channels, increased flood crests, degraded facilities for water-based recreation, increased industrial and domestic water treatment costs, premature aging of lakes by enrichment of the water with silt-carried fertilizer that promotes algae growth, destroys crops, and reduces productivity of flood plain soils.

Erosion and sedimentation are natural processes that are usually gentle actions releasing controlled amounts of silt from watersheds to receiving streams. Surface mining activities accelerate these natural processes and short duration, high intensity storms can become a violent force moving thousands of tons of soil in a brief period of time. Cover is a very important factor. With the removal of ground cover, water moves across the denuded area on its own terms picking up soil particles as it flows and leaving gullies behind. The susceptibility of strip-mined land to erosion depends on physical characteristics of the overburden, degree of slope, length of slope, climate, amount and rate of rainfall, type and percent of vegetative ground cover.

By development of erosion and sedimentation control plans before disturbance of the area, many of the detrimental effects of strip mining can be prevented.

CONTROL MEASURES

The key to minimizing erosion and sediment problems is in the control of water flowing into, within, and from the surface mining area. Control measures must be designed according to sound engineering principles that will fit the topography,

Sediment and Erosion Control

soils, rainfall, climate, and land use of areas they are to protect. These controls include a wide variety of measures and facilities that are either vegetative or mechanical in nature. Their objective is to prevent accelerated erosion and sedimentation. To be effective, controls must be properly installed and maintained. To insure that the detailed erosion and sediment control procedures are implemented by the surface mine operator, these plans are inegrated with the mining and reclamation sequences.

Sediment yield from a mined watershed is the result of erosion from the disturbed area and the movement of this eroded material from the watershed. Therefore, sediment yield varies, not only with the extent of disturbance within the watershed, but also with the proximity of the disturbed area to the natural stream channel.

Thus, surface mines and access roads, where the outer fill slopes approach the stream channel, will yield greater quantities of sediment than those separated from the channel by a filter zone. Experience has shown that a protective, vegetated strip of undisturbed soil between the toe of the fill and natural drainways usually prevents muddy water from reaching streams. This filter zone must be wide enough to absorb all the muddy water that runs off outslopes. Sediment is collected on the undisturbed litter and soil, and only clear water enters the stream.

The required filter-zone width varies with the steepness and length of the outslope between the toe and the drainage channel. A minimum distance of 100 ft between the strip mine operation and any stream is required by Kentucky law. Often, steep slopes will require that the filter zone be more than 100 ft wide.

The salvage and stockpiling of topsoil for later use as topdressing is another parameter that must be considered. Stockpile areas must be located far enough from water courses so that they will not provide a source of sediment during storm runoff. Critical slopes on stockpiles must be avoided, especially if the material is easily eroded. Temporary soil stabilization measures should be established immediately after the stockpile operation is completed.

If the stockpiling is a continuing operation, then temporary stabilization methods are implemented in stages as stockpiling progresses. Stabilization can be accomplished with a vegetative cover or by chemical means. Chemical soil stabilizers penetrate the surface and bind soil particles into a coherent mass that reduces erosion by wind and rain. A quick growing cover of herbaceous species will also provide temporary protection against erosion.

One of the primary rules for good erosion and sediment control is that all earthmoving activities be planned in such a manner that the minimum amount of disturbed area will be exposed for the minimum amount of time. This objective can be accomplished by developing the area in stages with progressive backfilling and reclamation. Consideration must be given to critical areas such as steep slopes and high soil erodibility. Climatic factors as they relate to vegetative stabilization must also be considered; for example, topsoil should not be placed while in a frozen or muddy condition or when the subgrade is excessively wet.

Sediment Control Basins

Since sediment causes more off-site damage than any other aspect of strip mining, it is essential that steps be taken for the control of sedimentation. Sediment retention basins can be effectively utilized for collection and holding of eroded material before it reaches the main streams, thus preventing damage to areas downstream. By detaining storm water, sediment basins also reduce peak flows. Since most of the settleable solids drop out of suspension quickly in quiet water, it is necessary that the basins remain filled with water.

Sediment basins should be located on all drainways carrying concentrated flows from the disturbed areas. They should be located as close to the sediment source as possible and before the drainageways reach the main stream. Maps that delineate the various phases of mining and reclamation should also show the location of all sediment control retention structures. The drainage plan must indicate the sequence of construction, with all necessary structures being built in a specific area before the initiation of clearing and grubbing operations.

Sediment control structures are created by the construction of a barrier or a dam across a drainway, or by excavation of a basin, or by combination of these methods to trap and store eroded material. These catch basins are nearly always temporary sturctures. However, they can be designed as permanent structures if there is a need for them, if they will not endanger life and property in case of failure, and if a responsible party will continue maintenance. Where maximum storage can be obtained with a basin of planned size, it should be constructed adjacent to the drainageway and be of the diversion type. After the mining is completed and the area stabilized, diversions can be closed with the collected sediment isolated from further flows. Sediment control basins are classified as either primary or secondary, according to their design, location, and intended use.

Primary Basins: There are three basic types of primary basins which are described below.

Excavation or Dugout — This type of basin is a water impoundment made by excavating a pit or dougout. An earth embankment is sometimes used with the dugout to increase its capacity.

Earth Embankment — This basin is a water-retention-type structure constructed across a waterway or other suitable location to form a sediment catch basin. Where topography of the site restricts storage requirements, structures may be built in series so that the cumulative total of the sediment storage capacity will equal the storage requirements for each acre of disturbed area in that watershed.

Leaky Dam — This basin is a rock-french-drain-type structure that is used to momentarily stop runoff water so that it can deposit its sediment load before leaking through the dam. It has been used very successfully on small watersheds of less than 150 acres and on larger areas in combination with earth embankments to catch initial sediment loads.

Secondary Basins: Secondary basins consist of facilities that are not adequate for sediment control when used alone. They are used to catch initial sediment loads near the disturbed area, and thus they lengthen the time between cleanouts for downstream primary structures. Types of secondary structures are as follows.

Sediment and Erosion Control

Gabions — This structure is made up of large, multicelled, rectangular, wire-mesh baskets that are rock-filled. They are used as building blocks in the construction of sediment control structures. Gabions are used mainly in building small check dams across drainways; however, they are very versatile, and under favorable conditions, they can be used as primary structures. During construction, foundations must be properly prepared and the gabions securely keyed into the foundation and abutment surfaces. Rock used in filling the baskets must be durable and adequately sized.

Log and Pole — This check dam structure is strictly a stop-gap measure for collecting sediment in small drainways. They have not proven to be very successful in field use and should only be used in an emergency situation. They must be replaced at the first opportunity, after the emergency no longer exists.

Rock Dam — This type of check dam is a barrier built across a drainageway to retard storm runoff and form a small sediment collection basin to assist in sediment control. Such dams are not substitutes for primary structures. Rock check dams are usually used where small localized sedimentation problems exist and the drainage area is less than 50 acres.

Primary Basin Design Criteria

Technical information and design criteria for primary sediment control basins are as follows.

Excavation or Dugout: An example of standards for the design and construction of excavated sediment ponds may be found in *Drainage Handbook for Surface Mining,* Department of Natural Resources, Division of Reclamation, Charleston, West Virginia, January 1, 1972. Under "capacity requirements," the excavated sediment pond must have a minimum capacity to store 0.125 acre-feet per acre of disturbed area in the watershed.

Stanford Research Institute, in its 1972 report *A Study of Surface Coal Mining in West Virginia,* stated that as this storage capacity is only for a type II storm of 24 hours duration with a 10 year frequency, greater safety would be provided by sediment storage requirements of 0.28 acre-feet per acre. 0.125 acre-feet dams need maintenance and clean-out when they are 60% filled. This level could be reached in less than a year.

Actual operational experience has shown that some of these basins fill up after only one moderate storm, depending mainly on the soil type. Greater than 60% filling reduced the basin's capacity to retain runoff long enough for sediment to be deposited. Removed materials should not be permitted to reenter the drainage system.

Recent studies by the U.S. Forest Service, in which they have measured silt buildup in sediment ponds, revealed that 0.2 acre-feet of storage per acre of disturbed land was a reasonable figure.

Earth Embankment: These sediment retention basins are constructed to detain water long enough to allow soil particles that the water is carrying to settle out by natural gravitation. It must be recognized that basins of the size that will normally be constructed will not retain the runoff long enough to settle out colloidal material. Location of the embankment is critical to successful installation and operation of the sediment basin. Topography of the watershed will play an important part in selecting the dam site. Construction material must be readily available and the site should provide maximum storage of silt behind the structure. Failure of the structure should not result in the loss of life or damage to property.

All earth embankment type sediment retention basins must be designed and constructed according to sound engineering principles. Practically all failures of this type of structure can be traced to faulty engineering, i.e., inadequate emergency spillway, or inadequate capacity for the area drained, which can result in filling of the sediment storage area after only one storm, insufficient retention time to settle out suspended solids, or a larger-than-anticipated volume of water that sweeps the dam away. Sediment control basins must be installed before disturbance within the immediate watershed. If the proposed mining area is located in several watersheds, basin construction is scheduled in advance of mining so that the affected watershed will be protected before disturbance.

There is no substitute for good planning, design, construction, and maintenance of sediment control basins if they are to provide effective sediment control of runoff water and prevent off-site damages. However, it must be recognized that there are locations where the physical characteristics of the terrain are such that effective sediment control basins cannot be constructed. If these conditions exist, then surface mining should be prohibited.

Technical assistance regarding sedimentation problems can generally be obtained from the local county soil and water conservation district and the Soil Conservation Service, U.S. Department of Agriculture. In many cases, the District will schedule the assistance of qualified personnel to conduct surveys in the proposed mining area with the operator to gather data for development of plans to control erosion and sediment, including location and design of needed sediment control structures. A publication entitled *Engineering Standard for Debris Basin for Control of Sediment from Surface Mining Operations in Eastern Kentucky*, has been prepared by the U.S. Department of Agriculture Soil Conservation Service. The specifications presented therein may be considered a representative example.

As stated before, sediment control basins work on the theory that by reducing the velocity of the runoff water, natural gravitational settling will clear the water before discharge into receiving streams. Experience in the field with this type of structure strongly indicates that gravitational settling alone would not be sufficient to clarify muddy water. Some clay soils are of colloidal nature and may stay in suspension for weeks, causing turbid water conditions. Material such as lime, alum, and organic polyelectrolyte added to the muddy water will cause flocculation of the suspended particles into an agglomeration of particles; then gravitational settling out will occur and clarification of the treated water takes place.

A test project near Centralia, Washington demonstrated that natural gravitation alone would not lower the suspended solids to acceptable levels. The high clay content of the soil caused the runoff water to be turbid. These colloidal particles were found to carry a negative electrical charge, repelling each other and resisting flocculation and settling. Tests indicated that addition of an organic polyelectrolyte with a positive electrical charge would effectively cause flocculation of the suspended particles, and then gravitational settling would occur. Two settling ponds

Sediment and Erosion Control

were constructed in series, and the organic polyelectrolyte was added at the discharge point of the first pond. Flocculation, settling, and clarification takes place in the second pond. The State of Washington waste discharge permit requires that the discharge not be more than 5 Jackson's Turbidity Unit above normal background. Turbidity range of the receiving stream during the rainy season was found to be between 20 and 55 JTU's. Three-times-daily water testing showed that the turbidity of the water at the pond discharge weir would remain within the range of 85 to 120 JTU's without chemical treatment. After treatment with an organic polyelectrolyte, the decantate from the second pond was clear and had a range of 4 to 15 JTU's.

Leaky Dam (Rock-French Drain): Generally, a formal design is not required; however, criteria for the use and construction of this type of dam have been developed by the Stanford Research Institute and are as follows:

   The height of the rock dams shall not exceed 20 ft measured from the flow line of the channel to the top of the dam.

   All materials used in rock dams shall be end-dumped and dozer-placed in lifts not to exceed 5 ft.

   The downstream portion of rock dams from the downstream toe to the upstream shoulder shall be constructed of boulders not smaller than ½ yd^3 nor larger than 2 yd^3.

   The upstream portion of the rock dam from the upstream toe to the upstream shoulder shall be constructed of shot rock not larger than ¼ yd^3.

   The side slopes of the dam shall be less than 1½:1.

   The top of the dam shall be level in both grade and template.

   The width of the top of the dam shall be a minimum of 10 ft.

   No emergency spillway or principal spillway will be erected in this type of dam.

PIT DRAINAGE

Pit drainage is the control of water that is being removed from the pit area during actual mining operations so that it will not provide sediment for receiving streams. Surface runoff, rainfall, and seepage water often collects in the working pit areas and must be removed. If this accumulated water is in an area where equipment is moving back and forth, large quantities of spoil are churned up and put into suspension. A much too common practice is to bulldoze a cut through the bench crest and discharge the accumulated water onto the outslope.

Observation indicates that serious erosion occurs on spoil outslopes when pit water is caused or permitted to flow onto or over these areas. The deleterious effect on the environment can be very great, and in some cases, entire streams have been destroyed by this eroded spoil and sediment from pit areas. If the water comes in contact with toxic materials, another problem is added that can be as bad or worse than the sedimentation problem.

Pit water should be released slowly through the use of siphons or pumps with outlets below the toe of the outslope. Pumping or siphoning can be regulated to control the flow and to prevent overloading of the natural drainage ways or holding ponds. Holding ponds may also act as settling basins for sediment and/or be a part of the chemical treatment facilities for toxic water.

BENCH DRAINAGE

Bench drainage is associated only with contour mining on steep slopes and involves removing water from the bench area. This is accomplished by making waterways draining to an outlet in the direction of the bench slope. In no instance should water be discharged over the bench crest without the use of structural means to protect against erosion. Lowering of water from the bench to the receiving stream should be done by using the natural drainways available. When natural drainways are not available, then grassed waterways or rocklined chutes, flumes, ditches, or pipes are used.

The method of controlling erosion and sedimentation from the bench area and outer slopes will vary, depending on local conditions. Sediment originating on the bench should be confined there and not be released in the discharge water. This objective can be accomplished by the proper use of check dams, shallow ponds, or swales on the solid bench. Ponds should be constructed so as to be dry between runoff periods. On virtually all sites, it will be necessary to construct a diversion ditch above the highwall to divert water away from surface mining areas. This device should be constructed in such a manner that it can remain as a permanent part of the water disposal system. Proper outlets to such diversions are an essential part of the plan, and in most instances will require a sediment pond or debris basin.

REVEGETATION

With all mechanical measures, it is still imperative that the spoil be revegetated as rapidly as possible. Immediate cover should be obtained, regardless of long-range revegetation plans. Chemical amelioration of the spoil in the form of lime and nutrient fertilizers is generally necessary if revegetation practices are to be successful.

For the purposes of sediment and erosion control, roughness and scarification can be utilized to reduce the production of sediment and to aid in the establishement of other erosion control practices, particularly revegetation efforts. If grading is up and down the slope, runoff and erosion are encouraged by the grouser bar marks left by crawler tractors. If, however, the grading is accomplished on the contour, or across the grade, the grouser bar marks will tend to retain moisture.

On a seedbed, the marks trap and retain seed and moisture. This seed is often covered by soil being carried downslope by runoff and the bar marks may be the only areas in which seed remains after a rather severe storm. If the seed thus trapped is a turf-forming grass, it may be sufficient to establish an acceptable vegetative cover without requiring a reseeding program.

Another example is the slope that is to receive a mulch (woodchips) to protect it from excessive erosion between seeding seasons. If the slope has been scarified, the woodchips will adhere to the soil surface with greater tenacity than they will to a smooth-graded surface. Infiltration of rainfall is enhanced when a surface is left in a rough condition. This factor is also important when erosion, sediment, and storm runoff controls are planned and implemented together in a total conservation program.

COAL-HAUL ROADS

Coal-haul and mine access roads are defined as any road constructed, improved or used by the operator (except public roads) that ends at the pit or bench. These roads constitute approximately 10% of the total area directly disturbed by the surface mining operation. In some cases, the land disturbed by haul roads exceeds the area included in the mining operation.

Studies of the U.S. Forest Service in eastern Kentucky show that typical contour mining roads exhibit poor alignment, excessive grades, insufficient strength and durability, and poor drainage. Mine roads in other Appalachian States, particularly where contour mining is practiced, appear much the same as those in Kentucky. Access roads were also found to be a large source of sediment. It is possible that of the sediment that finds its way into the streams as much (or even more) originates from the haul roads as from the mining operation.

Most roads are built as cheaply as possible, and good road-building design and practice are often ignored. Maintenance schedules are generally inadequate, and upon completion of mining, haul roads are usually abandoned, with little or no attempt made to bed them down. Such roads deteriorate very rapidly.

Area mining, which is practiced in flat to gently rolling terrain, presents fewer haul road related environmental problems. These roads through necessity are generally well engineered because of the heavy equipment using them, such as 240 ton haul trucks. They must have wide beds, good alignment, and adequate drainage to permit coal haulers to run at top speed during all seasons of the year. Excessive dust can be a public nuisance and a driving hazard, and it is hard on equipment. Calcium chloride and sodium chloride have proven to be effective materials for controlling dust. Two applications during the summer when the ground surface is moist at the rate of ½ lb/yd^2 have been suggested. The most common procedure is to keep the roads wet by using water trucks. Sediment that reaches the streams can be traced to one or more of the following five basic phases of haul-road life.

Designing and Planning the Haul and Access Road System: It is important to plan the access without damaging other resources such as streams, timber, etc. Careful planning can minimize the amount of land in roads, thus reducing the amount of acreage disturbed. Design criteria should include acceptable grades, widths, strength, durability, drainage and filter strips. Factors affecting design criteria that must be considered in the planning include:

The expected traffic volume per unit of time that will be generated by all probable users.

The weight per axle or tire that those users will exert on the travelway.

The time duration through which each user can be expected to use the road.

The speed at which traffic should flow during periods of maximum traffic volume.

The expected ratio of available engine power to gross vehicle weight for the primary haulage vehicle using the road.

The bed width of the haulage vehicle that will be the primary road user.

The ability of the forest floor below the road to act as a sediment filter or trap.

Location: Based upon the design standards, several alternate roadways should be located and evaluated. The routes are selected and plotted on topographic maps or aerial photos. From the maps and photos it is easy to determine the slope, aspect, grade and pinpoint obstacles that must be avoided (such as rock outcrops, natural scenic formations, property lines, and wet areas.)

After the roads have been tentatively located on maps, they are walked on the ground and the centerline flagged. Adjustments in grade or alignment are made by the locating party instead of the construction crew. Road locations may be changed several times before the final route is selected. All flagging except that marking the final route should be removed in order to avoid confusion during construction.

Construction and Drainage: Actual construction should always be performed in dry weather. Wet materials in the subbase and base of the road will not dry out and may leave if the material freezes. Trees and brush should be windrowed at the toe of the fill to act as a sediment filter and add support for the fill section. Organic material should never be buried in the fill section, as it cannot be compacted and upon decaying will serve as passageways for water. Water entering the fill will result in a saturated condition causing slips and slides.

Six feet beyond the cut bank and 3 feet beyond the toe of the fill should be cleared to help the roadbed dry out faster after a rain. Cutting rather than bulldozing is recommended, because the ground litter is not disturbed and erosion is reduced.

Experience has shown that a protective strip of absorbent undisturbed forest soil between the road and stream usually prevents muddy road water from reaching streams. This strip, often called a filter strip, should be wide enough to absorb all the muddy water that runs off road surfaces. A minimum distance of 100 ft has been suggested between the road and stream. Seeding of the overcast soil and road shoulders immediately after construction will help minimize erosion and stream sedimentation. If this cannot be done or is not effective, sediment catch basins can be installed.

Roads for all weather use and high speed with heavy equipment need a surface or wearing course in addition to the subbase and base course. A variety of materials can be used for surfacing: slag, crushed stone, reddog, stream gravel and many others. The material chosen should be sound, durable, and not contain acid producing or toxic elements that could cause stream pollution. Unburned coal refuse and waste should never be used for surfacing.

The usefulness and permanence of roads depends on how well they are drained. It is poor economy to skimp on drainage. Uncontrolled water will erode and break up road surfaces, thus destroying their usefulness and increasing maintenance costs. Drainage control structures are one of the most important items on any roadway. Their design depends on the length of time that the road will be used and the hydrologic data for the area. During the field reconnaissance, the location, type, and size of drainage structures are noted. Most states have sizing charts, and techniques used by their highway departments for culvert and drainage structures design. These charts and techniques are easily adapted for use on coal-haul roads.

Maintenance: If a road is to be kept serviceable and properly drained and prevented from having an undesirable effect on stream water quality, then maintenance is required. Basically, maintenance is keeping the drainage system functioning properly and grading the road to its original shape. Maintenance costs can be minimized if the road was designed and constructed according to good engineering principles and if timely repairs are made in a proper manner. In most cases, maintenance is applied only to smoothing of the road surface, and drainage facilities receive little attention until their failure damages the travel-way itself.

All ditches, culverts, and bridges must be inspected on a regular schedule and repaired or cleaned whenever damaged or obstructed. At no time should grading leave a berm between the roadbed and the ditch line. When pulling ditches, the backslope should not be undercut because this will cause sloughing into the ditch and result in washout and bank erosion. Daylighting heavily shaded roads by cutting away overhanging trees so that the road will dry quickly from exposure to sun and wind is good preventive maintenance.

Abandonment and Bedding Down: When a haul road is abandoned, steps must be taken to minimize erosion and establish a vegetative cover. For complete abandonment, culverts and other structures are removed, and the natural drainage pattern is restored. Side ditches should be obliterated and properly spaced grade dips or water bars should be constructed to handle roadway cross drainage. A water bar must be placed at the head of all pitch grades, regardless of spacing. All road surfaces must be ripped, treated with soil amendments, seeded with grasses, legumes, and trees, and mulched. Seeding will help stabilize the abandoned roads, provide food for wildlife, and improve the aesthetics.

An effective program to check erosion from haul roads must consider all phases, and specific procedures must be established for each one during the planning stage. Knowledge is currently available on all phases of haul-road life. Applicable criteria have been developed by the U.S. Forest Service through years of engineering study and experience. The U.S. Environmental Protection Agency, Region 10, Seattle, Washington, has developed guidelines for the construction of logging roads that have been modified for mine-access roads. Many states include haul-road standards in their surface mine regulations and require these roads to be bonded. West Virginia Surface Mining Regulation 20-6, Series VII, Section 5, is a representative example of State access road controls.

REVEGETATION STUDIES

The material in this chapter is excerpted from the following reports:

PB 187 738
PB 191 360
PB 231 559

REVEGETATION STUDIES AT THREE PENNSYLVANIA SITES

As part of an experiment in the restoration of lands strip-mined for coal, revegetation studies were conducted at three backfilled strip-mine sites in north-central Pennsylvania. The sites are located in Clinton, Elk and Clearfield Counties near the villages of Westport, Benezette, and Penfield.

Since the sites upon which these studies were conducted are located in the same general area and at approximately the same elevation, it was felt that climatic conditions such as temperature and rainfall could be considered constant among them, and that the data and conclusions of this report are applicable to other sites having similar climatic conditions.

Preparation of Study Areas

Varied amounts of lime and fertilizer were applied to the spoil. In all cases the spoil was thoroughly disked both before and after the lime and fertilizer were applied with a tractor drawn spreader. The fertilizer used contained 10% nitrogen, 10% phosphate, and 10% potash.

When certain pH levels were desired the amounts of lime required were determined through soil-sampling and analysis. At least 24 evenly spaced soil samples were taken in each area in which a different pH level was desired. These samples were titrated and analyzed, and from the results of the analyses the amounts of pure lime required were calculated. The quantities of lime applied were corrected to account for the actual analysis of the liming material used.

Mixtures of grasses and legumes were recommended by the committee because the grasses supply a quick ground cover and the legumes are a continuing source of nitrogen for the grass and contribute to the ground cover. Quick ground cover is desirable as a control for early erosion damage.

The nitrogen supplied by the legumes is important because the nitrogen of the fertilizer would be exhausted within a few years. The seeds were also applied to the treated area with the tractor drawn spreader. The entire area was disked lightly after the application of the seed. The moisture content of the spoil was high at the time of planting, but severe drought conditions prevailed during the following two summers.

5,786 trees and shrubs were planted on about 6-foot centers by two-man teams. The trees and shrubs were planted in holes about 4 inches in diameter and about 8 inches deep. The seedlings were not watered, and drought conditions prevailed during most of the first growing season. Lack of moisture may have had a more adverse effect on some species than on others. The seedlings were received in good condition and were kept in cold storage until planted.

Effect of Lime and Fertilizer on Grass-Legume Mixtures

For the studies on the effect of lime and fertilizer on grass-legume mixtures, a finely and evenly divided spoil having an initial pH of 4.5 was selected. This spoil was divided into three primary plots, and the primary plots were subdivided into three secondary plots.

Two primary plots were limed to a desired pH level and the other was not treated. Of the three secondary plots, within each primary plot, one received no fertilizer and the other two received 250 and 500 pounds per acre respectively. Seven different grass-legume mixtures were planted across the nine plots so that 63 samples were obtained.

These were evaluated in terms of ground-cover percentage which was that percent of the earth that appeared covered with vegetation when viewed from eye level. The data from the individual samples did not show a consistent trend. The reported data was obtained by averaging the individual ground-cover percentages both within and across the treatment plots.

Table 4.1 gives the pH levels, the amount of fertilizer applied, and the average ground-cover percentage in each secondary plot and each primary plot. Table 4.2 gives the seeding rate and the average ground-cover percentage for each grass-legume mixture across the treatment plots.

TABLE 4.1: AVERAGE GROUND-COVER PERCENTAGE FOR THREE pH LEVELS

pH Level for Primary Plot	Fertilizer lb/acre	Average Ground Cover for Secondary Plot, percent	Average Ground Cover for Primary Plot, percent
4.5 (not treated)	0	20	
	250	25	25
	500	30	

(continued)

pH Level for Primary Plot	Fertilizer lb/acre	Average Ground Cover for Secondary Plot, percent	Average Ground Cover for Primary Plot, percent
5.5	0	30	
	250	35	35
	500	40	
6.5	0	30	
	250	35	35
	500	40	

Source: PB 187 738

Table 4.1 shows that, within each lime treatment, there was an increase in the ground-cover percentage as the amount of fertilizer applied was increased. When the pH of the spoil was increased to 5.5, the average ground-cover percentage increased by 10%. No further increase in the average ground-cover percentage was obtained when the pH of the spoil was increased to 6.5. The same ground-cover percentage was obtained in the area limed to a pH of 5.5 with no fertilizer as in the untreated area with 500 pounds of fertilizer per acre.

TABLE 4.2: GRASS AND LEGUME GROUND-COVER PERCENTAGE

Mixture	Seeding Rate, lb/acre	Average Ground Cover, percent
Creeping red fescue Serecia lespedeza	30 8	45
Creeping red fescue	40	40
Kentucky 31 tall fescue Crownvetch	30 10	40
Kentucky 31 tall fescue Serecia lespedeza	30 8	35
Timothy Birdsfoot trefoil	6 8	25
Kentucky 31 tall fescue Birdsfoot trefoil	30 8	22
Perennial rye grass Rye grain	50 2.5*	20

*bu/acre

Source: PB 187 738

Table 4.2 shows that the mixtures containing a fescue did much better than those without. One exception to this is the mixture containing Kentucky 31 tall fescue and birdsfoot trefoil. It is felt that this exception was due to the development, after the planting had been completed, of several acid water seepage points in one of the areas in which this mixture was planted. This acid water reduced the ground-cover percentage in that area to zero, thereby lowering the average for the mixture.

The legumes accounted for only a small part of the average ground-cover percentage. Since these legumes start slowly by spread, their contribution to the total ground cover should increase.

Effect of Lime and Fertilizer on Trees

For the studies on the effect of lime and fertilizer on trees, two 6 acre areas were selected. One had an initial pH above 4.5, the other was below 3.5. Both of the areas were divided into six equal plots, and a different amount of lime and fertilizer was applied to each. Seven species of trees were planted across these treated plots. Tables 4.3 and 4.4 show the amounts of lime and fertilizer applied, the species and number of trees planted, and the second year survival percentage for each species in each plot. Survival percentage was determined by actual count of the surviving trees at the end of the second growing season.

TABLE 4.3: SECOND-YEAR SURVIVAL PERCENTAGE WITH INITIAL SPOIL pH ABOVE 4.5[1]

	Survival percentage					
	plot 1	plot 2	plot 3	plot 4	plot 5	plot 6
Lime, tons/acre.....	0	1	2	0	2	4
Fertilizer, lb/acre.	0	150	0	300	300	600
Japanese larch......	82	72	70	82	62	46
Jack pine...........	89	56	86	80	73	84
White pine..........	73	82	58	66	86	76
Austrian pine.......	78	40	48	44	64	60
Norway spruce.......	94	97	93	93	86	92
Pitch pine..........	90	100	87	96	74	95
European alder......	87	38	56	64	40	58

[1]/ 88 European alders were planted; and 176 of each of the other trees were planted.

Source: PB 187 738

No appreciable effect on the survival percentage of any species is apparent as a result of varying the amounts of fertilizer. Table 4.3 shows that where the initial pH was above 4.5 there was no appreciable effect on the survival percentage of any species by varying the amounts of lime applied. The strip not limed or fertilized shows generally as high a survival percentage as the strip where 4 tons of lime and 600 tons of fertilizer were applied. Table 4.4 shows that where the initial pH was below 3.5 the amount of lime applied affected survival percentages of all species, but that very large amounts of lime are required.

From the data it can be concluded that: (1) on spoils having an initial pH above 4.5 the application of lime is likely to have no effect on the survival percentage, but on spoils having a pH below 4.5 the survival percentage should increase as the amount of lime applied increases until the pH of the spoil reaches 4.5; (2) on spoils having an initial pH below 3.5 a very large amount of lime is required to obtain any survival at all; and (3) fertilizer does not affect the second-year survival percentage.

TABLE 4.4: SECOND-YEAR SURVIVAL PERCENTAGE WITH INITIAL SPOIL pH BELOW 3.5[1]

	Survival percentage					
	Plot 1	Plot 2	Plot 3	Plot 4	Plot 5	Plot 6
Lime, ton/acre	0	4	8	0	8	16
Fertilizer, lb/acre	0	250	0	500	500	1000
Japanese larch	0	22	40	0	25	56
Jack pine	0	0	14	0	16	22
White pine	0	14	10	8	25	33
Austrian pine	0	0	0	0	0	2
Norway pine	0	20	10	4	25	27
Pitch pine	14	50	69	14	68	72
European alder	0	16	14	4	11	36

[1]/ 186 trees of each species were planted.

Source: PB 187 738

Relative Hardiness of 14 Species of Trees and Shrubs

Fourteen species of trees and shrubs were planted in a spoil having an initial pH above 4.5. Table 4.5 lists the species of trees and shrubs planted, the number planted, and the survival percentage. The species are listed in order of decreasing hardiness based on the average survival percentage.

Five species exceeded 70% survival, which would be considered good. These five included a spruce, a locust, a larch, and two pines, and provided a wide variety of types of cover. All other species except the Sawtooth oak, the Austrian pine, and the European alder approached 50% which would be considered adequate cover.

TABLE 4.5: AVERAGE SURVIVAL PERCENTAGE OF TREES AND SHRUBS WITH INITIAL SPOIL pH ABOVE 4.5

Species	Number planted	Survival percentage
Norway spruce	220	82.5
Black locust	244	80.5
Japanese larch	238	77.0
Pitch pine	238	75.5

(continued)

TABLE 4.5: (continued)

White pine	222	74.0
Jack pine	236	68.5
Lespedeza natob	251	66.5
Bristly locust	260	55.0
Autumn olive	251	54.5
Scotch pine	222	48.0
Tat honeysuckle	212	47.5
Sawtooth oak	260	42.5
Austrian pine	238	42.0
European alder	248	34.5

Source: PB 187 738

PLANTING OF 14 TREE SPECIES ON PENNSYLVANIA ANTHRACITE STRIP MINE SPOILS

Materials and Methods

Four strip mine spoil types were selected as experimental planting sites. These types are characterized by distinct differences in the predominant parent material as follows:

Type I — Black carbonaceous shale
Type II — Gray to yellow shales
Type III — Sandstones and conglomerates
Type IV — Glacial till and surface deposits

Each spoil type was represented in the experimental plantings by both graded and ungraded conditions. The spoils of the ungraded sites varied from 10 to 15 years old (since stripping) and slope angles were approximately 40°. The graded sites were 1 to 3 years old (since grading), and most slope angles were between 5° and 10°. All planting sites were sampled systematically to a depth of 12 inches for analysis of selected physical and chemical properties.

The soil fraction (material passing through a 2 mm sieve) varied considerably among the spoil types, from less than 30% in Type I to 70% in Type IV. Moisture retention percentages also varied. So individual spoil types and grading conditions varied considerably in their capacities to store and to supply moisture for tree survival and growth. However, under average rainfall conditions in the anthracite region, retention capacities are adequate for normal tree survival and growth. Although pH values varied considerably between spoil types and grading conditions, acidity itself was not a problem. In spoil Types I and III, pH values were appreciably higher in graded spoils than in ungraded spoils. The reasons for this are not clear. The lower pH values may be due partially to weathering processes, leaching of bases, and surface erosion on the ungraded spoils. On the other hand, grading of spoils exposed unweathered and unleached material.

Exchangeable bases, extractable phosphorus, and total nitrogen varied from one spoil type and one grading condition to another. The overall natural supply of these essential plant nutrients was low; nitrogen, in particular was probably too low for normal tree growth.

The total bases (Ca + Mg + K) in graded spoil Types I, II, and III were considerably higher than in their ungraded counterparts. Ca accounted for 79 to 83% of the total exchangeable bases in the graded spoils and 53 to 64% in ungraded spoils. There were no appreciable differences in chemical or physical characteristics between graded and ungraded sites on spoil Type IV.

The 14 species used in the experiment (Table 4.6) were selected to include those that had shown the most promise in earlier mine spoil plantings in both the anthracite and bituminous regions of Pennsylvania. The field layout consisted of 32 blocks; 8 on each spoil type, with 4 replicates each on graded and ungraded spoils. Blocks were 84 by 50 feet, with the longer dimension on the contour, thus providing for 14 rows (6 feet apart and 50 feet long) running up and down slope.

TABLE 4.6: TREE SPECIES AND AGE OF THE PLANTING STOCK

Species	Age
Hybrid poplar, *P. maximowiczii x P. trichocarpa NE-388*	(1)
European alder, *Alnus glutinosa* (L.) Gaertn.	1-0
Black locust, *Robinia pseudoacacia* L.	1-0
Jack pine, *Pinus banksiana* Lamb.	2-0
Red pine, *Pinus resinosa* Ait.	2-0
Scotch pine, *Pinus silvestris* L.	2-0
Pitch pine, *Pinus rigida* Mill.	2-0
White pine, *Pinus strobus* L.	2-0
Virginia pine, *Pinus virginiana* Mill.	1-0
Austrian pine, *Pinus nigra* Arnold	2-0
Japanese larch, *Larix leptolepis* (Sieb. and Zucc.) Gord.	2-0
European larch, *Larix decidua* Mill.	2-0
Norway spruce, *Picea abies* (L.) Karst.	2-0
White spruce, *Picea glauca* (Moench) Voss.	2-0

[1] Cuttings

Source: PB 191 360

In each block, 11 rows as a group were assigned to the 11 coniferous species and 3 rows were assigned to the hardwoods. The species were randomized separately, by rows, within these groups, The conifers were planted with 2 foot spacing, or 17 trees per row. Thus, in the 32 blocks, 800 seedlings of each coniferous species and 544 seedlings (or cuttings, in the case of hybrid poplar) of each hardwood species were planted. These added up to 8,800 conifers and 1,632 hardwoods for a grand total of 10,432 seedlings and cuttings.

Planting was done by crews of experienced planters, using planting bars, during typical spring weather when spoil moisture conditions were favorable. Survival and height growth were recorded at the end of each growing season and were summarized by species, spoil type, and grading condition. During the 5 year period, particularly in 1966 and 1967, a few blocks or parts of blocks were destroyed by renewed stripping or grading operations that were not anticipated when the sites were selected. Consequently, only limited use could be made of statistical methods for analysis of data.

Results

Tree survival on graded sites was better than on ungraded sites. Overall survival was 70% for hardwoods and 60% for conifers on graded sites, compared to 40% for hardwoods and 32% for conifers on ungraded sites. The overall growth of conifers was poor to fair. Only three species—jack pine, Virginia pine, and Japanese larch—exceeded an average height of 3 ft. Growth of hardwoods was better: average heights on graded sites ranged from 6.4 to 13.0 ft. The first three growing seasons were extremely dry, accounting for some of the mortality and retarded growth.

Much variation occurred between species, spoil types, and replicates. A number of instances of 90% or higher survival, and growth considerably better than average, were recorded. The 5 year results in terms of survival and total height are discussed below in relation to the three independent variables—spoil type, grading condition, and species.

Spoil Type: The most marked spoil type effect was the poor performance on Type IV (glacial till and surface deposits). This was unexpected in view of the relatively high content of "soil"-size particles and the near neutral reaction of Type IV spoil. The reason for this is not clear. However, the surface of this spoil becomes readily sealed, thus reducing rainfall infiltration, increasing water runoff, and causing intense surface erosion that resulted in tree mortality. This was not so pronounced in the other spoil types. Also, pH was possibly somewhat above the optimum for growth of essential mycorrhizal fungi and other symbiotic root organisms.

No strongly consistent differences were found among the other three spoil types. Type I usually was somewhat less favorable than Types II and III, but this did not hold for all replicates and species. Because of its relatively low content of "soil"-size particles and its generally darker color, spoil Type I, and spoils like it, can be expected to be the least favorable of these three types.

Spoil types frequently vary within themselves and intergrade with each other to a degree that differences affecting plant performance are not clearly or consistently evident, at least during the early stages of growth.

The tendency of rocks to roll and slide, particularly on spoil Types I and II, is another critical feature that adversely affects survival and growth. Although no attempt was made to evaluate the relative contribution of erosion and rock rolling and sliding to mortality, there is strong evidence that considerable mortality and damage was caused by rock rolling and sliding on spoil Types I and II and by erosion on spoil Types III and IV.

Erosion and rock sliding, or a combination of both, reduced tree growth by injuring and partially covering trees. Partially covered trees frequently bend, and subsequently their growth slows.

Grading Condition: The effect of grading on tree performance was much more clearly defined than the effect of spoil type. Survival on graded spoils averaged almost twice as high as on ungraded spoils. This effect was most strongly expressed on spoil Type IV, but was remarkably consistent for all species and on all spoil types. Growth was consistently better on the graded spoils, although in some instances the differences were relatively small.

These results disagree with work reported from the Central States, where grading was found to have adverse effects. However, the adverse effects there were most evident on spoils derived from limestone and having a high content of clay. Such material, when graded, becomes very compact and impervious to air and water. None of the Pennsylvania anthracite spoils are of limestone origin, nor do they have such high proportions of clay. The coarser-textured anthracite spoils obviously were improved by the degree of compaction resulting from grading operations. Besides its compaction effects, grading also improves anthracite spoils as planting sites by reducing steepness of slopes. Thus, damage from erosion and rock slidings are lessened.

Species: As expected, apart from spoil type and grading condition effects, individual species performed differently. In the following paragraphs an attempt is made to present the results, based on survival and growth data and supplemented by subjective judgement, by pointing out the better (or poorer) performers and by relating their performance to site factors, planting stock used, and inherited characteristics of early growth.

Hardwoods — The three hardwoods—hybrid poplar, black locust, and black alder—varied appreciably in their performance. Their survival and height growth appeared to be affected primarily by grading and spoil type. All three species had excellent survival and good growth on graded spoil Types I and III, and poor survival and growth only on ungraded spoil Type IV. Hybrid poplar, with an overall average survival of 77 and 55% and a total height of 13.0 and 3.2 ft for graded and ungraded spoils respectively, performed best among the three hardwoods.

Pines — The performance of the seven pine species varied appreciably between grading condition and between species. In general, jack, red, Scotch, and pitch pines had better survival than white, Virginia, and Austrian pines. Virginia pine had the best overall height growth, followed by jack and Scotch pines. The growth of Virginia pine deserves special attention since 1-0 nursery stock was used, compared to 2-0 nursery stock used for the other pines.

Larches — Japanese and European larches behaved similarly with respect to spoil type and grading condition. Both species survived adequately (45 to 50%) and grew relatively well (2.3 to 3.3 ft). There was no drastic difference in survival or height growth as in the hardwoods and pines. Their best performance was on spoil Type II, and their poorest on spoil Type IV. Average height growth of both species under all conditions was comparable to that of jack, Virginia, and Scotch pines.

Spruces — Norway and white spruces had similar survival and growth rates, but were less successful than all other species. Adequate survival for both species was found only on graded spoil Types I and III and also, for white spruce, on ungraded spoil Type II. On the other spoil types and grading conditions, survival was poor to negligible. Average heights varied between 0.5 and 1.0 ft. Since slow growth during the early years is normal for these species at this age, this does not necessarily show whether the species is succeeding or failing. The low number of surviving trees on certain spoil types and grading conditions prohibited firm conclusions about height growth.

Discussion and Conclusions

Results obtained in this study confirm and supplement previous findings and provide evidence that many forest plantings on anthracite strip mine spoils can survive and grow well. Certain species may even thrive. However, one can raise many questions about the cause of variability in performance between spoil types, and graded versus ungraded conditions, and about the suitability of individual species and genera.

Since neither the causes of mortality nor spoil characteristics affecting survival and growth were studied in detail and correlated, this study does not fully explain why such differences in performance of many species were found. However, there is no doubt that the causes of variability in survival and growth lie in obscure differences, primarily in physical make up, of the planting media.

A few factors, based on yearly survival counts and growth data and on visual observations made during visits to the experimental sites, became apparent. Most mortality occurred within 2 years after planting. It also appeared evident that erosion and rock sliding and rolling were the main factors contributing to mortality and stunted growth.

Because the rate of erosion is related to the soil fraction content and the slope angle, ungraded spoil Type III and, in particular, Type IV are susceptible to the most severe erosion. This makes the environment for tree establishment and growth very critical on these types, especially during heavy rains.

Further study of erosion control on these spoils warrants consideration. It was observed that the seedlings of species having reasonable height, stem diameter, and well-developed root systems had a better chance to survive. Jack, red, Scotch, and pitch pines, and occasionally Austrian pine, the larches, and hybrid poplar, apparently are superior in withstanding the adverse effects of strip mine spoils. On the other hand, Virginia and white pines, both spruces, and under certain conditions, even black locust had difficulty in becoming established, especially when planted on steep ungraded spoils.

Perhaps most remarkable is the strong evidence that plantings on graded sites had considerably higher survival and better height growth. For the sake of objectivity, it should be mentioned that the year of planting, and the following two growing seasons, were extremely dry. This, without question contributed to mortality. These experimental plantings were too young to provide firm conclusions, especially about height growth. However, the following generalizations can be made:

After 5 years all four anthracite strip mine spoil types supported planted forest trees.

The degree of performance varied from poor to excellent, depending on tree species, spoil type, and grading conditions.

Hardwoods survived somewhat better and had much better height growth than conifers. Hybrid poplar clone NE-388 performed best.

Jack, red, Scotch and pitch pines showed superior survival and attained good height growth among conifers. Virginia pine was better only in height growth.

Larches survived well under a wide range of conditions and attained good growth on ungraded sites.

Survival of spruces was poor to adequate. Their early slow growth makes them a poor planting choice, especially on ungraded sites where erosion and rock sliding occur continuously.

Erosion and rock sliding are prime factors of tree mortality, especially on ungraded sites. So, graded sites, regardless of spoil type, are superior tree growth media.

VEGETATIVE ANALYSIS OF 81 STRIP MINE AREAS OF PENNSYLVANIA

Description of Study Area

Mercer County is located in northwestern Pennsylvania, approximately midway between Pittsburgh and Erie. Strip mining began in Mercer County after the second World War, reaching a peak between 1950 and 1964. The total area stripped involved 4,470 acres or approximately 1% of the area of the county.

There were four active surface mining operations in the county at the time of the study. The strip mine spoil in Mercer County has been classified as either gentle sloping or moderately steep slope where some areas have been leveled but the majority of the area remains as rough terrain, ranging from 10 to 100 acres in size (1,040 acres). Steep spoil is characterized as having many short, steep slopes and some with water at the base and the acreage ranging from 10 to 200 acres in size (2,910 acres). The abandoned strip mine areas in Mercer County are characterized by a spoil consisting of fragments of limestone, shale, and impure coal.

The majority of the coal mined extensively in Mercer County was the Brookville coal of the Clarion formation. The Clarion formation is composed of shale, sandstone and at least two coals and their underclay. The base of the formation is defined as the base of the Brookville coal or the base of its underclay, if underclay is present.

The Scrubgrass coal occurs near the top of the formation. The Clarion coal of the Clarion formation is either absent or is combined with the underlying Brookville layer. The Brookville coal is overlayed with 30 to 90 feet of overburden consisting of glacial till, sandstone, shale and limestone. The Brookville coal seam reaches its greatest thickness in the southern portion of the study area. Pyrite is present in the Brookville coal throughout most of the area. In the southern portion of Mercer County, the lower Kittanning coal was mined extensively. The Kittanning coal consists of clay shale, sandy shale and sandstone and represents disposition under a variety of conditions.

Vegetative Analysis

A preliminary examination has been completed on 81 strip mine areas in Mercer County, varying in age from 1 to 30 years. The mines were located on aerial photographs. The vegetative types and the area coverage was determined from a site examination as well as from aerial photographs of the area. The vegetative associations were classified as to the dominant vegetation according to coverage area. These vegetative associations were then determined as to whether they were natural revegetation of the area or whether these areas were planted during reclamation. Further analysis was completed on the species utilized by the U.S. Soil Conservation Service in reclaiming strip mines. These species were rated as poor, fair, good or excellent according to their survival on the site.

The pH varied from 4 to 8 and indicated that a considerable variation exists between, as well as within, strip mine areas. This variation is probably the result of the methods employed during the actual mining and back-filling at the completion of the strip operation. The moisture content and organic matter remained relatively constant to a depth of 0.9 meter and they were related to the strip operation.

Vegetative analysis of 81 strip mine areas indicated that an equal distribution occurred between the pine, deciduous, mixed deciduous and grassland associations. The pine association is found less frequently than the other three associations. The dominant species in the pine association is white pine *(Pinus strobus)* with an even distribution between red pine *(Pinus resinosa)*, Virginia pine *(Pinus virginiana)*, larch *(Larix laricana)* and red and black spruce *(Picea rubers* and *Picea mariana)*. The mixed deciduous association consisted of quaking aspen *(Populus tremuloides)* as the dominant species with an even distribution between red maple *(Acer rebrum)* and black cherry *(Prunus serotina)*.

The mixed deciduous and pine associations are characterized by a mixture of the previous two associations with the addition of black locust *(Robina hispida)*. Grassland associations are characterized by a variety of grasses with switch grass *(Ranicum virgatum)* and sweet clover *(Melilotus* spp) as two of the dominant species. There was also a mixture of other varieties of grasses present in the grassland associations. For example, birdsfoot trefoil *(Lotus americanus)* has been used extensively in reclamation of strip mines.

Pine associations are a result of direct planning by the mine operator following the completion of the surface mining operation. The deciduous associations are a result of natural revegetation with the possible exception of black locust which was planted on strip mine sites among pines. The pine association provided the least amount of cover, while the grassland association provided the maximum amount of coverage within the shortest length of time.

On the other hand, the mixed pine-deciduous association provided excellent cover on mines after 15 or 20 years following the mining operation. Therefore, after a strip mine operation grasses could be planted on the site followed 3 to 4 years later by a planting of a mixed pine-hardwood association for both beautification and food and cover for wildlife. In addition to these vegetative associations, an additional 16 species of grasses and 30 species of trees and shrubs have been planted by the U.S. Soil Conservation Service with a varied degree of success (Tables 4.7 and 4.8).

TABLE 4.7: ESTIMATED SURVIVAL OF GRASSLAND SPECIES

Common Name	Species	Survival in Strip Mines*
Switch grass	Ranicum virgatum	Excellent
Birdsfoot trefoil	Lotus americanus	Excellent
Crown vetch	Coronilla varia	Good
Sand reed grass	Ammophila breviligulata	Poor
Alfalfa	Medicago sativa	Good
Redtop	Agrotis ālba	Fair
Sand love grass	Eragrostis spp.	Poor
Tall oats grass	Arrbenatherum elātus	Fair
Rig blue stem #145	Solidago caēsia	Fair
Tall fescue	Festuca elātum	Fair
Indian grass	Hierochloe odorāta	Poor
Wagner flat pea	Lathyrus sylvestris	Good
Deer tongue grass	Paspalum xalapaese	Excellent
Weeping love grass	Eragrostis curvulae	Good
Canada blue grass	Póa compressa	Fair
Hop clover	Tribilium agrárum	Poor

*Poor - 0-25% survival
Fair - 25-50% survival
Good - 50-75% survival
Excellent - 75-100% survival

Source: PB 231 559

TABLE 4.8: ESTIMATED SURVIVAL OF TREES AND SHRUB SPECIES

Common Name	Species	Survival in Strip Mines*
Buck thorn	Rhamnus spp.	Excellent
Cardinal autumn olive	Elaeagnus curvulae	Good
Bristly locust	Robinia hispida	Excellent
Black locust	Robinia pseudo-acácia	Excellent
Common bittersweet	Celastrus scandens	Fair
Pfitzer juniper	Juniperus chinensis pfitzerana	Good
Red-oser dogwood	Cornus stohonifera	Poor
Amur honeysuckle	Lonicera amoena	Fair
Amur privet	Ligrustrum amurense	Fair
Mugo pine	Pinus mugo	Poor
White pine	Pinus strobus	Good
Virginia pine	Pinus virginiana	Fair
Red pine	Pinus resonsa	Good
Eastern red cedar	Juniperus virginilána	Fair
Douglas fir	Pseudotsuga taxifolia	Fair
White birch	Betula papyrifera	Excellent
Black birch	Betula leuta	Fair
Yellow crabapple	Malus sieboldii	Excellent
Crabapple	Malus baccata	Excellent
Hall deer apple	Malus halliana	Excellent
Hawthorne	Crataegus succulenta	Excellent
Sawtooth oak	Quercus aculueina	Fair

(continued)

TABLE 4.8: (continued)

Common Name	Species	Survival in Strip Mines*
Forsythia	*Forsythia* spp.	Fair
Shrub honeysuckle	*Lonigera maakii*	Fair
Rose of Sharon	*Hibiscus syriacus*	Poor
Mountain ash	*Pyrus americana*	Good
European alder	*Alnus incana* (acid soil)	Good
Arbor vitae	*Thuja occientalis*	Fair
Chinese chestnut	*Castanea mollissima*	Fair
Hybrid poplar	*Poulus balsamifera x P. gilladensis*	Fair

*Poor - 0-25% survival
 Fair - 25-50% survival
 Good - 50-75% survival
 Excellent - 75-100% survival

Source: PB 231 559

SPOIL AMENDMENT STUDIES

The material in this chapter is excerpted from the following reports:

COM-72-10623
PB 226 905
PB 232 069

USE OF SEWAGE EFFLUENT AND LIQUID SLUDGE

Many thousands of acres stripped in Pennsylvania before passage of the Surface Mining Conservation and Reclamation Act remain void of vegetation due to the extremely acid and toxic nature of the spoil material. These sites are subject to serious erosion problems and continue to produce large quantities of acid runoff. The specific objectives of the study described in PB 232 069 were:

To determine effects of irrigation with municipal sewage and liquid sludge on the germination, survival and growth of grass and legume species;
To study the changes in the chemical composition of the spoil leachate after effluent and sludge irrigation; and
To determine the effect of effluent and sludge irrigation on the chemical characteristics of the spoil, and on spoil temperature.

Methods

Box Construction and Spoil Material: Six large boxes were constructed during the summer of 1971 at the Waste Water Renovation and Conservation Project Facility of the Pennsylvania State University. Each box measured 4 feet wide, 4 feet deep, and 32 feet long, and was constructed of plywood, with the bottoms and tops open. The inside walls of the boxes were lined with a reinforced plastic sheeting, to prevent moisture from moving through the box. The outside of the boxes were painted white. No lining was installed in the bottoms which rested on topsoil, thus allowing free passage of leachate through the spoil and into the soil.

Spoil Amendment Studies

The bottom of each box was filled to a depth of six inches with white sand to provide a medium in which to seat tension lysimeters. Three tension lysimeters were installed in each box for collection of leachate samples, the porous ceramic ends of which penetrated into the sand layer. The remaining three and one-half feet of each box was filled to the top with approximately 25 tons of spoil per box.

The spoil material was obtained from a spoil bank over the Lower Kittanning coal seam in Clearfield County in Pennsylvania's bituminous coal region. The spoil was extremely toxic, having a pH in the range 2.0 to 3.0, and was largely composed of shale. This spoil bank had supported no vegetation at all for 23 years.

Vegetation and Seeding: On May 22, 1972 the six boxes were seeded with eight grass and eight legume species. Each two-box treatment replicate was seeded with both vegetation types, thus, three boxes were seeded with grasses and three boxes were seeded with legumes. Eight, four foot sections were delineated along the 32 foot length of each box and separated with wooden dividers. Each 4 foot by 4 foot plot of each box received 2 ounces of seed of one species, equivalent to a seeding rate of 340 pounds of seed per acre. The grass and legume species used and the seeding design are given in Figure 5.1.

FIGURE 5.1: GRASS AND LEGUME SEEDING DESIGN*

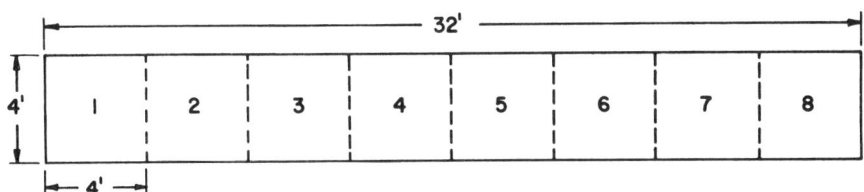

One of each of the three treatment rates was applied to a box planted with grasses, and one of each was applied to a box planted with legumes.

Grasses	Legumes
1 Garrison creeping foxtail	1 Sericea lespedeza
2 Saratoga smoothbrome	2 Chemung crownvetch
3 Reed canarygrass	3 Lathro flatpea
4 Blackwell switchgrass	4 Iroquois alfalfa
5 Weeping lovegrass	5 Pennscott red clover
6 Redtop	6 Sweet clover
7 Deertongue	7 Ladino clover
8 Climax timothy	8 Arnot bristly locust

*The numbered box compartments correspond with the numbered species.

Source: PB 232 069

Effluent and Sludge Treatment and Design: The treated municipal sewage effluent and liquid digested sludge used in the treatment of the spoil material was obtained from the Pennsylvania State University Sewage Treatment Plant which serves the University and the borough of State College. Effluent applications were made weekly. Effluent was applied to the spoil through an irrigation system with full- and part-circle sprinklers.

Applications of sludge were also made weekly for a total of seven weeks. When possible the sludge was applied on the day prior to effluent irrigation. This system of sludge application followed by effluent treatments made it possible to wash sludge particulates from the vegetation. Sludge was applied to the spoil by pumping it from a holding lagoon into the irrigation system main line with a trash pump, and finally to the boxes through a ¾" garden hose. Samples of sludge and effluent were taken during each irrigation and stored under refrigeration for chemical analysis.

The treatments were set up to irrigate one box of grasses and one box of legumes with two inches of effluent plus two inches of sludge per week, referred to as the two-inch combination treatment (2E + 2S). The second pair of boxes, one planted with grasses and one with legumes, was irrigated with one inch of effluent plus one inch of sludge per week, the one-inch combination treatment (1E + 1S). The grass and legume control boxes received no treatments.

The two-inch combination treatment boxes received a total of 33 inches of effluent and 14 inches of sludge through the growing season, while the one-inch combination treatment boxes received 16.5 inches of effluent and 7 inches of sludge.

Leachate Sampling: Samples of leachate were collected weekly, generally on the fourth day after effluent irrigation, and placed under refrigeration for chemical analysis with the effluent and sludge samples at a later date. Leachate samples were obtained by placing a vacuum in the tension lysimeters with a vacuum pump prior to the wastewater applications. The samples obtained from the three lysimeters in each box were combined to give an average sample of leachate for that box.

Chemical Analysis of Effluent, Sludge and Leachate: Refrigerated samples of effluent and leachate were analyzed for pH, total acidity, sulfates, nitrate nitrogen, organic nitrogen, ammonium, orthophosphate phosphorus, total phosphorus, potassium, calcium, magnesium, manganese, iron, copper, boron, aluminum, zinc and sodium. Sludge samples were analyzed for all but sulfates, orthophosphate phosphorus, and total acidity.

Temperature Measurements: The maximum and minimum daily and mean monthly air temperatures were obtained from the Pennsylvania State University Weather Station, and corrected for determinations at the spoil box site.

The correction procedure was based on data from previous summers establishing the monthly relationship between University Weather Station recorded air temperatures and air temperatures recorded in the field. No recorded data on air temperatures in the field was available for the term of the study, thus necessitating use of the corrected data. Subsurface spoil temperatures for the three and six inch depths were recorded on each box from June 14 to August 11 on 14 days of moderate to intense solar radiation.

Also, surface temperature data was available, collected during the summer of 1971 on the spoil boxes used in this study. Surface temperatures were recorded on days of good to intense solar radiation.

Vegetation Responses: Following the termination of effluent and sludge irrigation in September 1972, grass and legume dry matter production, areal cover, and height growth were determined. In addition, plant samples were extracted from the spoil with their root systems intact for comparison of species development between treatments.

Areal cover and height growth measurements were made for each species on each box on September 21, 1972. Cover determinations were made by visually dividing each four foot by four foot box plot into 16 one square foot subplots and estimating the percentage of vegetation cover for each subplot. Summation of subplot percentages gave the total percent cover for each 16 square foot plot. Height growth was measured to the nearest 0.01 foot, several measurements being taken for each plot and averaged to get the height growth of each species on each treatment.

On September 21 and 22, 1972, the grasses and legumes were clipped and collected for dry matter determinations. Bagged vegetation was oven dried at 80°C for 48 hours, cooled, and weighed to the nearest 0.1 gram. The weight in grams for each species was converted to pounds per acre.

Foliar Analysis: Subsequent to dry matter production measurements, samples of grasses and legumes from each treatment were used for foliar analyses. Oven dried vegetation was ground in a Wiley Mill to pass through a 1.0 millimeter sieve, ashed and analyzed on an emission spectrometer. Foliar samples were analyzed for phosphorus, potassium, calcium, magnesium, manganese, iron, copper, boron, aluminum, zinc and sodium. A separate plant sample was analyzed for nitrogen by a micro-Kjeldahl procedure.

Spoil Analysis: On September 30, 1972, samples of the spoil material were taken from three depths in each box. Separate samples were taken from the 6 inch, 18 inch and 30 inch depths in the spoil beneath four plots from each box. The spoil samples were dried, sieved to pass a 1.0 millimeter screen, and extracted with ammonium acetate solution. The ammonium acetate extracts were then analyzed on the emission spectrometer for the same elements as those in the foliar analyses except nitrogen.

In addition, the sieved spoil samples taken from the six inch depth were analyzed for pH by mixing spoil material and distilled water in a one-to-one ratio, allowing the slurry to stand 30 minutes, remixing and reading the pH on a pH meter.

Results and Discussion

Chemical Quality of the Sewage Effluent and Sludge: The treated sewage effluent and liquid digested sludge were analyzed to determine their chemical composition. The range and mean values of the constituents in the effluent and sludge are given in Tables 5.1 and 5.2. One of the most noticeable properties of the effluent is its slightly alkaline pH. The irrigation of acid spoils with two inches of alkaline effluent per week has a considerable flushing effect on acid spoil leachate in the upper rooting zone of the spoil.

This flushing of acid leachate with alkaline effluent should improve the chances for survival of grass and legume species susceptible to the toxic acid conditions found in many spoils. Effluent total nitrogen averages 20.1 mg/l; very low compared to that of the sludge, but each two-inch application supplied about 10 pounds of N per acre. Concentrations of manganese, iron, boron and aluminum were very low, all less than 1 mg/l and well below levels toxic to most plants.

The sludge was slightly more alkaline than the effluent, and would have a similar neutralizing effect on acid conditions in the spoil. In addition, mean values for total nitrogen reach nearly 500 mg/l with a considerable amount in the organic form, valuable in soil building processes. The greatest amount of sludge nitrogen was ammoniacal.

Since the sludge was anaerobically digested, one would not expect to find much nitrate nitrogen present. The small amount found was probably formed after delivery to the storage pond. However, the reduced acidity in the surface spoil layers resulting from the alkaline effluent and sludge would create a more favorable environment for such autotrophs as Nitrosomonas and Nitrobacter. Under these conditions, the process of nitrification may take place resulting in the production of nitrates.

TABLE 5.1: CHEMICAL COMPOSITION OF SEWAGE EFFLUENT

Constituent	Range		Mean
	Minimum	Maximum	
pH	6.70	7.99	7.33
	mg/l	mg/l	mg/l
Organic Nitrogen[a]	1.2	3.8	1.7
Ammonium-N	2.8	7.1	5.3
Nitrate-N	6.8	17.5	13.1
Orthophosphate-P	4.8	6.2	5.5
Total acidity[b]	0.1	0.3	0.2
Sulfate Sulfur	11.5	18.5	16.0
Potassium	2.0	68.7	23.6
Calcium	23.8	46.5	34.2
Magnesium	12.6	21.2	17.9
Manganese	0.0	0.1	0.1
Iron	0.5	2.0	0.9
Copper	0.1	0.3	0.2
Boron	0.1	0.3	0.2
Aluminum	0.3	1.0	0.7
Zinc	0.2	0.5	0.4
Sodium	33.4	44.5	38.2

[a] Does not include ammoniacal nitrogen.

[b] Expressed in meq/l as hydrogen.

Source: PB 232 069

TABLE 5.2: CHEMICAL COMPOSITION OF LIQUID DIGESTED SLUDGE

Constituent	Range Minimum	Range Maximum	Mean
pH	7.30	7.52	7.41
	mg/l	mg/l	mg/l
Organic Nitrogen[a]	69.0	644.0	195.1
Ammonium-N	127.0	500.0	300.9
Nitrate-N	2.3	4.4	3.1
Phosphorus[b]	69.0	272.1	118.3
Potassium	75.6	191.6	136.1
Calcium	33.2	290.8	82.2
Magnesium	10.2	72.5	27.1
Manganese	0.3	8.9	1.9
Iron	1.5	90.2	17.7
Copper	0.4	8.9	2.4
Boron	0.2	2.1	1.0
Aluminum	2.9	95.4	18.7
Zinc	0.8	11.7	3.5
Sodium	25.6	74.3	48.5
Total Dry Solids	850	8190	2220
Volatile Solids	440	5470	1390

[a] Does not include ammoniacal nitrogen.

[b] Total P by dry ashing plus 6N HCl.

Source: PB 232 069

Concentrations of manganese, iron, copper, boron, aluminum and zinc in the effluent were well below levels toxic to most plants. Concentrations of these constituents were substantially higher in the sludge than in the effluent, but the values represent total amounts rather than water-soluble concentrations.

The fertilizer values of the sludge and effluent applied on the spoil are given in Table 5.3, including the total amounts of nitrogen, phosphorus and potassium applied over the period of the study. The fertilizer value of the sludge, compared to that of the effluent, is much greater.

The one inch effluent applications supplied the fertilizer equivalent of 750 pounds of a 10-5-14 fertilizer while the one inch sludge applications supplied the equivalent of 3,955 pounds of a 20-11-7 fertilizer. The sludge had roughly 25 times the fertilizer value of the effluent in nitrogen and P_2O_5, on an equal volume application basis.

Weather Conditions from April Through September, 1972: The weather conditions experienced during the spring and summer of 1972 were not typical of normal conditions.

TABLE 5.3: FERTILIZER EQUIVALENT FOR THE EFFLUENT AND SLUDGE

Treatment	Amounts of Nutrients Applied in the Effluent and Sludge (pounds per acre)			Fertilizer Equivalent			
				Amount (pounds per acre)	Fertilizer Formula (percent)		
	N	P	K		N	P_2O_5	K_2O
1-inch effluent	75	16	88	750	10	5	14
plus							
1-inch sludge	791	188	216	3955	20	11	7
Total	866	204	304	3464	25	14	11
2-inch effluent	150	32	176	1500	10	5	14
plus							
2-inch sludge	1583	375	432	7910	20	11	7
Total	1733	407	608	6928	25	14	11

Source: PB 232 069

With minor exceptions in the months of May and September, mean monthly temperatures were lower than the normal values computed for the previous 30 year period. During the month of June, the deviation in the mean monthly air temperature from the normal for that month was nearly 8°F. The patterns of precipitation were also quite unusual.

There was a considerable surplus in each of the first three months of record, with an 8.72 inch surplus in June alone. From July through September, however, there was a deficit in each of the three months. The deficit was most severe in August when precipitation was 3.04 inches below normal. Thus the first three months of record were unusually wet, and the last three somewhat dry.

On an overall basis, the weather conditions during the term of the study could be characterized as being unusually cool and wet, with a final surplus of 6.82 inches of precipitation for the six months recorded. The more moderate temperatures and greater amounts of precipitation most likely had a beneficial effect on grass and legume survival and growth.

Spoil Temperatures: Surface temperatures of the spoil material were measured on seven days between August 20 and September 29 during the summer of 1971 on the same treated boxes of spoil used in the 1972 study. The spoil had been seeded with grasses and legumes and had received weekly applications of effluent and sludge starting in the last week of July, at treatment rates identical to those used in 1972. Measurements of surface temperatures were made one day after treatment on spoil void of vegetation, however, in some box compartments vegetation surrounded the point of measurement. Averages of treatment mean observations show that the 1E + 1S and 2E + 2S treatment rates and the control differed significantly from each other in their effect on surface temperatures, as did the two cover types. It appears inconsistent that one of the highest treatment rates (2E + 2S legume) also had the second highest average surface temperatures. At the 0.95 level of confidence, the grass control and the 2E + 2S legume treatment differed significantly from the 1E + 1S and 2E + 2S grass treatments, while the legume control differed only from the 1E + 1S grass treatment. Thus the effect of treatment on spoil surface temperature one day after treatment application appears to be negligible, on the order of 1° to 3°F, if any.

The highest surface temperature observed throughout this segment of the study was 119.3°F on the 1E + 1S legume treatment, and the highest treatment mean observation for any one day was 115.2°F on the 2E + 2S legume treatment. Thus, none of the observed temperatures fell within the range that is lethal to plants. There are several possible explanations for this.

First, measurements were not made continuously and were partially dependent on weather conditions and cloud cover. Second, periods of rain prior to measurement may have had a cooling effect on the control boxes, as well as the cooling effect of irrigation on treated boxes. And third, surface temperatures were measured relatively late in the summer, starting on August 20, two months after the period of maximum solar radiation. These factors most likely combined to lower surface temperatures below those generally attainable.

Subsurface temperature measurements were made at the three and six inch spoil depths on the four treated boxes and the two controls during a two month period of the summer of 1972. At the three and six inch depths, the grass and legume controls had significantly higher average temperatures than any of the treated boxes. The cover type had no significant effect on subsurface temperature at either depth.

The highest single temperatures observed in the spoil material were on the grass control. Temperatures of 97.7° and 90.5°F were recorded at the three and six inch depths, respectively. The 2E + 2S grass and legume treatments had the lowest average temperatures at both depths. At the 0.95 level of confidence, the grass control was significantly different from the 1E + 1S grass and the 2E + 2S legume and grass treatments at the three inch depth. At the six inch depth, the grass control differed significantly from all four treated boxes.

Grass and Legume Growth Responses and Foliar Analysis: At the end of the growing season (October 1972), the grasses and legumes were measured to determine the percentage areal cover of the plots on which they were seeded and the average height growth. The vegetation was subsequently harvested for dry matter production determinations and foliar analysis. No grass or legume germination occurred on the controls.

The species have been ranked in descending order by the amount of dry matter production on the 2E + 2S treatment. With the exception of Garrison creeping foxtail, the grasses follow the same descending order for dry matter production on the 1E + 1S treatment. The first five legume species show an inverse relationship in dry matter production on the 1E + 1S treatment compared to that obtained on the 2E + 2S treatment. The grasses which showed the best growth response were weeping love grass, Blackwell switch grass and deertongue. Reed canary grass also showed an impressive yield of 3,215 pounds per acre on the 2E + 2S treatment.

Weeping love grass, one of the most acid tolerant species of grasses, produced 136% more dry matter (11,067 pounds per acre) than Blackwell switch grass (4,689 pounds per acre), its closest competitor on the 2E + 2S treatment. The relationship was nearly the same on the 1E + 1S treatment where weeping love grass produced 126% more dry matter than Blackwell switch grass. Saratoga smooth brome (930 pounds per acre) and Garrison creeping foxtail (0 pounds per acre) gave the poorest yields on the 2E + 2S and 1E + 1S treatments, respectively.

The legumes that yielded best were ladino clover, Iroquois alfalfa, sericea lespedeza and Pennscott red clover on the 2E + 2S treatment, with production ranging from 3,078 to 2,718 pounds per acre. The yield of bristly black locust was fifth best on the 2E + 2S treatment with 2,432 pounds per acre, but was the highest of all legumes tested (1,193 pounds per acre) on the 1E + 1S treatment.

Pennscott red clover gave the second highest yield on the 1E + 1S treatment with 519 pounds per acre, 44% of that of the bristly black locust. Lathro flat pea (761 pounds per acre) gave the lowest yield on the 2E + 2S treatment, and sweet clover (0 pounds per acre) gave the lowest yield on the 1E + 1S treatment.

The 2E + 2S treatment resulted in greater yields than the 1E + 1S treatment for both grasses and legumes. The yields on the 1E + 1S treatment of the most productive grasses, relative to their corresponding yields on the 2E + 2S treatment, were as follows: weeping love grass, 55%; Blackwell switch grass, 58%; deertongue, 48%; and reed canary grass, 20%. The ratio of yields for the legumes was: ladino clover, 7%; Iroquois alfalfa, 13%; sericea lespedeza, 17%; Pennscott red clover, 19%; and bristly black locust, 49%.

The height and growth of the grasses and legumes is more indicative of the natural stature of the species than of successful establishment and growth on strip mine spoils. For example, deertongue has a prostrate growth form and Blackwell switch grass has an erect form which is three to four times as high, but both had essentially the same total dry matter yield. It was evident, however, that the height attained by a particular species increased with the higher rate of effluent and sludge treatment. For every species but one, that being weeping love grass, average height growth was greater on the 2E + 2S treatment.

The results of the areal cover determinations were arranged with species ranked in descending order for the 2E + 2S treatment. As with the dry matter production, weeping love grass, Blackwell switch grass, feed canary grass and deertongue had the highest percentage cover on both the 1E + 1S and 2E + 2S treatments. Weeping love grass ranked first overall with an average of 97.5% cover of the spoil on the 2E + 2S and 1E + 1S treatments.

Blackwell switch grass ranked second with an average of 88% cover. Redtop also did very well on the 2E + 2S treatment with 95% cover. Garrison creeping foxtail gave the poorest percent cover on both treatments.

Four of the five most productive legumes (ladino clover, bristly black locust, Iroquois alfalfa and sericea lespedeza), also achieved the highest percent cover of spoil on the 2E + 2S treatment, with sweet clover replacing Pennscott red clover as the fifth. The areal cover values for these five species ranged from 99 to 90%. On the 1E + 1S treatment, bristly black locust was the only legume achieving high (81%) coverage, while Iroquois alfalfa was second with 30%.

It is interesting to note that sweet clover had good cover on the 2E + 2S treatment, but none at all on the 1E + 1S treatment. Lathro flat pea had the poorest cover on the 2E + 2S treatment with 40%. For every species tested, percent cover was greater on the 2E + 2S treatment than the 1E + 1S treatment. The macronutrient and micronutrient contents of the grass and legume foliar samples are given in Tables 5.4 and 5.5. Macronutrients are expressed as a percent of the dry weight, and micronutrients are expressed in micrograms/gram of dry matter.

TABLE 5.4: MACRONUTRIENT CONTENT OF GRASS AND LEGUME DRY MATTER

Treatment	Species	Macronutrients (percent of dry weight)				
		Nitrogen	Phosphorus	Potassium	Calcium	Magnesium
	Grass					
1 + 1	Garrison Creeping Foxtail	--[a]	--	--	--	--
	Saratoga Smoothbrome	3.54	0.29	1.26	0.28	0.25
	Reed Canarygrass	3.10	0.26	0.70	0.26	0.41
	Blackwell Switchgrass	1.78	0.12	0.55	0.19	0.38
	Weeping Lovegrass	1.83	0.14	0.72	0.20	0.16
	Redtop	3.87	0.33	1.30	0.43	0.40
	Deertongue	3.80	0.20	0.63	0.21	0.60
	Climax Timothy	3.65	0.29	0.89	0.49	0.22
2 + 2	Garrison Creeping Foxtail	2.63	0.18	1.37	0.29	0.21
	Saratoga Smoothbrome	3.79	0.31	1.52	0.40	0.33
	Reed Canarygrass	3.33	0.25	1.04	0.31	0.33
	Blackwell Switchgrass	2.04	0.14	1.00	0.26	0.29
	Weeping Lovegrass	2.31	0.16	0.78	0.31	0.21
	Redtop	3.91	0.36	1.60	0.50	0.39
	Deertongue	4.48	0.36	1.77	0.31	0.55
	Climax Timothy	3.72	0.35	1.00	0.66	0.27

Treatment	Species	Macronutrients (percent of dry weight)				
		Nitrogen	Phosphorus	Potassium	Calcium	Magnesium
	Legume					
1 + 1	Sericea Lespedeza	3.76	0.19	0.74	0.29	0.30
	Chemung Crownvetch	5.11	0.34	1.27	0.73	1.01
	Lathro Flatpea	5.32	0.34	2.33	0.81	0.76
	Iroquois Alfalfa	3.91	0.44	2.04	0.71	0.77
	Pennscott Red Clover	3.84	0.35	1.88	0.97	0.91
	Sweet Clover	--[a]	--	--	--	--
	Ladino Clover	4.24	0.33	1.97	0.67	0.73
	Bristly Black Locust	4.92	0.21	0.91	0.48	0.64
2 + 2	Sericea Lespedeza	3.10	0.21	0.74	0.48	0.30
	Chemung Crownvetch	4.31	0.32	1.73	0.72	0.85
	Lathro Flatpea	4.35	0.30	2.34	0.97	0.73
	Iroquois Alfalfa	3.75	0.45	1.55	0.69	0.36
	Pennscott Red Clover	3.94	0.35	2.05	1.04	0.57
	Sweet Clover	4.31	0.57	1.86	1.01	0.87
	Ladino Clover	4.44	0.39	0.99	0.98	0.47
	Bristly Black Locust	5.94	0.28	1.50	0.62	0.49

Source: PB 232 069   [a]No seed germinated.

TABLE 5.5: MICRONUTRIENT CONTENT OF GRASS AND LEGUME DRY MATTER

Treatment	Species	Micronutrients (micrograms per gram)						
		Manganese	Iron[b]	Copper	Boron	Aluminum[b]	Zinc	Sodium[c]
	Grass							
1 + 1	Garrison Creeping Foxtail	--[a]	--	--	--	--	--	--
	Saratoga Smoothbrome	150	300	15	13	152	49	622
	Reed Canarygrass	195	500	15	10	500	85	464
	Blackwell Switchgrass	112	106	11	7	51	44	121
	Weeping Lovegrass	198	212	12	6	89	72	261
	Redtop	318	488	18	14	185	82	530
	Deertongue	339	412	25	15	227	78	140
	Climax Timothy	157	500	19	18	500	110	391
2 + 2	Garrison Creeping Foxtail	236	486	13	11	170	81	491
	Saratoga Smoothbrome	169	300	15	13	110	66	625
	Reed Canarygrass	146	211	10	13	106	71	561
	Blackwell Switchgrass	162	158	14	10	62	58	281
	Weeping Lovegrass	189	167	12	7	85	66	225
	Redtop	313	247	19	15	120	119	436
	Deertongue	375	217	17	15	86	105	235
	Climax Timothy	194	500	21	19	500	165	364

Treatment	Species	Micronutrients (micrograms per gram)						
		Manganese	Iron[b]	Copper	Boron	Aluminum[b]	Zinc	Sodium[c]
	Legume							
1 + 1	Sericea Lespedeza	117	265	7	27	167	52	105
	Chemung Crownvetch	1200	388	17	78	143	237	138
	Lathro Flatpea	673	423	16	32	165	158	505
	Iroquois Alfalfa	334	401	15	70	302	203	1000
	Pennscott Red Clover	397	280	18	48	117	138	454
	Sweet Clover	--[a]	--	--	--	--	--	--
	Ladino Clover	293	427	14	43	165	99	1000
	Bristly Black Locust	924	391	19	59	244	94	303
2 + 2	Sericea Lespedeza	155	139	13	23	89	60	206
	Chemung Crownvetch	654	383	25	50	60	226	388
	Lathro Flatpea	521	300	20	31	80	192	591
	Iroquois Alfalfa	147	217	20	50	60	129	819
	Pennscott Red Clover	125	260	24	37	107	103	561
	Sweet Clover	270	187	17	80	56	119	644
	Ladino Clover	300	500	25	40	430	169	1000
	Bristly Black Locust	886	371	30	58	115	119	499

[a] No seed germinated. [b] Values over 500 are indicated as 500. [c] Values over 1000 are indicated as 1000.

Source: PB 232 069

The grasses harvested from the treated strip mine spoils used in this study contained higher concentrations of the macronutrients than those reported as normal in important feeding stuffs with the exception of the element potassium, the concentrations of which were somewhat lower. Concentrations of nitrogen and magnesium were higher in the legumes than values reported as normal.

Phosphorus concentrations were somewhat higher for certain legumes, calcium was lower, and potassium showed no definite relationship to those values reported as normal for the legumes. Low concentrations of potassium in the grasses and calcium in the legumes appears inconsistent with the fact that considerable amounts of all macronutrients were supplied in the sludge.

Comparing the 1E + 1S and 2E + 2S treatments shows that for the grasses, concentrations of all macronutrients but magnesium were higher in the vegetation receiving the higher treatment rate. Magnesium was higher for some species of grasses and lower for others with the 2E + 2S treatment. The legumes gave different responses than grasses to the increased irrigation of the 2E + 2S treatment. Nitrogen, phosphorus and potassium concentrations showed no definite trend, with some species of legumes having higher concentrations with the higher treatment rate, and some species having lower concentrations. Calcium concentrations were higher in the legumes receiving the 2E + 2S treatment, and magnesium concentrations were higher with the 1E + 1S treatment.

For the most part, concentrations of manganese in the grasses and legumes tested were below those levels considered highly toxic (400 to 500 ppm). Redtop and deertongue were the only two species on either treatment whose concentrations of manganese exceeded 300 micrograms per gram. Redtop did not produce much dry matter, but deertongue did quite well on both treatments.

Garrison creeping foxtail gave poor yields on the 2E + 2S treatment with 236 micrograms of manganese per gram of dry matter, and produced nothing at all on the 1E + 1S treatment. But climax timothy and Saratoga smooth brome, the two grasses giving the poorest yields, had relatively low concentrations of manganese. Weeping love grass and Blackwell switch grass, which yielded well, had concentrations of manganese less than 200 micrograms per gram.

In general, the legumes contained much higher concentrations of manganese than the grasses. Chemung crown vetch, lathro flat pea and bristly black locust all had extremely high concentrations of manganese on both treatments. The crown vetch and flat pea did quite poorly; the bristly black locust did quite well. Although no definite trend appeared to exist for the grasses, the 1E + 1S treatment resulted in higher concentrations of manganese in most of the legumes than the 2E + 2S treatment.

The range of boron encountered in the grasses tested was between 6 and 19 micrograms per gram, certainly well below toxic levels (>59, 99 and 68 µg/g for red clover, alfalfa and bird's-foot trefoil respectively). The two species of grasses that gave the best yields contained the least boron on both treatments. Climax timothy contained the most boron.

The legumes contained considerably more boron than the grasses, ranging between 23 and 80 micrograms per gram. All values appeared to fall within nontoxic levels, possibly with the exception of Chemung crown vetch and sweet clover.

No clear relationship was observed between legume yield and boron concentrations, however, concentrations in the legumes were lower for the 2E + 2S treatment than for the 1E + 1S treatment even though the former received more boron. Certain species that did well exhibited higher boron concentrations than others that did quite poorly, such as bristly black locust with 59 micrograms of boron per gram of foliage compared to lathro flat pea with 32 on the 1E + 1S treatment. The degree of tolerance to boron is very much species dependent.

The toxic effects of aluminum may be even more severe than those of manganese. An earlier study showed no aluminum toxicity in alfalfa at less than 200 micrograms per gram in the stems and leaves, light toxicity from 200 to 325, medium toxicity from 325 to 450, and severe toxicity at over 450 micrograms per gram.

Of the eight species of grass tested, two had concentrations of aluminum in the dry matter exceeding 500 micrograms per gram, Reed canary grass with the 1E + 1S treatment and climax timothy on both the 1E + 1S and 2E + 2S treatments. Reed canary grass gave a somewhat lower yield on the 1E + 1S treatment, but did fairly well on the 2E + 2S treatment where aluminum concentrations in the foliage decreased to 106 micrograms per gram.

The two grass species performing best, weeping love grass and Blackwell switch grass, also had the lowest concentrations of aluminum in their dry matter, less than 90 micrograms per gram for both species on both treatments. Saratoga smooth brome gave the lowest yield on the 2E + 2S treatment and second lowest on the 1E + 1S treatment with aluminum concentrations of 110 and 152 micrograms per gram, respectively. Except for Blackwell switch grass and climax timothy, all aluminum concentrations were lower with the 2E + 2S treatment than the 1E + 1S treatment.

Contrary to the results for manganese where the concentrations in the legumes exceeded those in the grasses, average concentrations of aluminum in the legumes were considerably less than those in the grasses for both treatments. On the 1E + 1S and 2E + 2S treatments, concentrations of aluminum in the grasses averaged better than 243 and 155 micrograms per gram, respectively, while corresponding concentrations in the legumes averaged 186 and 125 micrograms per gram.

Again, concentrations of aluminum in the legume foliage decreased with the increasing treatment rate except for ladino clover, which showed a rise in dry matter aluminum by 2.6 times at the highest rate. The extremely high concentration of aluminum in ladino clover (430 micrograms per gram) on the two inch treatment appears inconsistent with the fact that ladino clover produced the greatest yield (3,078 pounds per acre), especially since the aluminum content appears to be well within the potentially toxic range.

There appears to be a possibility of foliar contamination by the sludge. Most other legumes tested were well below toxic levels in their aluminum content, especially on the 2E + 2S treatment.

Iron is an essential element in plant nutrition, although the high concentrations commonly encountered in acid spoil leachate may pose a toxicity problem. Plant iron is generally in the range of 50 to 120 ppm on a dry weight basis. The iron contents of the grasses and legumes under consideration were, with one exception, in excess of the concentrations reported above, the exception being Blackwell switch grass with 106 micrograms per gram. Sericea lespedeza contained 139 micrograms of iron per gram of dry matter on the 2E + 2S treatment, and all other grasses and legumes had higher concentrations. Thus a strong potential for iron toxicity existed, although no symptoms of toxic conditions were evident.

The two most productive grasses contained the lowest concentrations of iron on both treatments. However, deertongue also gave good yields with iron concentrations of 412 and 217 micrograms per gram on the 1E + 1S and 2E + 2S treatments, respectively. Garrison creeping foxtail, Saratoga smooth brome, redtop and climax timothy, the four grasses giving the poorest yields, contained higher concentrations of iron.

The two most productive legumes contained considerably more iron than their counterparts of the grasses, although on the average, there was almost no difference between grass and legume iron contents by treatments. Averaged over species, the grasses contained 360 and 286 micrograms per gram on the 1E + 1S and 2E + 2S treatments while the legumes contained 368 and 295 micrograms per gram. Ladino clover and bristly black locust showed high concentrations on both treatments. For most of the grass and legume species tested, iron concentrations were higher with the 1E + 1S than with the 2E + 2S treatment.

Concentrations of 20 to 30 ppm of copper in snapbean (*Phaseolus bulgaris* L.) tissues were accompanied by yield reductions, and severely reduced growth and high toxicity at 40 or more parts per million copper.

For the most part, copper concentrations in the grass and legume species tested appeared to fall within a safe range. Deertongue had the highest concentration of the grasses with 25 micrograms per gram on the 1E + 1S treatment. The legumes averaged considerably higher copper concentrations on the 2E + 2S treatment, six species having values of 20 or more micrograms per gram. Bristly black locust had the highest copper content with 30 micrograms per gram.

Leachate Quality: Samples of leachate taken from the 3.5 foot depth in the spoil were analyzed for certain chemical constituents to determine if significant differences existed between the various treatments. The results of the analyses for chemical constituents are shown in Table 5.6. Statistical tests for significant differences were applied. Weekly trends of leachate pH, sulfates, nitrates, ammonium and phosphates were graphed. The results are summarized in Table 5.6.

The highly acid soil solution and the toxicity of certain metals made soluble by the acid conditions is one of the primary reasons for vegetation failure on the spoil bank environment. The pH of the leachate taken from the 3.5 foot depth in the spoil remained quite low for both the treated spoil and the controls. Averaged by treatments, there was no significant difference between the 1E + 1S treatment and the control, both having a pH of about 2.7.

The spoil receiving the 2E + 2S treatment produced leachate of a significantly higher pH, averaging 2.9 on the 2E + 2S legume treatment and 3.1 on the 2E + 2S grass treatment. At the 95% confidence level, the grass covered spoil had a significantly higher leachate pH than did the legume covered spoil. The legume and grass controls and the 1E + 1S grass treatment differed significantly from the other three treated boxes.

TABLE 5.6: MEAN CHEMICAL COMPOSITION OF LEACHATE SAMPLES

Treatment	Vegetation Type	pH	Mean Concentrations of Leachate					
			Sulfates (SO_4)	Nitrates (NO_3-N)	Ammonium (NH_4-N)	Organic Nitrogen	Orthophosphate P	Total Acidity
			mg/l	mg/l	mg/l	mg/l	mg/l	(meq/l) as H
1 + 1	Grass	2.64	312	2.2	17.6	14.9	0.13	15.8
	Legume	2.77	179	3.2	27.2	15.5	0.07	6.9
2 + 2	Grass	3.08	95	6.9	29.9	16.3	0.17	3.3
	Legume	2.91	107	5.5	26.4	15.5	0.07	4.0
Control	Grass	2.71	442	1.2	1.9	5.4	0.11	18.1
	Legume	2.68	400	1.3	2.5	7.6	0.17	17.7

Source: PB 232 069

The extremely low pH of the one-inch grass treatment is unexplained, although localized differences in the spoil material surrounding the tension tubes, which collected the water, may have been the cause. The leachate pH was relatively stable through the study for all treatments but the two-inch grass and the legume control, which showed a general increase in the latter weeks of the summer.

Although the pH of leachate at lower depths remains highly acid, evidence from another experiment indicates that the pH of leachate at the six-inch depth (in the rooting zone) can be raised considerably through irrigation with effluent and sludge. Data collected during the summer of 1971 on similarly treated spoils shows that leachate pH at the six-inch depth can be raised to within the range of from 5.0 to as high as 8.0. Untreated spoils maintained a rooting zone pH of below 3.0.

Total acidity, in milliequivalents per liter expressed as hydrogen, generally was inversely related to the pH. Both treatments and the control, as well as both cover types differed significantly from each other. The 2E + 2S treatment resulted in significantly lower total acidity, averaging 4.0 and 3.3 on the legume and grass treatments, respectively. The legume and grass controls had the highest total acidity of 17.7 and 18.1, respectively, which were significantly higher than all but that of the 1E + 1S grass treatment.

The concentrations of sulfates in the leachate range from 95 to 442 milligrams per liter. The 2E + 2S grass and legume treatments averaged lowest with 95 and 107 milligrams SO_4 per liter, respectively, while the controls averaged highest with 400 (legume) and 442 (grass) milligrams per liter. Both controls differed significantly at both confidence levels from the treatments receiving effluent and sludge.

The 2E + 2S treatments were significantly lower in sulfate concentrations than the 1E + 1S treatments. Sulfate concentrations in the leachate were significantly higher with the grass cover.

Spoil Amendment Studies   113

In general, sulfate concentrations followed a downward trend through the summer on both controls and the 1E + 1S grass treatment. The other treatments maintained relatively stable sulfate concentrations. Sulfate concentrations in spoil water may be no more toxic to vegetation than chlorides, and probably just reflect the acid-forming properties of sulfides.

The nitrogen relationships in the leachate were quite indicative of the treatment of the spoil. Nitrate, ammonium and organic nitrogen concentrations were highest in the leachate from the 2E + 2S treatments and lowest in that of the controls. Mean nitrate values ranged from 1.2 milligrams per liter on the grass control to 6.9 on the 2E + 2S grass treatment. These mean values of nitrate were well below the limit of 10 milligrams per liter set for potable water by the U.S. Public Health Service.

Individual values for nitrates ranged only as high as 12.0 mg/l on the 2E + 2S grass treatment. At both confidence levels, the control had significantly lower concentrations of nitrate nitrogen in the leachate than the irrigated spoils. Thus, more nitrate nitrogen was available for plant uptake in the treated spoil than in the controls. The 2E + 2S treatment resulted in significantly higher concentrations of nitrates in the leachate, while cover type had no significant effect.

There were stable, depressed values of nitrate in leachate from the controls, slightly higher values showing a general upward trend for the 1E + 1S treatments, and even higher nitrate content of the leachate from the 2E + 2S treatments. The rising leachate concentrations of nitrate through the summer in the treated spoil indicates probable nitrification of organic and ammoniacal nitrogen added to the upper spoil layers.

Ammonium and organic nitrogen concentrations were much higher in the leachate from the irrigated spoil than the controls. Leachate ammonium averaged 2.2 milligrams per liter for the controls, 22.6 for the 1E + 1S treatments, and 28.1 for the 2E + 2S treatments. All mean values exceeded the potable water limit of 0.5 milligrams per liter. There were low concentrations of leachate ammonium in the controls and much higher ammonium concentrations for the treated boxes with gradually increasing values through the summer (with the exception of the 1E + 1S legume treatment).

Organic nitrogen content in the leachate averaged 6.5, 15.2 and 15.9 milligrams per liter for the control, 1E + 1S and 2E + 2S treatments, respectively. The leachate concentrations of ammonium and organic nitrogen are significantly lower for the controls than for any of the treatments. The legume cover was associated with higher mean concentrations for both constituents; however, the effect was significant only for ammonium-N.

Although these concentrations of nitrogen in the leachate appear to be rather high, and the potential for eutrophication exists, it must be remembered that very large amounts of nitrogen were applied to the spoil in the sludge applications, nearly 500 milligrams N per liter. Considering this, good renovation of the leachate nitrogen was achieved by the spoil, despite the relatively low number of available exchange sites on colloidal particles in the coarse spoil material. Another component of the leachate with the potential for eutrophication is the phosphate ion. Orthophosphate phosphorus concentrations in the spoil water ranged from 0.07 to 0.17 milligrams per liter, and averaged highest on the controls, although no statistically significant differences existed between the controls and treatments. The fact that the controls averaged higher concentrations of leachate orthophosphate seems to contradict that an average of 5.5 milligrams per liter of orthophosphate was supplied by the effluent in the treatments. The liquid sludge contained 118.3 milligrams P per liter.

Considering this high rate of application, good reduction in orthophosphate content of the leachate was achieved, concentrations being reduced by some 98% on the average. Orthophosphate concentrations remained relatively stable through the summer on all but the 2E + 2S grass and legume treatments, which had somewhat rising levels. Both controls experienced one week of very high values toward the end of the summer, possibly due to sample contamination.

The elemental constituents of the leachate are given in Table 5.7. Concentrations are averaged for the first nine weeks and the second nine weeks of the study, and the final mean leachate concentrations for the full term of the study are given. The phosphorus values reported are total phosphorus. Statistical tests for determination of significant differences between treatments for the elements manganese, iron, aluminum, copper, boron and zinc were applied. Trends in leachate concentrations of manganese, iron, aluminum and boron were graphed. The results are summarized below.

TABLE 5.7: MEAN ELEMENTAL CONSTITUENTS IN THE LEACHATE SAMPLES

Treatment	Vegetation Type	Time Period	P^a	K	Ca	Mg	Mn	Fe	Cu	B	Al	Zn	Na
1 + 1	Grass	first 9 weeks	10.3	60.7	30.6	41.9	11.6	109.3	1.1	0.5	77.0	2.3	13.9
		second 9 weeks	5.5	47.7	19.8	28.6	5.9	62.7	0.7	0.2	39.0	1.4	18.4
		full term	7.7	53.8	24.9	34.9	8.6	84.6	0.9	0.3	56.9	1.8	16.3
	Legume	first 9 weeks	7.0	44.8	22.1	19.6	4.9	42.1	0.6	0.2	28.5	1.2	20.7
		second 9 weeks	5.3	45.4	16.0	23.1	3.6	20.7	0.5	0.2	20.0	1.0	28.3
		full term	6.2	45.1	19.0	21.4	4.2	31.4	0.6	0.2	24.2	1.1	24.5
2 + 2	Grass	first 9 weeks	6.3	39.9	24.7	17.5	4.0	6.0	0.5	0.1	16.7	1.1	24.3
		second 9 weeks	3.5	39.8	16.8	13.4	2.6	2.6	0.8	0.1	10.6	0.8	25.8
		full term	4.8	39.8	20.5	15.3	3.3	4.2	0.7	0.1	13.5	1.0	25.1
	Legume	first 9 weeks	8.0	53.1	13.9	13.3	2.7	12.7	0.5	0.2	13.2	0.9	31.3
		second 9 weeks	4.9	39.8	12.2	14.0	2.3	7.0	0.5	0.1	11.6	0.8	28.2
		full term	6.4	46.5	13.0	13.6	2.5	9.9	0.5	0.2	12.4	0.8	29.7
Control	Grass	first 9 weeks	8.6	64.3	31.5	50.8	15.8	91.4	1.1	0.4	98.9	3.1	3.2
		second 9 weeks	4.4	34.9	30.7	43.1	14.0	49.2	0.9	0.2	100.0	2.4	3.4
		full term	6.5	49.6	31.1	46.9	14.9	70.3	1.0	0.3	99.5	2.8	3.3
	Legume	first 9 weeks	10.5	60.1	27.8	44.1	14.8	107.7	1.0	0.4	92.0	2.6	3.7
		second 9 weeks	7.0	43.1	30.3	42.7	16.1	69.2	0.9	0.3	93.1	2.2	5.0
		full term	8.6	51.1	29.1	43.3	15.5	87.3	0.9	0.3	92.6	2.4	4.4

[a] Total P.

Source: PB 232 069

There were no great differences between treatments and controls for either the phosphorus or potassium leachate concentrations, although the 2E + 2S treatment had the lowest average concentrations. Considerably more calcium and magnesium were found, however, in the leachate from the controls, and both were found to have the lowest average concentrations with the 2E + 2S treatment.

High concentrations of soluble manganese have been shown to be toxic to vegetation. Concentrations of from 1 to 10 milligrams per liter were found to have toxic effects on five legumes tested in an earlier study. Lespedeza and sweet clover were most seriously affected.

All leachate manganese concentrations appeared to exceed those values considered potentially toxic to plants. The lowest values were in the leachate from the 2E + 2S treatment, averaging 2.5 and 3.2 milligrams per liter, while the controls averaged highest with 14.9 and 15.5. Both the grass and legume controls were significantly higher in leachate manganese than the two treatments, while the 2E + 2S treatment resulted in the significantly lowest concentrations. Manganese concentrations with the grass cover were significantly higher than those with the legume cover. The 1E + 1S grass treatment was significantly higher than the other irrigated spoils, consistent with its lower pH.

The 1E + 1S and 2E + 2S grass treatments showed a downward trend in manganese concentration through the summer, while the 1E + 1S and 2E + 2S legume treatments decreased slowly and then increased again. Both controls showed a definite decrease and then increase in manganese content of the leachate, probably due in part to the patterns of precipitation through the spring and summer months.

Leachate iron and aluminum concentrations showed great similarity. Averaged by treatments, leachate iron concentrations were 7.1, 57.3 and 78.6 milligrams per liter for the 2E + 2S and 1E + 1S treatments and the controls, respectively. Averaged leachate aluminum concentrations were 12.9, 40.1 and 96.1 mg/l for the respective treatments. Concentrations were significantly lowest with the 2E + 2S treatment and significantly highest with the control for both elements. The grass covered spoil had significantly higher concentrations of both elements than the legume cover.

Although the relationship of iron toxicity to plant growth has not been conclusively determined in the spoil bank environment, it has been reported that aluminum in solution is the most common element toxic to plants in acid spoils. Concentrations of as low as 1.0 milligram aluminum per liter in culture solutions have severely reduced growth of some plant species. Thus, all concentrations of leachate aluminum were well above the potentially toxic levels.

At the 99% confidence level, both controls and the 1E + 1S grass treatment were significantly higher in iron than the remaining three treatments. The high concentration of leachate iron on the 1E + 1S grass treatment is not explained, but may again be due to extreme localized variability within the spoil material on that treatment. The 2E + 2S grass treatment gave significantly lowest iron concentrations. Aluminum content in the leachate was significantly higher on both controls than for any treatments. There were generally decreasing concentrations of leachate iron through the summer for both controls and treated spoils, with the exception of one weekly value on the 1E + 1S legume treatment.

As for trends of leachate aluminum, the controls remained high and relatively stable through the study, with aluminum concentrations in the legume control decreasing below 100 milligrams per liter for a time through the mid-summer period. The 1E + 1S and 2E + 2S grass treatments experienced decreasing values of aluminum. Leachate aluminum content fell and then rose again on the 1E + 1S legume treatment, but remained stable on the 2E + 2S legume treatment. Increasing concentrations toward the end of the summer may have been due to precipitation patterns.

In addition, concentrations of copper, boron and zinc in the leachate were analyzed. The significantly highest concentrations of copper occurred in the leachate from the control treatment, averaging 0.960 milligrams per liter. The lowest mean copper content was found in the 2E + 2S legume treatment with 0.467 milligrams per liter, although no significant differences existed between the 1E + 1S and 2E + 2S treatments. Significantly higher copper concentrations accompanied the grass cover as compared to the legume cover. These concentrations of copper would not be regarded as toxic.

Averaged by treatments, leachate boron concentrations were significantly lower with the 2E + 2S treatment, while the 1E + 1S treatment and the control showed no significant differences. Cover type had no significant effect on boron concentrations. Concentrations were highest for the 1E + 1S grass treatment (possibly due to increased solubility at the lower pH), and both controls, all of which were significantly higher than the other three treatments.

The highest mean concentration was achieved in the leachate from the 1E + 1S grass treatment, 0.347 milligrams per liter. The 2E + 2S treatments averaged 0.149 milligrams per liter between them. It has been reported that optimum growth of Kentucky bluegrass was obtained with from 1 to 5 milligrams per liter of boron in the soil solution. Best growth of alfalfa and sweet clover was achieved with 10 and 5 milligrams per liter in solution. All concentrations observed in this study appear to be less than toxic on this basis.

Trends of boron concentrations are as follows. The 2E + 2S treatments remained fairly stable at a very low concentration, fluctuating between 0.1 and 0.3 milligrams per liter. Both the grass control and 1E + 1S treatment showed decreasing values through the summer, while the legume control and 1E + 1S treatment showed considerable fluctuation and a decrease toward the end of the summer.

Zinc concentrations in the leachate followed the same pattern as manganese and aluminum: significantly higher on the control (averaging 2.58 milligrams per liter), and significantly lower on the 2E + 2S treatment (averaging 0.88 milligrams per liter). The grass cover had significantly higher zinc concentrations in the leachate than the legume cover.

One study found development and growth of oat plants to be normal with 10 milligrams per liter of zinc in solution. However, another study found 50% dry weight reduction in several crops of fruits and cruciferous vegetables when grown in nutrient solutions of from 1 to 10 milligrams per liter of zinc. Toxic levels of zinc for the grass and legume species tested in this study may be quite different; however, there is a possibility of toxicity, especially on the control treatments.

Spoil Analysis: Following the last applications of effluent and sludge to the spoil material, samples of the spoil were taken from the six-inch depth from four compartments of each treatment and one of the control boxes. Analyses were made to determine if there were significant differences in the chemical constituents of the spoil between treatments and control as a result of irrigation.

An exact analysis of variance was performed for spoil pH and the elements manganese, iron, aluminum and zinc. The factors in the analysis were treatment rate and cover type. The analysis was performed by reordering all terms last, with one of the two degrees of freedom associated with the treatment-cover interaction having a value of zero.

The results of this analysis showed that the cover effect on spoil properties was significant only in the case of manganese, and that the treatment-cover interaction was significant for only manganese and zinc. Furthermore, since in a practical sense the cover type will not have a meaningful effect on the spoil elemental properties in the first growing season, the few instances of significance for cover and the interaction are most likely due to inherent variability in the spoil itself.

Adopting this assumption (that the cover type has no significant effect on the spoil elemental constituents in the first year), the analysis of variance was performed again with the treatment rate as the only factor in the analysis. The results of this second analysis are summarized below.

The two treatments and the control each had a significantly different effect on spoil pH: the 2E + 2S treatment resulted in higher mean values (3.34), and the control had the lowest (2.71). At the 95% confidence level the pH of the spoil taken from both 2E + 2S treatments (3.39 and 3.30), were significantly higher than the pH of all but the 1E + 1S legume treatment (3.15), which did not differ significantly from the 2E + 2S legume treatment.

These differences show that the capacity of the spoil in the rooting zone for producing acid leachate has been reduced by treatment with effluent and sludge. Most likely, the mechanism of ion exchange is, in part, responsible for this increase in spoil pH. Ammonium and certain elements found in the effluent and sludge in ionic form may be available for replacement of the hydrogen ion at exchange sites on the spoil particles, thus having the effect of increasing spoil pH.

The results of the analysis for spoil elemental constituents at six-inch depth of even numbered plots of the treatment boxes and the legume control after termination of treatments are presented in Table 5.8, which gives mean concentrations of extractable elements in the spoil.

The ammonium acetate extractant used was buffered to a pH of 7.0. This is well above the range of maximum solubility for many cations such as manganese, iron, aluminum and zinc. The low solubility of such metals at this neutral pH is the probable reason for the low concentrations of these extractable spoil ions.

In addition, spoil phosphorus was determined using the same procedure. This method of determining phosphorus does not appear to be satisfactory, as the ammonium acetate does not bring phosphorus into solution. Concentrations reported are no better than water-soluble phosphorus. Concentrations of potassium, calcium and magnesium were lowest in the spoil from the control and highest in the spoil from the 2E + 2S treatments.

TABLE 5.8: MEAN ELEMENTAL CONSTITUENTS OF THE SPOIL MATERIAL

Treatment	Vegetation Type	Elements (micrograms per gram of spoil)			
		Phosphorus	Potassium	Calcium	Magnesium
1 + 1	Grass	3.5	26.0	68.4	41.0
	Legume	4.5	39.6	90.0	52.0
2 + 2	Grass	1.0	78.0	125.4	59.0
	Legume	3.5	97.0	136.4	65.6
Control	Legume	1.0	20.0	15.0	32.0
		Manganese	Iron	Copper	Boron	Aluminum	Zinc	Sodium
1 + 1	Grass	5.25	3.25	0.64	0.20	2.50	1.40	44.0
	Legume	4.05	2.15	0.64	0.24	2.55	0.90	58.0
2 + 2	Grass	7.30	1.95	0.54	0.20	3.20	0.40	73.0
	Legume	3.65	2.60	0.60	0.30	3.00	0.65	87.0
Control	Legume	4.70	1.10	0.84	0.20	2.25	0.40	49.6

Source: PB 232 069

Sodium content was lowest on the 1E + 1S grass treatment; however, averaged by treatments, sodium was lowest in the spoil from the control and highest in the spoil from the 2E + 2S treatments. The sum of the exchangeable bases was, therefore, much greater for the spoil receiving the highest treatment rate and lowest for the spoil from the control.

The results of the LSD tests for spoil manganese, iron, aluminum and zinc showed no definite patterns of significant differences between treatments for any of the four cations, which may be due in part to the neutral buffering of the extract. However, except for spoil manganese, concentrations of these spoil cations were lowest in spoil from the control.

No significant differences existed between the three treatment rates for spoil manganese. Spoil aluminum was significantly highest in spoil from the 2E + 2S treatment (at the 95% confidence level), while zinc was significantly highest for the 1E + 1S treatment. This would appear to be the opposite of the expected condition, although the contribution of cations by the effluent and sludge in exchange processes may be the reason for the highest concentrations in the irrigated spoil.

In addition to the chemical analyses, a nitrogen mass balance calculation was done to determine the fate of the great quantities of nitrogen applied to the spoil in the effluent and sludge.

The results show that considerable nitrogen is added to and retained by the spoil, most of which would quite likely be held in the upper spoil zone and thus be readily available for vegetation establishment and growth in subsequent growing seasons. The nitrogen stored in the spoil receiving the 1E + 1S treatments averaged 603 pounds per acre, 69% of that applied in the effluent and sludge. The spoil receiving the 2E + 2S treatments stored an average of 1,071 pounds of nitrogen per acre, almost 62% of that applied. Spoil from the controls experienced a slight loss of nitrogen due to leaching, averaging about 5.6 pounds of nitrogen per acre.

Conclusions

(1) Treated sewage effluent and liquid digested sludge pose no problems for vegetation in toxicity of heavy metals. They do, however, supply valuable amounts of certain nutrients, especially nitrogen, and organic matter.

(2) The effect of irrigation with effluent and sludge on spoil surface temperatures is negligible, on the order of a few degrees Fahrenheit one day after treatment, if any. Similarly, subsurface temperatures in the upper foot of spoil material may be depressed by a few degrees following treatment.

(3) Effluent and sludge irrigation makes possible the establishment and growth of certain grass and legume species on highly acid and toxic strip mine spoil material.

(4) On the average, the grasses did much better than the legumes in both dry matter production and percent areal cover of the spoil. Weeping love grass and Blackwell switch grass gave the best growth response of the grasses with 11,067 and 4,689 pounds of dry matter production per acre, respectively, at the 2E + 2S treatment rate. Over both treatment rates, bristly black locust gave the best growth of the legumes.

(5) The 2E + 2S treatment gave significantly better results in terms of grass and legume dry matter production and areal cover.

(6) Weeping love grass, Blackwell switch grass, Reed canary grass and deertongue gave the best percentage cover of the spoil among the grasses, especially on the 2E + 2S treatment. Ladino clover, bristly black locust, Iroquois alfalfa, sericea lespedeza and sweet clover also gave good cover on the 2E + 2S treatment. These species would be best suited for use in stabilization of soil and erosion control.

(7) For the most part, the grasses that achieved the greatest dry matter production (weeping love grass and Blackwell switch grass), had the lowest concentrations of toxic metals in the above-ground plant tissues. However, legumes having good production (ladino clover and bristly black locust), often had high concentrations of metals in their dry matter.

(8) Highest concentrations of metals and lowest concentrations of nitrogen were found in leachate from the control spoil. Leachate from the spoil receiving the 2E + 2S treatment had the lowest concentrations of metals and acidity, and the highest concentrations of nitrogen. Thus, spoil treated with effluent and sludge produces leachate more favorable and less toxic for grass and legume establishment.

(9) Concentrations of nitrates and pH in spoil leachate may show steadily increasing values with time through continued effluent and sludge irrigation. Steadily decreasing concentrations of sulfates, manganese, iron, aluminum and boron may be achieved in leachate from spoil receiving regular effluent and sludge treatment.

(10) Spoil pH may be increased by the process of ion exchange through irrigation with effluent and sludge.

(11) Considerable amounts of nitrogen may be retained by the spoil material following irrigation with effluent and sludge, especially in the organic form. Treated sludge also supplies significant amounts of organic matter to the spoil. Together, these improved spoil conditions greatly enhance the prospects for establishing vegetation on spoil banks.

The results of this study indicate that treated municipal sewage effluent and liquid digested sludge are a valuable means of amending harsh conditions which make spoil banks so unsuitable for establishment and growth of vegetation. In particular, the effluent and sludge have considerable nutrient value and soil building potential, and can aid in the reduction of toxic concentrations of spoil leachate.

USE OF SLUDGE AND DEHYDRATED SLUDGE PRODUCT

Gob (refuse) piles and slurry areas created from coal mining activities in Illinois remain devoid of vegetation for many years. The spoils which result from strip mining generally will become revegetated naturally, but there are some that also remain bare. A high sulfur content and the resulting low pH appears to be responsible for the lack of plant growth. The levels of available nitrogen, phosphorus, and potassium are frequently very low in these materials and should be supplied when attempts are made to establish vegetation.

The problem of the disposal of municipal sludge is becoming more acute. Its value as a source of plant nutrients is well established. Its use on agricultural land is frequently prohibited because of public objections. It has been proposed to use municipal sludge on gob, slurry and spoil areas for plant nutrients in attempts to grow vegetation. If use of sludge proves practical, the problems with municipal sludge disposal and land reclamation can be solved. The study reported in PB 226 905 was to determine if plants will grow on properly treated gob, slurry and spoil. The source of nutrients was municipal sludge and Milorganite, a dehydrated sludge product. Lime was included as a variable because sludge has little neutralizing effect, and gob, in particular, usually has a high sulfur content and a very low pH.

Materials and Methods

Soil material samples were taken at 0 to 6 inches and 6 to 12 inches at five sites in the gob area and five sites in the slurry area of a mine at Fiatt, Illinois. The soil was tested for pH, available phosphorus, potassium and total sulfur. The material was extremely variable in all four characteristics. The pH in 11 cases out of 20 was below 3.

The level of available phosphorus (P_1) was generally low, 14 cases out of 20 were below a reading of 20 pounds per acre. The available K was also low in most cases, 11 cases had a reading below 150 pounds per acre. The most significant value probably is total S. Some samples contained well over 7% S. It has been estimated that 30 tons per acre of lime is required to neutralize the effect of 1% S in the soil.

Spoil Amendment Studies

For greenhouse studies several hundred pounds of material were collected at site 2 (gob) and site 9 (slurry). These bulk samples were taken from 0" to 12". In addition, spoil material was received from a southern Illinois site. The materials were air dried. The soil test results were as follows.

	pH	P_1 (lb/acre)	K (lb/acre)	Total S (percent)
Gob	2.1	22	86	5.46
Slurry	5.3	16	130	3.73
Spoil	4.0	6	66	–

Four hundred and fifty grams of material were placed in 4" plastic pots after 100 ppm of K as KCl and the appropriate amount of lime and Milorganite had been thoroughly mixed into it. In the case of gob where sludge was used, the sludge was added to the surface and when it had dried sufficiently, the crust of sludge on the surface was mixed slightly into the material to a depth of 2".

Ten seeds of perennial ryegrass, *Lolium perenne,* were planted ¼" deep, and sufficient tap water was added for germination and subsequent growth. The ryegrass seedlings were counted sometime after establishment and the top growth was harvested 40 days after planting and dry weight was determined. The plants were allowed to recover and after 23 to 25 days the top growth was harvested again. The soil material was allowed to dry out.

Five sorghum *(Sorghum vulgare)* seeds were planted ½" deep and the pots were watered for germination and growth. The top growth was harvested after 45 days and the dry weight was determined. In the case of gob, ryegrass was again planted after the sorghum was harvested. Only one harvest was taken 50 days after planting In no case was the soil material in the pot removed and remixed after the first ryegrass planting. When a new crop was planted only the surface 1" was stirred enough to allow for soil-seed contact.

Soil samples were taken from each pot with a small probe after the sorghum was harvested. The pH and P_1 values were determined.

The levels of lime and sludge or Milorganite used were determined after studying the soil test results and are given in the data tables. There were 3 replications in the gob experiment and 4 in the slurry and spoil experiments. The data obtained was not analyzed statistically. The results are very graphic and determinations of experimental error would not aid appreciably in interpretation.

The experiments on the three materials were started at different times so the results cannot be compared directly. Greenhouse temperature ranged from 65°F at night to 90°F during the day. However, due to variations in light and daytime temperatures, growth conditions varied throughout the period.

Perennial ryegrass was selected as a test plant for these trials because it is small-seeded and grows rapidly. Previous experience had shown that it grows well under greenhouse conditions. Sorghum was selected for growing after two cuttings of ryegrass because it was expected that the greenhouse temperatures would be too high for ryegrass as summer approached. This was not the case and the sorghum plants turned purple, and did not grow normally presumably due to low night temperature (65°F).

Field trials may indicate that other species are better suited for field plantings than ryegrass. Sorghum probably will not have a place in the field since it is an annual.

Experimental Results

The complete data of the gob experiment are given in Table 5.9. Seedlings did not emerge from the gob unless lime and/or sludge were added. However, enough of one or the other permitted emergence. The effect of sludge without lime is remarkable because soil samples taken May 1 indicated that sludge had only a slight effect on pH. Perennial ryegrass and sorghum emerged from a soil which later showed a pH of 3.1.

Seedlings emerged where 40 or more parts per thousand of lime was added without sludge, but made little growth because of a lack of soil nutrients. A combination of lime and sludge was much better than either alone. Twenty parts per thousand of lime and sludge with 250 parts per million of N each permitted no (or only slight) growth. Applied together they supported nearly as much growth of ryegrass and sorghum as any other treatment combination.

The highest rate of sludge applied (375 parts per million of N or 5,250 pounds dry matter per acre) did not adversely affect the plants, and when the last crop of ryegrass was grown, it supported more growth than where only 250 parts per million of N had been applied. The soil test values show that where sludge was applied with 20 parts per thousand or less of lime the pH was raised 0.4 to 1.6 pH units by the application of sludge at the rate of 375 parts per million of N.

The maximum pH, 7.4, was reached with 80 parts per thousand of lime, an amount far less than enough to neutralize the estimated potential acidity. The results of the test for P_1 indicate that there still was free lime present where 160 and 320 parts per thousand of lime was applied. This is suggested by the low P_1 readings even where the highest rate of sludge was used. The free lime apparently was reacting with the acid in the test solution so that the available phosphorus was not detected. The two highest rates of sludge increased the level of P available to the plant substantially.

The complete data of the slurry experiment are given in Table 5.10. Perennial ryegrass seedlings emerged where neither lime nor Milorganite were applied. This was expected since the pH of the slurry (5.3) was high enough to permit germination. Without Milorganite, ryegrass and sorghum growth was very limited. After the first harvest of ryegrass the yields increased up to 375 parts per million of N per acre. Sorghum yields were increased up to 500 parts per million of N per acre.

The application of lime increased the pH only slightly and the maximum pH was about 7.0. It is of interest that the pH of the check increased from 5.3 to 6.9 during the course of the experiment. The corresponding increases in the gob and spoil were 0.3 and 0.9 pH units, respectively. The level of available phosphorus was raised to adequate levels with sludge at the rates of 160 and 320 parts per million of N per acre.

The complete data of the spoil experiment are given in Table 5.11. Perennial ryegrass and sorghum seedlings emerged where neither lime nor Milorganite were applied.

TABLE 5.9: PERENNIAL RYEGRASS AND SORGHUM GROWING ON GOB

Sludge, N,ppm	Rate of lime, pp 1000						
	0	10	20	40	80	160	320
No. seedlings,[1] 1 week							
0	0	0	3	27	25	29	27
125	0	1	15	29	28	27	26
250	0	20	26	27	28	25	26
375	21	28	29	28	28	28	29
Perennial ryegrass, dry wt. in grams, 0-40 days							
0	0	0	.02	.15	.15	.17	.22
125	0	.01	.12	.47	.43	.29	.54
250	0	.19	2.25	2.38	2.22	1.72	1.76
375	1.10	2.36	2.94	2.61	2.35	2.05	2.22
Perennial ryegrass, dry wt. in grams, 40-65 days							
0	0	0	.01	.29	.19	.26	.25
125	0	.01	.28	.81	.62	.44	.51
250	0	.48	2.86	2.52	2.07	1.64	1.58
375	1.39	2.74	3.20	2.71	2.71	2.43	2.57
Sorghum, No. of seedlings,[2] 21 days							
0	0	1	7	12	12	14	15
125	0	3	10	12	11	10	12
250	0	9	11	10	13	12	12
375	7	12	13	13	10	10	14
Sorghum, d.m. grams, 22 days							
0	0	.02	.12	.24	.30	.22	.23
125	0	.01	.14	.19	.23	.18	.17
250	0	.14	.26	.35	.40	.29	.23
375	.10	.28	.37	.34	.38	.32	.37
Ryegrass, d.m. grams, 50 days							
0	0	0	.06	.13	.15	.10	.09
125	0	.01	.14	.16	.10	.14	.13
250	0	.22	.60	.54	.78	.59	.48
375	.21	.82	.84	.88	1.00	.87	.80
Soil test values							
pH, May 1, 1973							
0	2.4	3.0	3.5	6.6	7.4	7.4	7.5
125	2.7	2.9	4.2	6.3	7.4	7.4	7.4
250	2.8	3.4	3.6	5.9	7.2	7.3	7.3
375	3.1	3.6	5.1	6.5	7.2	7.1	7.4
P_1, May 1, 1973							
0	15.0	5.7	7.3	4.3	4.3	7.0	3.7
125	13.3	8.3	7.7	6.0	5.0	5.0	2.3
250	107.7	84.7	22.3	45.3	40.3	12.0	4.0
375	99.0	53.0	82.3	46.7	76.0	5.0	2.3

[1] 30 seeds were planted.
[2] 15 seeds were planted. 5 seeds per pot. 2 seedlings per pot were allowed to grow.

Source: PB 226 905

TABLE 5.10: PERENNIAL RYEGRASS AND SORGHUM GROWING ON SLURRY

Milorganite, N, ppm	Rate of lime, pp 1000			
	0	4	8	16
	No. seedlings,[1] 13 days			
0	36	40	39	38
125	36	36	38	37
250	38	35	36	37
375	36	36	35	37
500	37	31	37	40
	Perennial ryegrass, dry wt. in grams, 0-40 days			
0	.36	.38	.35	.38
125	1.05	.97	.87	1.01
250	1.69	1.38	1.40	1.48
375	1.68	1.45	1.19	1.64
500	1.68	1.10	1.72	2.08
	Perennial ryegrass, dry wt. in grams, 40-65 days			
0	.43	.45	.43	.48
125	1.34	1.35	1.15	1.19
250	1.87	1.66	1.60	1.59
375	2.34	2.21	1.91	2.31
500	2.38	1.91	2.29	2.35
	Sorghum, No. seedlings,[2] 13 days			
0	17	16	17	15
125	19	19	18	18
250	17	16	18	16
375	11	16	16	14
500	16	17	17	19
	Sorghum, d.m. grams, 50 days			
0	.92	1.50	.82	.67
125	.96	.88	1.07	.80
250	.89	1.13	1.00	1.00
375	1.30	1.65	1.48	1.73
500	1.96	1.91	1.70	2.17
	Soil test values			
	pH, May 31, 1973			
0	6.9	6.8	7.0	7.0
125	6.6	6.8	7.0	7.0
250	6.8	6.8	7.0	7.0
375	6.6	6.8	7.0	7.1
500	6.7	6.8	6.9	7.0
	P_1, May 31, 1973			
0	14.5	14.5	15.5	15.2
125	23.5	24.0	25.2	22.0
250	33.0	29.2	34.5	32.8
375	45.5	34.0	37.5	35.8
500	54.2	43.8	46.0	44.2

[1] 40 seeds were planted.
[2] 20 seeds were planted. 5 seeds per pot. 2 seedlings were allowed to grow.

Source: PB 226 905

TABLE 5.11: PERENNIAL RYEGRASS AND SORGHUM GROWING ON SPOIL

Milorganite, N, ppm	Rate of lime, pp 1000					
	0	2.5	5.0	10.0	20.0	40.0
Perennial ryegrass, No. seedlings,[1] 24 days						
0	29	27	36	32	31	39
20	40	34	36	33	34	29
40	36	34	34	35	32	37
80	33	35	33	31	28	27
160	21	29	36	31	27	28
320	21	34	34	33	26	35
Perennial ryegrass, dry wt. in grams, 0-43 days						
0	.16	.23	.40	.28	.24	.33
20	.20	.36	.50	.44	.51	1.08
40	.24	.47	.55	.60	.53	.96
80	.15	.94	1.00	.75	1.02	.99
160	.45	1.19	1.55	1.43	.93	1.36
320	.58	1.63	2.35	2.06	1.31	1.59
Perennial ryegrass, dry wt. in grams, 43-65 days						
0	.10	.20	.27	.22	.17	.26
20	.36	.49	.75	.40	.45	.89
40	.47	.56	.84	.68	.62	.91
80	.11	1.14	1.22	.98	1.08	1.07
160	.80	1.58	2.30	2.04	1.54	1.62
320	.74	3.12	3.22	3.60	2.25	2.25
Sorghum, No. seedlings,[2] 50 days						
0	14	13	14	12	13	13
20	8	16	15	11	11	14
40	9	12	13	10	13	16
80	12	14	14	11	11	15
160	12	11	19	12	9	11
320	16	15	15	13	15	10
Sorghum, dry wt. in grams						
0	.57	1.24	1.60	1.14	1.23	1.13
20	.53	1.33	1.07	1.51	.85	1.14
40	.84	1.26	1.37	1.22	.99	1.43
80	.93	1.66	.98	1.21	1.67	1.19
160	.70	1.47	2.44	.85	1.35	1.86
320	1.06	3.94	2.73	5.70	3.83	2.26
Soil test values						
pH, June 28, 1973						
0	4.9	5.3	6.6	7.9	7.9	8.1
20	4.9	5.8	7.4	7.7	7.6	7.8
40	4.8	6.0	6.4	7.2	7.8	8.0
80	4.6	5.8	5.8	7.6	8.1	8.1
160	4.7	6.0	6.5	7.6	7.8	7.8
320	4.5	5.4	6.6	7.4	7.6	8.2
P_1, June 28, 1973						
0	3.5	5.0	6.8	5.8	5.8	3.8
20	7.0	8.0	5.0	5.2	8.8	13.2
40	5.5	7.8	9.2	10.2	8.2	9.2
80	6.0	9.8	9.0	9.8	13.2	10.2
160	25.3	24.0	20.8	20.2	14.2	18.2
320	32.7	22.2	44.8	35.8	37.2	25.2

[1] 40 seeds were planted.
[2] 20 seeds were planted. 5 seeds per pot. 2 seedlings were allowed to grow.

Source: PB 226 905

The pH of the check, 4.9, was high enough to permit germination. However, dry matter yields of ryegrass were increased by lime rates up to 5.0 parts per thousand on the second cutting and on the sorghum. The production of dry matter of all harvests was increased by applications of Milorganite up to 320 parts per million of N per acre. Applications of Milorganite did not raise the pH even where no lime was applied as it did in the case of the gob. The level of available phosphorus, P_1, was not raised to an adequate level until sludge at the rate of 160 parts per million of N was applied.

Discussion

Perennial ryegrass and presumably many other plants can be established on gob, slurry, and spoil materials similar to those used in these trials. To do so lime can be used to raise the soil material pH, if necessary, and sludge can be used as a source of nutrients. It may be necessary to supply some potassium because the level in sludge is low in relation to plant needs.

Sludge in the gob trial enabled ryegrass to germinate and grow at a very low pH, 3.2, but this trial was not long enough to determine how long this effect is. It appears that the combined effect of lime and sludge is much greater than the sum of the single effects.

The amount of lime required to adjust the pH, particularly of the gob, indicated that only a small proportion of the potential acidity was developed in the time of this experiment. In determining the amount of lime that should be used in the field it will be necessary to know the rate at which the various forms of sulfur are converted to an acid form.

If the rate is very low, an amount of lime necessary for establishment may be enough to last for many years. If the rate is very high, enough lime should be added to neutralize the potential acidity or about 30 tons per acre of lime for each 1% sulfur in the material. The actual situation is probably between these two extremes. It is important to know the rate of conversion so that lime can be used efficiently.

The results of these trials indicate that the soil material of similar gob piles, slurry areas and spoils can support plant growth if they are treated with lime and sludge. However, since conditions were ideal in the greenhouse for germination and growth it is not known if field plantings will be successful where moisture stress and temperature conditions are much more extreme.

It is known that there are extreme differences in these materials, not only between sites, but within sites. Since only one sample of each material was used in this trial it will be necessary to conduct similar studies with samples from numerous sites before generalizations can be made about the establishment of vegetation on all sites.

USE OF FLY ASH ON KANSAS STRIP MINE SOILS

Experimental Procedure

Three soils from strip mined land in southeast Kansas were collected, air dried, pulverized to pass an 8 mesh screen, and placed in pots. Two of the soils were from newly mined land in Crawford County (designated A and B) and the third soil (designated C) was from older mined land but recently leveled. The soil and fly ash chemical and physical characteristics measured are presented in Table 5.12. The fly ash was obtained from the settling basin of the Kansas Power and Light Plant at Lawrence, Kansas. Fly ash was applied at the rate of 0, 25, 50, 100, 200, 400, and 800 tons per acre.

In addition, a treatment with lime at the rate recommended by soil test was included for each soil. (The lime requirement on soil B exceeded the soil test limits and the applied rate was estimated.) All pots received a fertilizer application of N, P and K based on soil test determinations.

TABLE 5.12: CHEMICAL AND PHYSICAL CHARACTERISTICS OF SOIL AND FLY ASH

Sample	Organic Matter (percent)	pH	Effective $CaCo_3$ (pounds/acre)	Available P (pounds/acre)	Exchange K (pounds/acre)	CEC (meq/100 g)	Bulk Density (g/cc)	Field Capacity (percent)	Texture Sand (percent)	Silt (percent)	Clay (percent)
Soil A	1.7	4.8	7,500	50	257	12.7	1.18	20.6	28	72	0
Soil B	1.5	3.1	20,000+	190	93	14.1	1.19	27.5	22	74	4
Soil C	2.2	6.3	3,000	34	252	14.1	1.18	24.1	22	74	4
Fly ash	–	7.7	–	42	1,000+	–	1.48	17.0	46	52	2

Source: COM-72-10623

The fly ash was thoroughly incorporated with the soil and the containers brought to field capacity. The pots were allowed to incubate for a period of five weeks with core soil samples taken at weekly intervals for pH determinations. At the end of the five week incubation, oats were seeded on the pots and allowed to grow for 35 days after planting. Emergence notes were taken on number of seedlings emerged at one and two weeks after planting. The pots were then thinned to 10 plants per pot. At the end of five weeks, the plants were clipped at ground level, weighed, and then dried at 65°C for 24 hours before taking dry matter yields.

Results and Discussion

The fly ash and lime treatments had a favorable effect on the soil pH (Figure 5.2). In a test on the fly ash for its neutralizing capacity, it was found to be 1.5% as effective as pure calcium carbonate. This agrees well with the experimental results comparing pH change of fly ash treatments to lime application. On the extremely acid soil (B), no treatment was sufficient to bring the pH to neutrality. The estimated rate of lime (25,000 pounds ECC/A) was too low.

The neutralizing effect of the fly ash and lime were found to occur quite rapidly as noted by the relative small change in pH of the soils over the duration of the study (Figures 5.3, 5.4, 5.5 and 5.6). Under field conditions the rate of neutralization would be slower because the fly ash or lime would not be as thoroughly incorporated. The initial soil samples taken one day after mixing of the soil with the fly ash showed a pH close to that of samples pulled after 5 weeks of incubation.

FIGURE 5.2: EFFECT ON THE pH AFTER 10 WEEKS

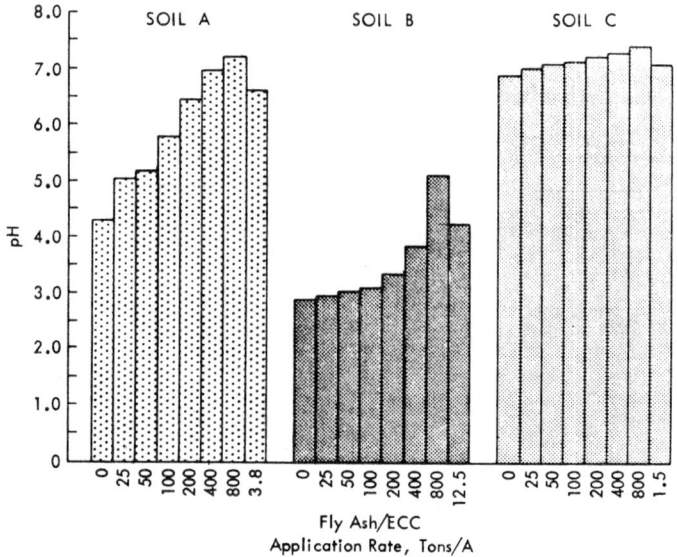

FIGURE 5.3: CHANGE IN pH AFTER ADDITION OF FLY ASH, CONTROL SAMPLES

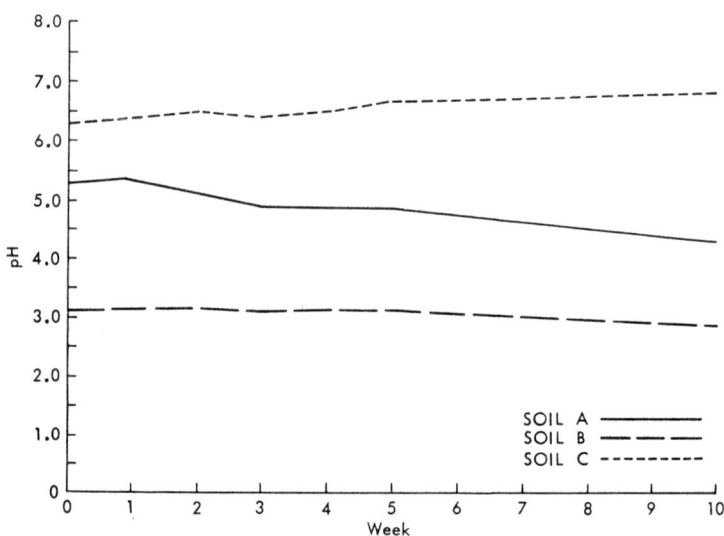

Source: COM-72-10623

FIGURE 5.4: CHANGE IN pH AFTER ADDITION OF FLY ASH, 50 TONS PER ACRE

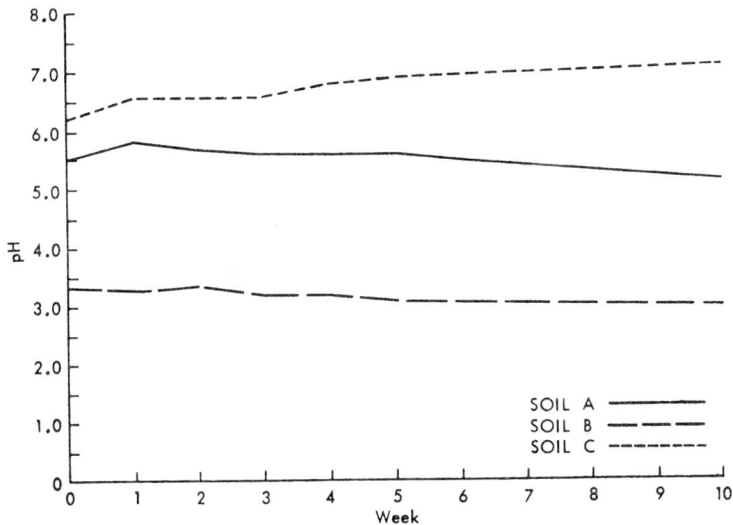

FIGURE 5.5: CHANGE IN pH AFTER ADDITION OF FLY ASH, 200 TONS PER ACRE

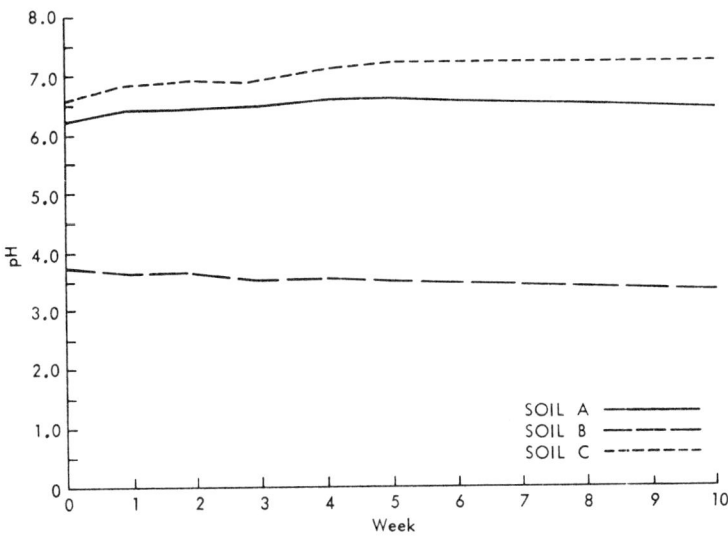

Source: COM-72-10623

FIGURE 5.6: CHANGE IN pH AFTER ADDITION OF FLY ASH, 800 TONS PER ACRE

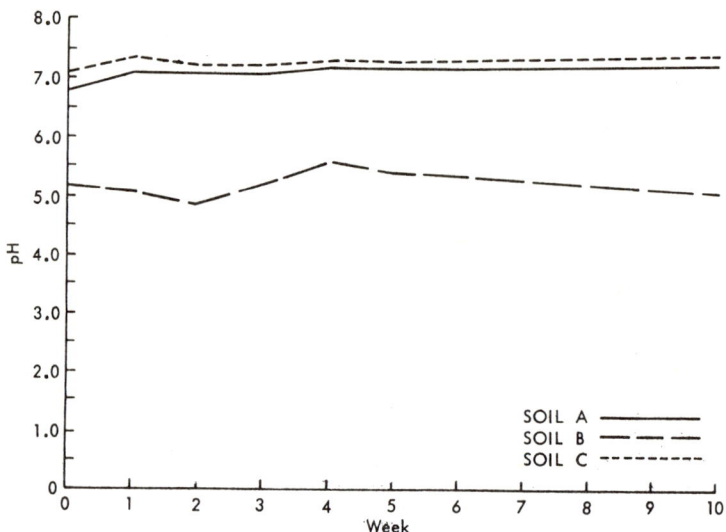

Source: COM-72-10623

A slight decline in pH over the 10 week period was noted at the 0 and 50 tons per acre fly ash rates for soil A, but not on the 200 and 800 tons per acre fly ash rates. Soil C indicated a slight rise in pH during the 10 weeks of the study at all rates of fly ash. Part of this pH change with time is probably due to incomplete equilibration of the soil-fly ash when the initial pH reading was taken.

Oat plant emergence was recorded at the end of one and of two weeks after planting. Good emergence was noted on all pots except those with none or low rates of fly ash on soil B. Not only was the number of seedlings reduced on these treatments, but there was several days' delay in emergence.

Although seedlings emerged on all pots, there was no growth of the seedlings on the extremely acid soil B until the pH had been brought above 4.0. The pH of the soil was the main factor influencing growth on soil B. Dry matter yields show a marked improvement in growth due to fly ash compared to the control. Figure 5.7 shows the effect of fly ash and lime on dry matter yield after 35 days' growth.

On soil A, part of the difference in growth is related to a pH change, but in addition the fly ash has contributed other factors for increased plant growth. This is evident by comparing fly ash to lime. Similar comparisons can be made for soil C which has a much higher initial pH. The beneficial effect of fly ash cannot be determined without additional soil and plant analyses. The higher rates of fly ash could cause some alteration in physical properties that could contribute to yield increases.

FIGURE 5.7: EFFECT OF FLY ASH AND LIME ON DRY MATTER YIELD

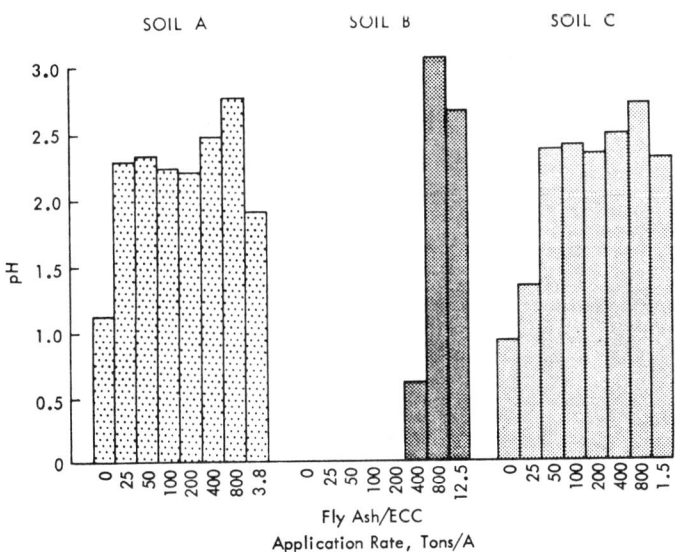

Source: COM-72-10623

RECLAMATION OF WEST VIRGINIA ACID SPOIL WITH FLY ASH

Specific objectives were to determine (1) if fly ash can sufficiently increase the pH of highly acid spoils to sustain the growth of grasses and legumes; (2) the best proportion of fly ash to spoil; and (3) if fly ash-spoil mixtures can support plants long enough to increase the organic matter in the mixture sufficiently to reduce the toxic potential of the ash. Overlying these objectives was the need to find economical and useful outlets for large tonnages of fly ash.

Materials and Methods

Physical and Chemical Characteristics of Sites: The field experiments were conducted in northern West Virginia on sites that are commonly known as orphaned areas because they were not reclaimed after the coal was removed.

The initial site had been surface mined for Sewickley-bed bituminous coal more than fifteen years before. It was virtually devoid of vegetation except for scattered brush consisting of black locust *(Robinia pseudoacadia)*, wild cherry *(Padus virginiana)*, blackberry briers *(Rubus allegheniensis)*, and broom sedge *(Andropogan virginieus)* around the perimeter.

Physical and chemical properties of the spoil at the start of the experiment were soil pH, 3.5; nitrogen, phosphorus, potassium, and organic matter, very low; cation exchange capacity (CEC), 12 milliequivalents per 100 grams; composition of −10 mesh fraction; clay, 21%; silt, 23%; sand, 56%; bulk density, 1.38 g/cc.

The surface was very rocky, with rock sizes ranging from pebbles to boulders weighing several hundred pounds. The chemical analysis is shown in Table 5.13.

TABLE 5.13: CHEMICAL ANALYSIS OF SPOIL AT SITE 1*

Constituent	Weight Percent
Aluminum (Al_2O_3)	18.0
Silicon (SiO_2)	69.3
Iron (Fe_2O_3)	7.9
Phosphorus (P_2O_5)	0.1
Titanium (TiO_2)	0.7
Calcium (CaO)	0.2
Magnesium (MgO)	0.2
Potassium (K_2O)	2.4
Sodium (Na_2O)	0.7
Sulfur (S)	0.3
Hydrogen (H_2)	0.8
Carbon (C)	1.7

*Soil sample cores were taken to a depth of six inches and composited for analysis.

Source: COM-72-10623

The second site was on a spoil area that resulted from the surface mining of the Bakerstown bituminous coal seam. Stripping had been completed in 1956, the area had been backfilled and seeded in 1958, and parts of the area had been reseeded in 1959. Both seedlings were nearly total failures. Adequate ground cover developed on only about 1% of the area.

At the start of the study, soil samples from the entire area were obtained to a depth of six inches, composited, and analyzed. Results were as follows: bulk density, 1.28 grams per cubic centimeter; nitrogen (very low), approximately 15 pounds per acre; phosphorus (low), 21 pounds per acre; potassium (low), 36 pounds per acre; organic matter, not detectable; pH, 2.6 to 3.3. Iron carbonate, shale, clay and acid slicks were scattered and typified the entire 117 acres.

Fly Ash Characteristics: Table 5.14 shows some of the physical and chemical properties of the fly ashes used in these experiments. Concentrations of the elements are about the same or somewhat higher than those in many natural soils. Many of the major and minor plant nutrients are included. However, fly ash generally contains very little phosphorus and is devoid of nitrogen, both of which are essential to plant growth; thus these elements must be added through fertilization.

Fly Ash Application: Fly ash application rates ranged from about 150 to 800 tons per acre. All the plots including the controls were plowed and harrowed with ordinary farm machinery to thoroughly mix the fly ash and spoil and to prepare the surface for seeding and fertilization. Plots were fertilized at the time of seeding with a granular dry fertilizer at the rate of 1,000 pounds per acre; analysis: 10% nitrogen, 10% P_2O_5, and 10% K_2O. Seed for the plots consisted of a mixture of grasses that had grown well in previous experiments.

TABLE 5.14: CHEMICAL AND SCREEN ANALYSIS OF FLY ASHES

	Fly ash 1	Fly ash 2	Fly ash 3
pH	11.4*	9.1-10.6	11.9
Bulk density, g/cc	1.12	0.93	1.15
Chemical analysis, wt pct			
Aluminum (Al_2O_3)	23.9	23.9	23.6
Silicon (SiO_2)	46.3	42.2	47.7
Iron (Fe_2O_3)	22.9	24.0	15.6
Phosphorus (P_2O_5)	.3	.2	.6
Titanium (TiO_2)	.9	.8	2.7
Calcium (CaO)	1.9	4.0	3.5
Magnesium (MgO)	.8	1.2	1.5
Potassium (K_2O)	2.2	2.2	2.2
Sodium (Na_2O)	.6	.6	1.9
Cobalt	.02	.02	ND**
Boron	.008	.02	ND
Manganese	.03	.05	ND
Copper	.02	.02	ND
Molybdenum	.007	NT***	ND
Carbon	5-7	12.4	1.54
Sulfur (total)	.24	.51	.34
Screen analysis, wt pct			
+60 mesh	2	2	1
-60+100 mesh	5	3	2
-100+150 mesh	4	4	2
-150+200 mesh	8	7	4
-200 mesh	81	84	91

*pH of the fly ash used at site 1 was 11.4; when used the next year at site 2 the pH had dropped to the range 4.4 to 9.5.
**Not determined.
***Not detected.

Source: COM-72-10623

The mixture contained Kentucky 31 fescue *(Festuca arundinacea* Schrebe), perennial rye grass *(Lolium perenne),* orchard grass *(Dactylis glomertas),* redtop grass *(Agrostis alba),* and bird's-foot trefoil *(Lotus corniculatus)* and was applied at the rate of 43 pounds per acre.

Results and Discussion

Physical Characteristics: Mixing fly ash with mine spoils produced chemical and physical changes that enhanced plant survival and growth. Bulk density of the mixtures was decreased up to 41% by addition of the amendment, while the bulk density of a limestone treated control plot decreased only 5% (Table 5.15). This is to be expected since the bulk density of the fly ash used ranged from 0.92 to 1.15 grams per cubic centimeter. Decreased bulk density values resulted in greater pore volume, greater moisture availability, and higher air capacity; better conditions for root penetration and growth.

TABLE 5.15: AVERAGE BULK DENSITIES OF SOILS, SITES 1 AND 2

Site	Plot	Bulk Density (g/cc)
1	Untreated control	1.38
	3	1.15
	Limestone	1.31
2	Untreated control	1.55
	6 (½ acre)	0.92

Source: COM-72-10623

Soils consist of sand and clay and silt sized mineral fractions, with or without rocks. The relative proportion of these various fractions defines a specific soil texture classification. For example, the plow layer (6 inches deep and weighing 1,000 tons per acre) of a typical clay loam has the following analysis per acre: clay, 350 tons; sand, 350 tons; and silt, 300 tons. A typical silt loam has the following per acre analysis: clay, 150 tons; sand, 200 tons; silt, 650 tons.

Fly ash, which is mostly in the silt size range, was applied at rates of 150 to 800 tons per acre, and this proved to be sufficient ash (particularly the higher rates) to produce a textural change from loam to silt loam. Spoil at site 1 was changed from a sandy clay loam to a silt loam by application of 600 tons of fly ash per acre. At site 2, the texture changed from a clay loam to a silt loam. Although soil scientists consider alteration of texture virtually impossible under ordinary circumstances, this study showed that adding fly ash to surface mine spoil changed the textural classification.

Moisture Availability and Infliltration: Available Water — It is generally accepted that medium textured soils (those predominated by the silt fraction) hold more available water than either coarse (sandy) or fine (clay) textured soils. Thus, fly ash-treated spoil should have more water available for plant use.

To confirm this, soil moisture tension and gravimetric soil moisture data were obtained, the latter being measured each week at both strip sites from May through October. To obtain data that would more truly represent the soil in situ, the soil-moisture tension procedure was modified by the use of larger undisturbed core or ring soil samples. Normal procedure calls for the soil to be screened through a 10 mesh or 2 millimeter screen before use on the tension table.

Spoil Amendment Studies

Thus, since strip mine spoil generally contains plate-like pieces of shale and/or other bits of rock, it was hoped that a more realistic moisture picture would be obtained by this modified procedure.

Moisture tension (pressure) values, expressed in atmospheres, were $1/3$, 1, 5, 9, and 15. The $1/3$-atmosphere value is commonly referred to as the field capacity tension value, and the 15-atmosphere tension value is referred to as the wilt point. Between these pressures is the available water, or water that is generally available to plant usage and growth; that is, water for photosynthesis, respiration, and other physiological and metabolic processes.

Duplicate samples were obtained to minimize errors, and the samples were placed on the pressure membrane and allowed adequate time to absorb moisture and become saturated, at least 24 hours. The appropriate pressure was applied, and the samples were allowed to come to equilibrium. In some cases, samples were left under pressure for as long as four days to obtain equilibrium, after which the samples were weighed and oven dried and the percentage moisture was calculated. From these data the moisture curves were plotted, of which Figure 5.8 is typical.

Preliminary results indicate that much water is held at the very low tensions ($1/3$-atmosphere), with the fly ash-treated spoil holding somewhat more moisture than untreated spoil. It is difficult to explain the very sharp drop in moisture content as the pressure is increased from $1/3$ to 1 atmosphere, but it may be due to the presence in the undisturbed core of roots and of rock and shale particles larger than two millimeters, which, as stated above, are ordinarily removed.

FIGURE 5.8: MOISTURE CURVES

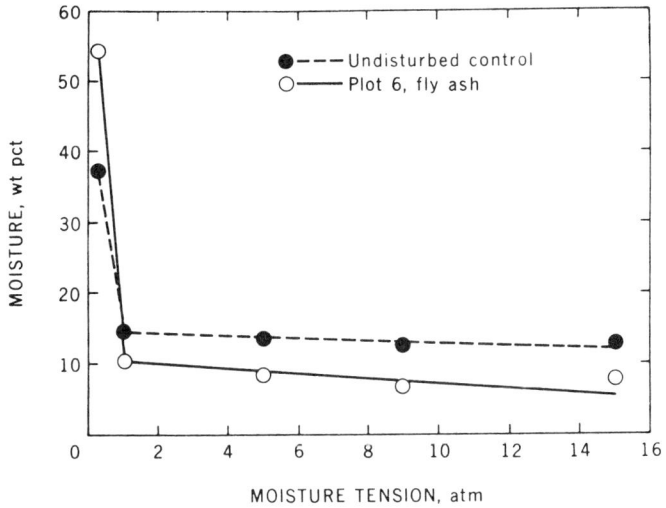

Source: COM-72-10623

Available water values (shown on Figure 5.8) for ⅓ and 15 atmospheres give an indication of the approximate range. Figures 5.9a, 5.9b and 5.9c gives the gravimetric moisture values for the plots indicated, along with precipitation for the site 2 area for each growing season.

FIGURE 5.9: MOISTURE CONTENT DETERMINATIONS FOR SPOIL ASH MIXTURES

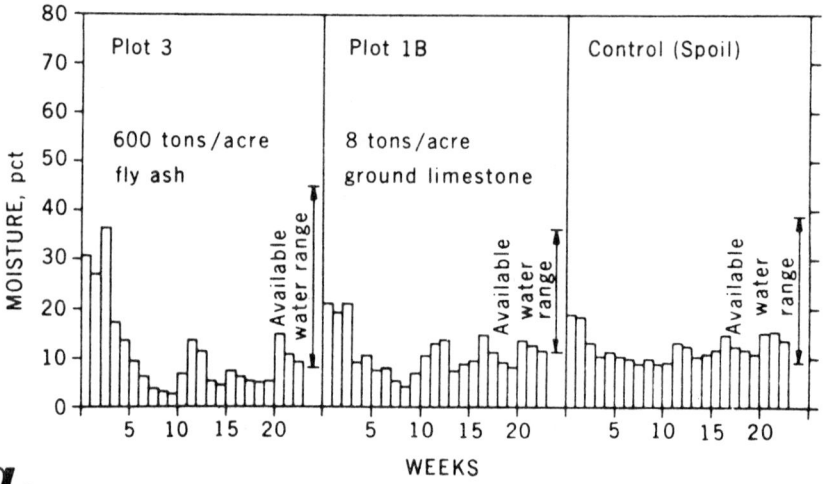

a.

Top Six Inches, Site 1, 3rd Year

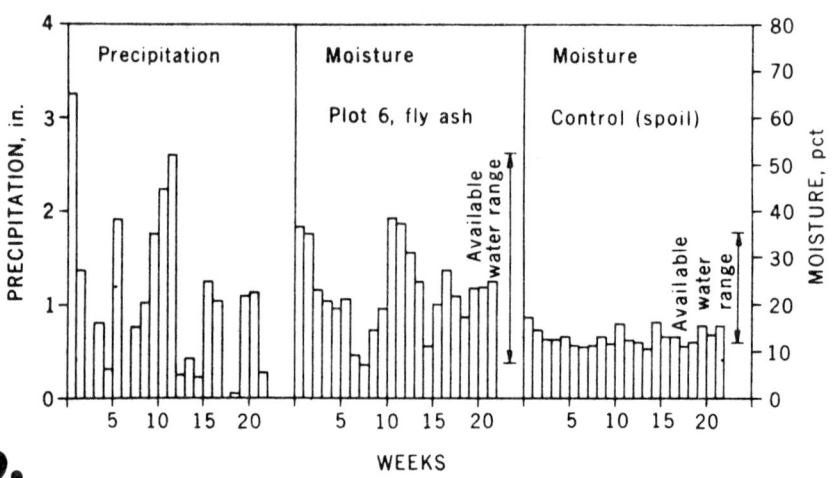

b.

Top Six Inches, Site 2, 2nd Year

(continued)

FIGURE 5.9: (continued)

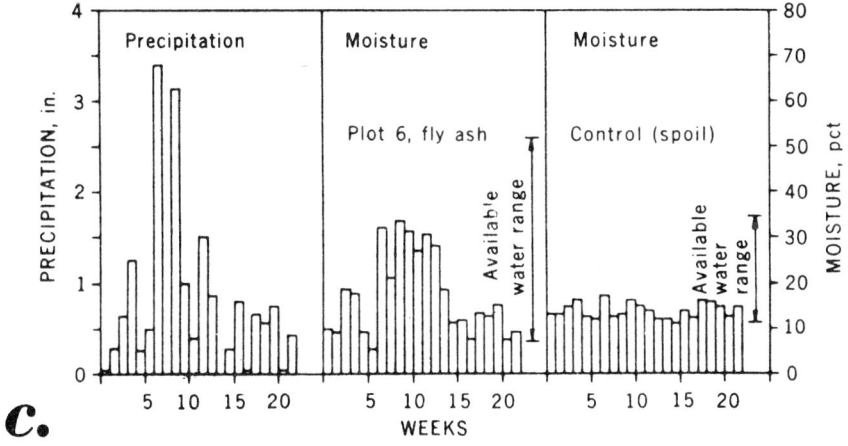

c.

Top Six Inches, Site 2, 3rd Year

Source: COM-72-10623

Undoubtedly fly ash can affect the moisture content, especially during wet periods (first two weeks, second growing season; Figure 5.9b). Furthermore, the available moisture range remains wider well into the dry periods (6th to 8th weeks and again from 20 to 22 weeks, site 2, third year; Figure 5.9c).

Water Infiltration — Moisture infiltration rates (determined by double-ring infiltrometer) at site 1 were moderate for one fly ash treated plot (600 tons per acre) and one limestone treated plot, while the control plot showed a moderately rapid infiltration rate (Table 5.16). (A nearby field soil was not available for testing.)

TABLE 5.16: WATER INFILTRATION AT FLY ASH-RECLAIMED SITES*

Site and Description	Accumulated Intake (inch)	Intake Rate (in/hr)	Classification**
1 Limestone treated	5.6	1.4	Moderate
Control	14.0	3.5	Moderately rapid
Fly ash treated	7.0	1.8	Moderate
2 Field soil	6.6	1.7	Moderate
Control	5.8	1.5	Moderate
Fly ash treated	6.7	1.7	***

*Average data at end of 4th hour (flow rate stabilized).
**Very rapid, 10.00 in/hr; rapid, 5.00 to 10.00; moderately rapid, 2.50 to 5.00; moderate 0.80 to 2.50; moderately slow, 0.20 to 0.80; slow, 0.05 to 0.20; very slow, 0.05 in/hr.
***Average of 8 plots. Range of accumulated intake after 4 hours, 2.9 to 14.4 inches; range of intake rate after 4 hours, 0.7 to 3.6 in/hr (moderately slow to rapid).

Source: COM-72-10623

At site 2, the undisturbed control had an average intake rate of 1.5 inches per hour compared with the nearby field soil value of 1.7 inches per hour; fly ash-treated plots had values both lesser and greater than these, 0.7 to 3.6 inches per hour (third footnote, Table 5.16). Such variable results may have been caused primarily by the shaly nature of the spoil.

Biological Properties: Evidence was also found that the microbiological environment of spoil-fly ash mixtures was improved. Nodules of active nitrogen-fixing bacteria were observed on the roots of bird's-foot trefoil plants, a legume that appears to thrive in fly ash-spoil mixtures.

Chemical Properties: Chemical analyses of the fly ashes and spoil show some differences in major and minor constituents. Of the major plant nutrients in spoil, nitrogen is negligible and available phosphorus and potassium are low. Although fly ash contains some phosphorus and potassium, it also is low in nitrogen. To assure an adequate and available supply of the major plant nutrients, the plots were treated with a commercial fertilizer containing 100 pounds each of N, P_2O_5, and K_2O.

Calcium and magnesium, also low in quantity, are necessary to sustain plant life. These elements in fly ash range from 1.9 to 4% for calcium and 0.8 to 1.5% for magnesium. A 600 ton application of fly ash may add as much as 48,000 pounds of calcium and 18,000 pounds of magnesium, which is equivalent to a total of 33 tons of liming material. Undoubtedly this large quantity of liming material accounts for the neutralization of acid in spoil sufficiently to increase pH to acceptable levels, as illustrated by the curves in Figure 5.10.

In humid regions or during periods of heavy precipitation, leaching may be as important as crop production in removing exchangeable bases and increasing soil acidity. Studies have shown, for example, that more calcium can be lost in drainage water than is removed in harvested crops. However, it is believed the calcium in fly ash is present in the glassy spheres portion of the solids, and, as a result, is likely to be released very slowly.

Other factors that affect leaching are slope and permeability. Both sites sloped about 3 to 10 degrees, but infiltration studies showed that precipitation entered the fly ash-treated spoil. This is important because adequate moisture is necessary to rapidly establish a vegetative cover to reduce erosion after fly ash treatment.

Lime Requirement: As can be seen from the pH curves in Figure 5.10, the pH of the untreated controls remained low throughout the study. The pH of the fly ash plots, initially increased by the addition of fly ash, declined gradually over a three year period, though not to a level that was detrimental to growth.

Liming materials present in the fly ash or added as limestone either were used by the plants or gradually leached from the soil. This, as expected, gradually lowered the pH. Application of fly ash in tonnage quantities therefore increased the calcium content and raised the pH, thus lowering the lime requirement.

Dry Matter Yields: During the third year at site 2, fertilizer (N, P, and K) rates were increased. Since nitrogen is the most easily lost by leaching, it was applied in equal portions of 70 pounds per acre as urea in the spring and after the first two cuttings.

FIGURE 5.10: pH OF SPOIL-FLY ASH MIXTURES

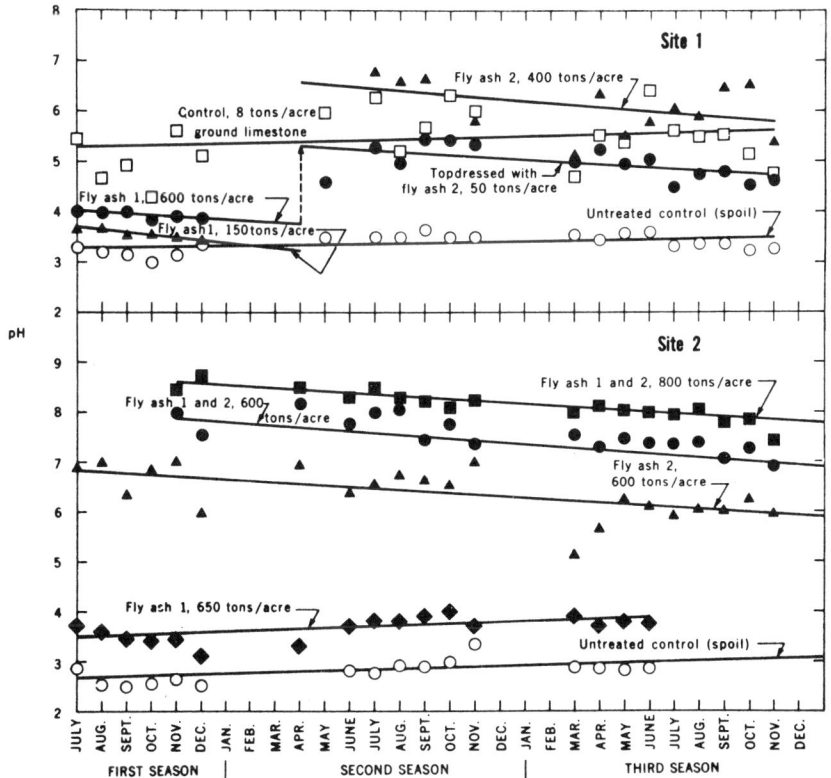

Source: COM-72-10623

Spring applications of P and K were 90 pounds per acre of P (as P_2O_5) and 180 pounds of K (as K_2O) from 0-15-30 analysis granulated fertilizer. This treatment increased yields and permitted three cuttings instead of the usual one or two per year. Average dry matter yields from all the plots for each cutting are given in Table 5.17).

TABLE 5.17: AVERAGE DRY MATTER YIELDS, SITE 2

	Dry Weight, Tons/Acre						
	1st Cutting		2nd Cutting		3rd Cutting		
Year	Date	Yield	Date	Yield	Date	Yield	Total Yield
First	June 1	1.6	—	None	—	None	1.6
Second	June 5	1.4	Sept. 5	0.9	—	None	2.3
Third	June 2	1.7	July 29	1.2	Sept. 30	0.9	3.8

Source: COM-72-10623

The overall average yields for the three cuttings are 44, 34 and 22% of the season total, which is comparable with those obtained at the West Virginia University Agronomy Farm, Reedsville, West Virginia, in high fertilization rate experiments. Furthermore, the total average yield of about 4 tons per acre for the third year at site 2 was typical of the yields obtained during West Virginia University's pasture experiments with perennial tall grasses.

To determine the nutritional value of dry forage taken from the plots, several bales were pulverized and fed to meadow voles, a rodent with the ability to digest forages. A control group of voles were fed pulverized dry forage that had grown on ordinary field soil. No significant differences were noted in weight gain or growth characteristics of the two groups at the end of the 38 day study. In addition, the landowner has fed hay from these plots to his cattle over a five year period, with no adverse effects.

ACID MINE DRAINAGE

The material in this chapter was excerpted from the following reports:

PB 217 872
PB 230 022
PB 238 538

COAL MINE DRAINAGE POLLUTION

Sources

All methods of surface and underground mining may result in some degree of mine drainage pollution. The quality and quantity of mine drainage pollutants produced from a mining operation depends upon such factors as the operating status of the mine, (i.e., active or inactive); hydrologic, geologic, and topographic features of the surrounding terrain; the type of mining method employed; and availability of air, water, and iron sulfide minerals.

In surface mines, the discharge of the pollutants is often intermittent, generally occurring during and immediately after periods of precipitation. Runoff in stripped areas may find its way to a surface stream or be trapped in inadequately restored trenches or pits formed during the stripping operation. When the runoff is trapped, pools which may contain high concentrations of mine drainage pollutants are formed. During subsequent periods of high runoff, these pools may overflow, releasing concentrated "slugs" of mine drainage pollution to receiving streams. Although streams that are only intermittently polluted may be of good quality much of the time, the aquatic life community of streams receiving slugs of acid mine drainage may be damaged for extensive periods of time.

Between flush-out periods, the pools in stripped areas often drain slowly into the backfill to emerge in the form of mine drainage seepages downslope from the stripping operation. They may also drain to underground mines underlying the stripped area, thus increasing the mine drainage flow from these mines. Mine drainage may continue to flow from inactive mines as long as air, water, and sulfide minerals are available.

Disposal of the refuse materials from mines and coal preparation plants is one of the most difficult problems associated with active operations. Refuse piles are a major source of acid drainage in some areas and washery residue spillage is a frequent source of the fine coal and silt pollution common in some streams.

For field inventory purposes, a coal mine drainage source is considered to be a surface or underground location resulting from the handling or extraction of coal, containing minerals whose solution by contact water is resulting in a highly mineralized drainage. A drainage of greater than one gallon per minute is considered large enough to be included in a source inventory.

Formation of Pollutants

Although the exact reaction process is still not fully understood, the formation of acid mine drainage is generally illustrated by the equations shown below. The initial reaction that occurs when iron sulfide minerals are exposed to air and water produces ferrous sulfate and sulfuric acid.

$$2FeS_2 + 7O_2 + 2H_2O \longrightarrow 2FeSO_4 + 2H_2SO_4$$

Subsequent oxidation of ferrous sulfate produces ferric sulfate.

$$4FeSO_4 + 2H_2SO_4 + O_2 \rightleftharpoons 2Fe_2(SO_4)_3 + 2H_2O$$

Depending on physical and chemical conditions, the reaction may then proceed to form ferric hydroxide or basic ferric sulfate.

$$Fe_2(SO_4)_3 + 6H_2O \rightleftharpoons 2Fe(OH)_3 + 3H_2SO_4$$

and/or

$$Fe_2(SO_4)_3 + 2H_2O \rightleftharpoons 2Fe(OH)(SO_4) + H_2SO_4$$

Pyrite can also be oxidized by ferric iron as shown below.

$$FeS_2 + 14Fe^{+3} + 8H_2O \longrightarrow 15Fe^{+2} + 2SO_4^{-2} + 16H^+$$

Other constituents found in mine drainage are produced by secondary reactions of sulfuric acid with minerals and organic compounds in the mine and along the stream valleys. Such secondary reactions produce concentrations of aluminum, manganese, calcium, sodium, and other constituents in the drainage water. These mine drainage constituents, along with iron and sulfate, are indicators of mine drainage pollution that may persist long after the acid in the drainage has been neutralized.

Crystalline forms of pyritic material are less subject to weathering and oxidation than amorphic forms. If the overburden also contains alkaline material such as limestone, acid water may not be discharged even though it is formed, because

Acid Mine Drainage

of inplace neutralization by the alkaline material. Discharges from this situation are usually high in sulfate.

Although there are conflicting opinions among researchers as to the importance of microorganisms in the productions of mine drainage pollution, there is evidence to indicate that microorganisms do contribute to pyrite oxidation. A number of bacterial species including *Thiobacillus thiooxidans*, *Thiobacillus ferrooxidans*, and *Ferrobacillus ferrooxidans* have been isolated from mine drainage waters.

Water Quality Evaluation

The intensity of coal mine drainage pollution is evaluated by measurement of various water quality parameters. The most common physical, chemical, and biological parameters used for evaluation are discussed below. The most commonly applied standard that reflects coal mine drainage pollution is pH. A pH of less than 6.0 or more than 8.5 is considered unacceptable. Other physical and chemical parameters and their respective range of values of concern are as follows: acidity, sufficient to lower alkalinity below 20 milligrams per liter; alkalinity, <20 milligrams per liter; sulfate, >250 milligrams per liter; hardness, >250 milligrams per liter; total iron, >1.0 milligram per liter; manganese, >1.0 milligram per liter; aluminum, >0.5 milligram per liter; suspended solids, >250 milligrams per liter; dissolved solids, >500 milligrams per liter.

Because the pollution of streams by coal mine drainage can be extremely damaging to aquatic life, biological observations and measurements are useful for evaluating the extent of pollution. Streams so polluted generally support only a few species of particularly tolerant plants and animals.

Damages to aquatic life from acid mine drainage are attributed usually to high concentrations of mineral acids, the ions of iron, sulfate, and the deposition of a smothering blanket of precipitated iron salts on the stream bed. In addition, zinc, copper, and aluminum have occurred at lethal concentrations in acid mine drainage; and arsenic and cadmium have been found at threshold concentrations. The toxicities of these elements are compounded by synergism among several of them: zinc with copper, zinc with cadmium, and copper with cadmium.

The toxicities of iron, copper, and zinc solutions are much greater in the acid waters polluted by coal mine drainage than in neutral or alkaline waters. Because of the complex chemical nature of coal mine drainage, it is impossible to assign its toxicity toward aquatic life to any single chemical constituent.

Toxic chemicals in acid mine drainage eliminate sensitive life forms; tolerant forms occasionally flourish to great numbers apparently unaffected by the pollutants. The specialized flora and fauna of acid mine drainage reflect harsh water quality conditions. Fish are usually not found when the pH of a stream is lower than 4.5. Conversely, populations of midge larvae (*Tendipes* sp.) may develop to nuisance proportions. A qualitative biological examination of a stream heavily polluted by acid mine drainage (pH 4.0 or lower) may reveal a community structure similar to the following.

Complex Plants: Cattails (*Typha* sp.) and some mosses; other vascular plant life is generally not found in acid mine drainage.

Algae: Dense flowing mats of species of the green alga *Ulothrix* are so common as to attract the attention of casual observers; gelatinous mats of chlorophyll-containing flagellates (*Euglena* spp.) often color stream beds dark green; microscopic examination commonly reveals other species of green algae, including *Microspora* spp., *Microthamnion* sp., the flagellate *Chlamydomonas* sp., great numbers of diatoms *Eunotia* sp., *Pinnularia* sp., and *Navicula* sp., and lesser numbers of *Surirella* sp.

Benthic Invertebrates: In severely polluted stream reaches, especially near the mine adits from which polluted water flows, no benthos will be found. In less severely polluted reaches, common inhabitants include midges (*Tendipes* sp. and others), alderflies (*Sialis* sp.), fishflies (*Chauloides* sp), craneflies (*Antocha* sp. and others), dytiscid beatles, and caddisflies (*Ptilostomis* sp.). Swampy areas polluted by coal mine drainage contain the above forms, plus water boatmen, dragonflies, damselflies and mosquitoes. Conspicuous by their absence are crayfish, blackflies, mayflies, stoneflies, and most species of caddisflies.

Absent Fish: Upstream reaches, not polluted by acid mine drainage, might support several species of rooted and floating vascular plants, twenty or thirty species of algae, fifteen or twenty species of benthic invertebrates, and a mixed community of fishes. Severely polluted stream reaches might support only three or four species of algae; in less severely polluted reaches, only one or two species of vascular plants, three or four species of algae, three or four species of benthic invertebrates, and no fish.

Damages

Discharge of acid coal mine drainage to surface waters changes the water quality by lowering the pH; reducing the natural alkalinity; increasing the total hardness; and adding undesirable amounts of iron, manganese, aluminum, sulfates, and other elements and suspended material.

Some tangible damages resulting from these quality changes that can be evaluated in monetary terms are the cost to municipal and industrial water treatment plants for the required additional treatment of polluted water and early replacement of equipment damaged by polluted water and the cost of early replacement of steel or iron structures and equipment such as culverts, bridges, locks, boat hulls, steel barges, pumps and condensers. Concrete structures may also be damaged.

Damages to recreational uses and aesthetic values are difficult to measure in economic terms, but are important. Such damages are; streams may be rendered less desirable or unusable for water-related recreational uses such as fishing, boating, water skiing, swimming, camping, and picnicking; the elimination or alteration of biological life; the lowering of property values along polluted streams.

PREVENTION OF ACID FORMATION

Since oxidation by oxygen is the primary reaction during early acid formation, the less time pyritic material is exposed to air, the less acid is formed. Thus, a positive preventive method is to cover pyritic materials as soon as possible with earth, which serves as an oxygen barrier. In terms of mining, this step is accomplished by current reclamation techniques and small open pits.

All techniques for preventing acid formation are based on the control of oxygen. There are two mechanisms by which oxygen can be transported to pyrite—convective transport and molecular diffusion.

The major convection transport source is wind currents that can easily supply the oxygen requirement for pyrite oxidation at the spoil surface. In addition, wind currents against the steep slope would provide sufficient pressure to drive oxygen deeper into the spoil mass. One factor in considering the degree of slope for regrading, especially on sides subject to prevailing winds, is that the wind pressure on the spoil surface increases as the slope increases. Thus the depth of oxygen movement into the spoil would increase as the slope increases.

Molecular diffusion occurs whenever there is an oxygen concentration gradient between two points, that is, the spoil surface and some point within the spoil. Molecular diffusion is applicable to any fluid system, either gaseous or liquid. Thus oxygen will move from the air near the surface of the spoil, where the concentration is higher, to the gases or liquid-filled pores within the spoil, where it is lower. The rate of oxygen transfer is strongly dependent on the fluid phases and is generally much higher in gases than in liquids. For example, the diffusion of oxygen through air is approximately 10,000 times as great as in water. Therefore, even a thin layer of water (several millimeters) serves as a good oxygen barrier.

The most positive method of preventing acid generation is the installation of an oxygen barrier. Artificial barriers such as plastic films, bituminous, and concrete would be effective, but these have high original and maintenance costs and would be used only in special situations.

Surface sealants such as lime, gypsum, sodium silicate, and latex have been tried, but they too suffer from high cost, require repeated application, and have only marginal effectiveness. The two most effective barrier materials are soil, including nonacid spoil and water. A 2-foot minimum thickness of soil or nonacid spoil that is required as a barrier is a function of the soil's physical characteristics, soil compaction, moisture content, and vegetative cover. Deeper layers of a sandy, dry granular material with large grain size and porosity would be required rather than a tightly packed saturated clay that is essentially impermeable.

Soil thickness should be designed on the basis of the worst situation—when the soil is dry and oxygen can move more readily through cracks and pore spaces devoid of water. A "safety factor" should be included to account for soil losses such as erosion. Vegetation not only serves as a barrier, but slows the rate of molecular diffusion because the pores are filled with water and not gases. As the vegetation dies, it becomes an oxygen user during the decomposing process and further aids the effectiveness of the barrier. The organic matter that is formed further aids in holding moisture in the soil.

Water is an extremely effective barrier when the pyritic material is permanently covered. Allowing the pyrite to pass through cycles where it is uncovered and then covered will worsen the AMD problem. Water barriers should be designed to account for water losses such as evaporation and include at least 30 centimeters of additional depth as a safety factor.

CORRECTIVE MEASURES

Acid Mine Drainage Control

Measures to control AMD are water control and in-place neutralization. Water serves not only as the transport media that carries the acid pollutants from the pyrite reaction sites and mine, but it also erodes soil and nonacid spoils to expose additional pyrite to oxidation. Facilities such as diversion ditches that prevent water from entering the mining area and/or carry the water quickly through the area can significantly reduce the amount of water available to transport the acid products.

These facilities are needed both during and following mining. Terraces, mulches, vegetation, etc. used to reduce the erosive forces of water are effective measures to prevent further pyrite exposure. These measures usually are performed during reclamation.

Alkaline overburden material and agricultural limestone can be blended with "hot" acidic material to cause in-place neutralization of the acid and assist in establishing vegetation. In some cases, alkaline overburden can be graded to cause acid seeps to drain through it. These techniques are more applicable to abandoned surface mines than to current mining, where proper overburden handling should prevent acid formation. The major exception may be those situations where an underground mine was breached and an acid discharge formed.

Acid Treatment

Where the formation of AMD cannot be prevented or the discharge controlled, treatment is necessary before the water can be discharged. The only method generally used for treating AMD is neutralization. The neutralization process provides the following benefits.

   Neutralization removes the acidity and adds alkalinity.

   Neutralization increases pH.

   Neutralization removes heavy metals. The solubility of heavy metals is dependent on pH up to a point: the higher the pH, the lower the solubility.

   Ferrous iron, which is often associated with AMD, oxidizes at a faster rate to ferric iron at higher pH's. Iron is usually removed in the ferric form.

   Sulfate can be removed if sufficient calcium ion is added to exceed the solubility of calcium sulfate; however, only in highly acidic AMD does this occur.

Some shortcomings of the neutralization process are:

   Hardness is not reduced and may be increased.

   Sulfate is not reduced to a low level; it usually exceeds 2,000 milligrams per liter.

The iron concentration usually is not reduced to less than 3 to 7 milligrams per liter.

A waste sludge is produced that must be disposed of.

Total dissolved solids concentration is increased.

A typical neutralization system would include adding an alkaline reagent, mixing, aerating, and removing the precipitate. Alkaline reagents that may be used are ammonia, sodium carbonate, sodium hydroxide, limestone, and lime.

Lime Treatment: Lime treatment is the most commonly used system. The lime reactions with AMD are as follows.

$$Ca(OH)_2 + H_2SO_4 \longrightarrow CaSO_4 + 2H_2O$$

$$Ca(OH)_2 + FeSO_4 \longrightarrow Fe(OH)_2 + CaSO_4$$

$$3Ca(OH)_2 + Fe_2(SO_4)_3 \longrightarrow 2Fe(OH)_3 + 3CaSO_4$$

AMD is discharged from the mine directly to a rapid mix chamber or to a holding/flow equalizing pond where it flows to the rapid mix chamber. Hydrated lime is fed to the rapid mix chamber either as a slurry or dry. If the ferrous iron concentration is low (less than 50 milligrams per liter), the water is treated to a pH of 6.5 to 8 and flows directly to a settling chamber. If the ferrous iron is high, the pH is usually raised to a higher level (8 to 10) and then passed to an aeration tank where the ferrous hydroxide precipitate is converted to ferric hydroxide.

Then the water flows to a settling chamber. The settling chamber may be a clarifier or pond. Here the iron, aluminum, calcium sulfate, and other heavy metals precipitate. The supernatant is the treated water. The precipitate or sludge is removed from the settling chamber and disposed of in a second pond, strip mine pit, underground mine, or landfill. In some cases, the pond serves as a settling chamber and permanent storage place for the sludge.

Except for large surface mines, lime systems are usually much less sophisticated than the one described above. They may be as simple as catching all the AMD in a small pond, then broadcasting by hand lime on the surface of the pond. This system is only effective when the pond is less than 1,000 square meters. Mixing of the lime and acid water is poor in this system, and excess lime is required. After the water is treated, it is pumped from the pond.

Water can also be treated as it is pumped from the pit by connecting a lime slurry tank to the suction end of the pump. As the water is pumped, the lime slurry is drawn into the AMD by the suction of the pump; the pump also serves to mix the lime and acid water. The discharge from the pump should pass through a settling pond to remove any precipitates. Commercial units which include automatic pH control are available on the market.

Limestone Treatment: The limestone reactions with AMD are as follows.

$$CaCO_3 + H_2SO_4 \longrightarrow CaSO_4 + H_2O + CO_2$$

$$3CaCO_3 + Fe_2(SO_4)_3 + 3H_2O \longrightarrow 3CaSO_4 + 2Fe(OH)_3 + 3CO_2$$

Although limestone is a cheaper reagent than lime and produces less and denser sludge, it has not received wide acceptance for several reasons: the carbon dioxide produced buffers the reaction, and it is difficult to raise the pH above 6 without using excessive material; limestone is ineffective with high ferrous iron water; the size, characteristics, and method of application of the limestone are critical; and the system is usually more complex than lime.

Several different treatment schemes have been utilized with limestone. Those most applicable to surface mine situations are streambed and ground limestone techniques. The simplest method is the placement of limestone in a streambed. The acid water is treated as the water flows through the bed. This method has proved to be ineffective in most cases because the limestone quickly becomes coated with iron, calcium sulfate, sediment, and biological growths that prevent the acid water from reacting with the limestone.

The method may have application for short-term temporary situations where any one installation will not be used for more than a month. A trench should be dug leading from the surface mine and filled with crushed limestone (2.5 centimeters). Basins to settle out any silt before the trench and a second to settle out the precipitate should be used. Surface water should be diverted away from the trench to prevent the limestone from being washed out during storms. If the limestone bed loses its effectiveness, the stone should be replaced or a new trench dug.

Pulverized limestone can be used in a manner similar to lime. The following factors should be considered in the selection of a limestone: high calcium carbonate content, low magnesium content, low amount of impurities, and large surface area, i.e., smallest particle size within economic bounds, –200 mesh or smaller is preferable. Methods for selecting limestone have been developed. The pulverized limestone can be fed as a slurry or dry. Two to three times the stoichiometric amount of limestone will probably be required, and even then a pH of only 6 to 6.5 will be reached. The reaction time of limestone is much slower than lime, and up to 30 minutes of mixing should be provided.

The split treatment of AMD with limestone and lime may offer some advantages in cost and improved sludge characteristics. It might also be used on ferrous iron AMD. A two-step process is required. First, the AMD is treated with limestone to a pH of 4.0 to 4.5 to take advantage of the pH range when limestone is most effective. The water then passes to a second reactor where lime is applied to raise the pH to the desired level. This process may have a cost advantage over lime alone and the desired sludge characteristics of the limestone process. With the proper combination of limestone and lime, a good effluent can be obtained even with ferrous AMD. This system is probably only applicable to large installations.

Anhydrous Ammonia: Anhydrous ammonia has been utilized for the neutralization of AMD. Such a system is attractive from the standpoint of ease of operation and maintenance. Usually, the only equipment used is a tank of anhydrous ammonia, a length of hose to discharge the material into the AMD, and a valve to control the flow of gas. Anhydrous ammonia is usually supplied by the dealer in pressurized tanks mounted on wheels. The user needs only the hose and valve. The tanks are easily moved from site to site and can be set up in a matter of minutes.

The disadvantages of anhydrous ammonia are: ammonia is lost to the atmosphere by diffusion or by air stripping where aeration is practiced; more sludge may be produced; the reagent cost is higher than lime or limestone; and ammonia-neutralized AMD may have a detrimental effect on a receiving stream because of the toxicity of ammonia to fish and aquatic life, the depression of dissolved oxygen levels as a result of nitrification, and nitrate enrichment, which may lead to accelerated eutrophication.

The detrimental effect on receiving streams is significant enough to warrant the recommendation that anhydrous ammonia not be used to treat AMD except under special conditions. Ammonia nitrogen levels as high as 1,625 milligrams per liter in AMD neutralized with anhydrous ammonia in laboratory studies have been reported. Nitrate N levels as high as 480 milligrams per liter in AMD being treated with anhydrous ammonia have been found in western Kentucky.

These levels of ammonia and nitrate are beyond desirable limits for stream. The only situation where anhydrous ammonia may be acceptable is where small volumes of AMD are to be treated, and all the treated water is applied to spoil banks as irrigation water and no runoff occurs. In this situation, the stream is not damaged, and the vegetation on the spoils receives the benefit of water and nitrogen.

Soda Ash: Sodium carbonate has been utilized for the treatment of AMD because of the simple feeders that have been developed. In most cases, soda ash briquettes have been used. A portion or all of the AMD is passed through a container holding the briquettes. The briquettes dissolve, neutralizing the water. These systems, which are usually used on small flows, are temporary and are easily moved. Their disadvantages are that good control of pH cannot be maintained, and at very high flows, they undertreat. Also, higher cost of soda ash militates against its use.

Sodium Hydroxide: One neutralizing system on the market uses sodium hydroxide. The addition of sodium hydroxide is controlled by the water level in a small flume. The device is considered suitable for remote location because it is easily moved, requires no electricity or power, and is simple to operate. The device is best suited for small flows. A baffle downstream of the device that ensures good mixing and a settling pond is desirable for best operation. The cost of sodium hydroxide is much higher than lime or limestone.

INFILTRATION OF WATER ON STRIP MINE SPOIL BANKS

The purpose of the research reported in PB 217 872 was to measure infiltration rates on some Ohio strip mine spoils, and to examine relationships between infiltration and soil variables such as bulk density, slope, texture, and moisture content. A sprinkling infiltrometer was used to apply rainfall of uniform intensity and raindrop characteristics to a number of spoil materials.

There is a large body of literature dealing with infiltration theory. A simple algebraic equation, $V_o = \frac{1}{2}St^{-\frac{1}{2}} + A$, describes the dynamics of infiltration. In this equation, V_o is infiltration rate, S is sorptivity, a measure of the capillary uptake or removal of water, and t is time. A equals permeability at a capillary potential of 0, and decreases with decreasing moisture content. S has the dimensions of length x time$^{-\frac{1}{2}}$ and A length x time^{-1}. As initial moisture content increases, S decreases while A increases.

Materials and Methods

The sprinkling infiltrometer was used in this study to produce artificial rainfall of uniform intensity (Figure 6.1). Water was pumped from a reservoir tank through a 7 LA nozzle at 0.422 kg/cm^2 pressure. With a nozzle height of 2.74 meters above ground, rainfall rate averages 11.2 centimeters per hour in the plot area.

At the end of each run on spoil, rainfall rate was measured during a 20-minute calibration run on a sheet-metal pan placed over the square plot frame. Rain gauges were set up at the center of each edge of the plot frame and read at 10-minute intervals during the spoil and calibration runs. A regression relationship for predicting rainfall on the plot from the rain gauge readings was developed from the calibration runs, and applied to the rain gauge readings taken during the runs on spoil. This allowed estimation of variability in rainfall rate during the run, and also corrected to some extent for variability of rainfall rate within the plot area.

FIGURE 6.1: WATER FLOW CHART FOR SPRAY INFILTROMETER

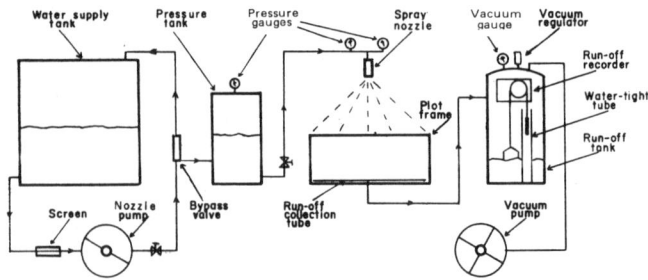

Source: PB 217 872

Runoff Collection and Measurement: Runoff was collected in a calibrated tank. In 1968 a water-recirculating pumping system was used to move the runoff from the plot to the tank, but in 1969 a vacuum collection system was installed. A Belfort water level recorder, employing a float-counterweight system, was used to measure the rate of change of water level in the tank. When the counter-weight moves from air to water during a run, an error results from the difference in density of the two media. Accordingly, in 1969 a watertight tube was mounted inside the calibrated tank so the counterweight would always operate in air.

Collection Tank Calibration: Known weights of water were added to the tank and height of rise measured with the water level recorder. Water temperature was noted and the volume of water computed from the temperature-density relationship.

Acid Mine Drainage

Computation of Infiltration: Runoff was determined by 5-minute intervals during the run on spoil, and subtracted from the observed rainfall rate to give infiltration. This computation assumes that all rainfall not appearing as runoff penetrates the soil when, in fact, some water is stored on the plot surface as puddles. If the dams forming these puddles break later in the run, runoff may increase so that apparent infiltration becomes negative. Erosion is a further source of error in measuring infiltration since space occupied by eroded soil in the calibrated tank is measured as runoff.

Characterization of Spoils

Slope: The 1.16 meter square plot frame was oriented with its upper and lower edges level. Slope of the plot was taken as the average of the slopes of the two sides as measured with an Abney level.

Soil Moisture: Immediately before each run, composite samples of the top 11.5 centimeters of spoil were taken adjacent the plot with a core sampler and placed in airtight soil moisture cans. Moisture content was determined gravimetrically.

Bulk Density: Following the second (wet) run on each plot, four undisturbed cores were taken from the plot area with a double-cylinder core sampler. These cores were dried to constant weight, weighed, and bulk density calculated from the relationship.

$$\text{Bulk Density} = \frac{\text{Oven Dry Weight (g)}}{\text{Volume of Core (cc)}}$$

Texture: Texture was measured on a composite spoil sample, using the hydrometer method.

Plant Cover: Plants on each plot were counted. Length of a representative sample was measured, and the total length of the plants on the plot was computed.

Cultural Practices: All spoil banks studied had been graded in accordance with the Ohio law. In addition, some areas had been disked, or subsoiled with a chisel to a depth of 45.5 centimeters. Number of diskings, number of subsoilings and time since last subsoiling were used as variables to assess the effect of these practices.

Water Supply: All runs in 1969 were on Unit 2 of the Eastern Ohio Resource Development Center near Caldwell, Ohio. Water came from a pond which gathered surface runoff. Work in 1968 was concentrated on Little Mill Creek watershed, adjacent the Coshocton Hydrologic station. Water for these runs came from a gravel aquifer near the Muskingum River in Coshocton. To determine changes in water quality, samples of both artificial rainfall and runoff were collected and analyzed for sulfate and manganese. Differences in concentration between rainfall and runoff was taken as the amount dissolved from the spoil. Admittedly, natural rainwater would have been better for this purpose, but these results provide some useful information.

Field Procedures

Plot sites included a range of slopes, textures, vegetative covers, and management practices. Work was concentrated on two areas, the watershed of Little Mill Creek near Coshocton, Ohio, where spoil banks resulted from mining the Middle Kittanning (No. 6) seam, and Unit 2 of the Eastern Ohio Resources Development Center, where the Meigs Creek (No. 9) seam had been mined.

The plot frame was driven deep enough to prevent movement of water out of or into the plot. The nozzle was centered 2.7 meters above the plot. Rain gauges were centered just outside the plot at the midpoint of the sides. The gauge on the uphill side was placed with its base flush with the ground and the other three gauges were placed at the same elevation. After soil moisture samples were taken, rainfall was begun, but diverted from the plot with a pipe until water pressure was at 0.422 kg/cm^2 and stable.

Statistical Methods

Infiltration rates observed for each run were fitted to the equation $V = A + \frac{1}{2}St^{-\frac{1}{2}}$. Goodness of fit, measured by R^2, ranged from 0.95 down to 0.009. The fit of the data to the above equation gave a significant correlation coefficient for 89% of the runs.

Whether an exponent of t other than $-\frac{1}{2}$ would give a better fit to the data was tested. The exponents -0.3, -0.4, -0.5, -0.6, -0.7, -0.8, -0.9 and -1.0 were used and the correlation coefficients for the respective equations compared. The exponent $-\frac{1}{2}$ results from theoretical treatment of infiltration into a rigid, porous medium. An exponent less than $-\frac{1}{2}$ implies a less stable medium, that is, the infiltration rate changes more rapidly with time as the exponent becomes more negative.

For those infiltrometer runs for which the algebraic equation gave a fit significant at the 5% level or lower, a stepwise regression analysis was made of the relationship between measured spoil characteristics and observed infiltration variables. Spoil characteristics tested in the stepwise procedure were volumetric soil moisture prior to each run, bulk density, percent clay, percent sand, percent slope, clay ratio, i.e.,

$$\frac{\% \text{ Clay}}{100 - \% \text{ Clay}}$$

and selected 2-factor interactions of these variables. Dependent variables were total infiltration per hour, infiltration rate at 2.5, 27.5, and 57.5 minutes, and exponent of t. Average rates of infiltration for each 5-minute time interval, and standard errors of these averages, were computed by coal seam (6 or 9) and for runs on dry or wet spoil.

Results

Average initial infiltration rates were practically identical on the Meigs Creek and Middle Kittanning spoils (Figure 6.2). Later rates showed more variability, but were not consistently higher on one spoil type. Average initial rates of infiltration were greater on dry spoils than on spoils that had received 11.2 centimeters of rain the previous day. After 25 minutes of rainfall, however, infiltration rates on wet or dry spoil were similar.

FIGURE 6.2: AVERAGE RATE OF INFILTRATION

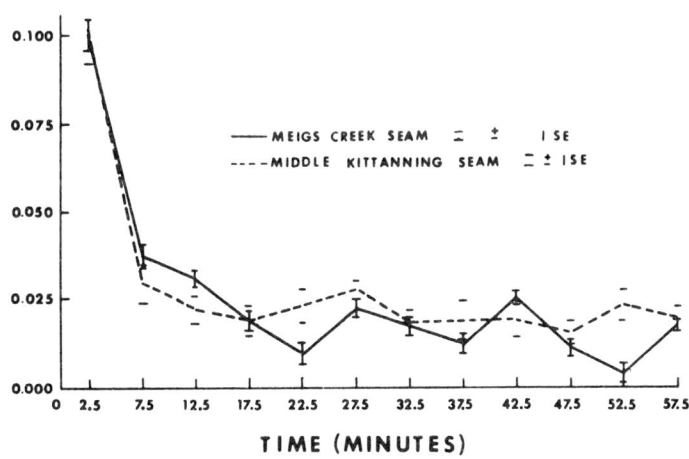

Source: PB 217 872

Observed infiltration rates at 5-minute intervals during each run were fitted to the equation $V = A + \frac{1}{2} St^{-C}$. In the range 0.3 to 1.0, a value of 1.0 for C gave the best fit most often, followed by 0.3 to 0.8.

As C changes from 0.3 to 1.0, infiltration rate diminishes more rapidly with increasing time. A larger value of C evidently indicates a less stable soil surface that changes more rapidly under the impact of raindrops. It is not surprising that these spoils should be unstable to raindrop impact since many of them lacked both protective vegetative cover and organic matter to stabilize aggregates.

In a stepwise regression procedure, independent variables related to the best value of C were investigated. Volumetric soil moisture was the only variable having a statistically significant relationship with C. The stepwise regression procedure includes variables in the regression equation in order of their reduction of the residual sums of squares. As each new variable is added, preceding variables are tested to see if they continue to make a significant contribution. Those that do not are deleted.

The spoils were classified in two groups, toxic and nontoxic, and the average increase in concentration of sulfate and manganese computed for the dry and wet runs on each type of spoil (Table 6.1). Average initial sulfate and manganese concentrations of the rainfall were 34.4 and 0.66 ppm, respectively.

TABLE 6.1: MEAN INCREASE IN CONCENTRATION OF SULFATE AND MANGANESE AFTER RAINFALL

Spoil Type	Sulfate (ppm)		Manganese (ppm)	
	Mean S.E. Dry Run	Mean S.E. Wet Run	Mean S.E. Dry Run	Mean S.E. Wet Run
Toxic	891.4 ± 254.9	465.0 ± 108.5	3.70 ± 0.85	2.21 ± 0.50
Non-toxic	83.0 ± 21.3	35.9 ± 9.79	1.14 ± 0.46	0.51 ± 0.17

Source: PB 217 812

On both types of spoils, runoff from the second, or wet, run had a lower concentration of manganese and sulfate than the dry run. The toxic spoils contributed a large amount of sulfate to the runoff. The recommended limiting concentrations of sulfate and manganese in drinking water are 250 and 0.05 part per million, respectively.

Discussion

The objective was to measure rates of infiltration on spoil banks and to identify spoil characteristics associated with variations in these rates. In general, infiltration was slow. Graded spoils from the Meigs Creek and Middle Kittanning seams did not differ noticeably in infiltration characteristics. Rates of infiltration were higher on dry spoil than on wet spoil during the first 25 minutes of the run but later rates were similar.

The empirical equation for infiltration, $V_o = \frac{1}{2}St^{-\frac{1}{2}} + A$, fits the data quite well in most cases, and when the exponent of t was changed to -1, the fit was improved. This improvement indicates that the spoil surface was less stable than a rigid, porous medium.

Concentration of sulfate and manganese in the runoff was considerably higher than in the artificial rain, especially on the toxic spoils. This demonstrates the potential of spoil banks for pollution of streams. It also shows that all spoil banks are not alike, so that mining and reclamation measures need to be tailored to the pollution potential of each particular kind of overburden.

Bulk density appeared as a variable in three of the four regression equations relating infiltration to spoil characteristics. The proportion of variation in infiltration accounted for by the regression equations was not large. One explanation may be the rather narrow range in infiltration rates encountered. Seventy-one percent of the plots had total infiltration between 0.508 and 2.03 centimeters per hour; distribution within this range was: 0.508 to 0.99 centimeter per hour, 18%; 1.0 to 1.49 centimeters per hour, 24%; 1.50 to 2.03 centimeters per hour, 29%.

A second explanation may be failure to measure some important variables governing infiltration. Recently a channel system concept of infiltration has

Acid Mine Drainage

been proposed. Infiltration rate is governed by the number of large pores in the soil, their continuity, presence of surface roughness for air exhaust, and presence or absence of a surface cover to prevent plugging of the channels by raindrop impact. Research results indicate that the channel system controls infiltration much more than soil texture, existing vegetation, or soil moisture regime.

Subsoiling helps create a channel system. A preliminary study was made of the effect of subsoiling in two mutually perpendicular directions on infiltration into spoil (Figure 6.3). The initial run on dry spoil gave quite high infiltration rates, but during the following run on wet spoil, infiltration declined sharply to rates similar to spoils without subsoiling.

FIGURE 6.3: INFILTRATION RATE ON SUBSOILED DRY AND WET SPOIL

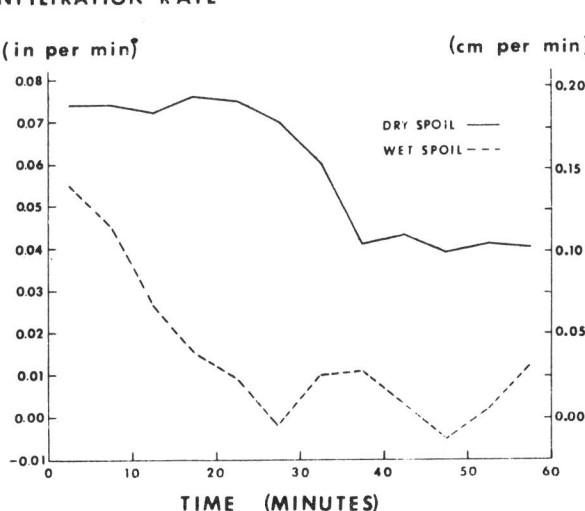

Source: PB 217 872

Conclusions

Infiltration of artificial rain on spoil banks was related to volumetric soil moisture, bulk density, clay percent, sand percent, and slope. Most of the variation in infiltration remained unaccounted for in stepwise regressions involving these variables.

MINE SPOIL POTENTIALS
FOR WATER QUALITY AND EROSION

The material in this chapter was excerpted from:

PB 208 817

The general goal of the research reported in PB 208 817 was to provide sufficient information about coal overburdens and spoils to enable operators to place, treat, and manage variable spoils in the most favorable manner to assure water and soils of good quality during surface mining and thereafter.

The area of principal concentration during the initial phase of study was selected to include the rock section exposed in Preston and parts of Monongalia, Taylor, Mineral, Grant, Tucker and Upshur Counties, W. Va., or Surface Mining Province 2. Some additional study of mine spoils was conducted on the eastern edge of Surface Mining Province 3 and correlative rocks exposed locally on high knobs of Surface Mining Province 2.

Samples of rock strata between the soil and a surface mineable coal seam were obtained by collecting expelled material from rotary drilling of blast holes. Cooperation of company drilling crews allowed acquisition of "rock chips" in 32 centimeter increments as drilling proceeded. The rock fragments from each interval were collected on a shovel and then poured into a one-pint, waxed paper carton.

The drilling equipment normally was located on a bench a few feet below the natural land surface, the soil and highly weathered rock having been removed with a bulldozer. Sampling from the drilling bench up to the adjacent soil surface was done by hand, in graduated increments. The soil profiles were sampled by morphological horizons according to accepted methods. At several sites overburden samples were obtained by hand collection at intervals on a fresh high wall using an extension ladder.

Preston County sites where rock chip, highwall, spoil and soil samples were obtained and/or studied in place were designated A through Z. Details concerning

Mine Spoil Potentials for Water Quality and Erosion 157

locations specifically mentioned are given below.

A - One mile south and one-half mile west of Valley Point; Upper Freeport coal; drainage into lower Glade Run, active 1969 1971.

C - Three-fourths mile south and one mile west of Valley Point; Upper Freeport coal; drainage into west fork of Glade Run; active 1969-1971.

E - Two and one-half miles south and one and one quarter miles east of Brandonville; Upper Freeport coal; drainage into lower first order tributary of Little Sandy Creek; active 1969-1971.

H - Two and one quarter miles west of Valley Point; Upper Freeport coal; drainage into Sovern Run; active 1969-1970, grading started 1971.

L - One-half mile south and one-half mile west of Valley Point; Upper Freeport coal; drainage into Glade Run; active 1969-1970, graded and planted to forage mixture 1971.

T - One-half mile north and one and one-half miles east of Albright; Bakerstown coal; drainage into Roaring Creek; active 1971.

SULFUR AND POTENTIAL ACIDITY

It has been well documented that the oxidation of pyrite is the major concern in considering water pollution and lack of vegetative cover on surface mined lands. The immediate concern of this investigation was to determine the amount of pyrite in rock strata between a strippable coal seam and the land surface.

It was assumed that this information would help explain why certain spoils were easily revegetated and produced little acid drainage water, while others were so acid that revegetation was considered impossible. In northern West Virginia, the spoil resulting from surface mining the Upper Freeport seam has been placed in the latter category, therefore, much of the effort of this investigation was toward characterization of the strata overlying the Upper Freeport in this area.

Sulfur Distribution in Overburden Material

Because the relatively humid climate in northeastern U.S. causes intense leaching within the upper few meters at the land surface, it was postulated that sulfate forms of sulfur, being soluble, would approach being negligible in occurrence. Under this assumption, also the fact that sandstones are low in organic materials except for coal veins, Upper Freeport overburden rock was analyzed for total sulfur. The distribution of total sulfur between the land surface and the Upper Freeport coal at sixteen borings from nine different locations within the Preston County area of investigation was determined. The data showed the consistent occurrence of a zone, essentially free of sulfur, penetrating about six meters from the land surface.

Changes in chemical and physical properties of geologic materials due to natural weathering action have long been recognized. The depth from the land surface to which these forces penetrate has been shown to range from tens of centimeters in the arctic to hundreds of meters in the tropics. Data obtained in the course of this project indicate a uniform depth range of four to eight meters in northern West Virginia; however, along local fractures oxidized minerals are found to depths of sixteen meters. Except in steep slopes where erosion is severe, the weathered, pyritic sulfur-free zone is a bountiful source of spoil material that will not produce significant quantities of acid when exposed as spoil.

The Upper Freeport coal is directly overlain by dark-colored shale, infrequently reaching a few meters thickness in the area of this study. Above this shale, or infrequently directly over the coal, the sandstone, extending to the horizon of the soils, consists of two zones: the lower, unweathered, dull gray portion; and the upper, weathered zone characterized by the brighter colors of oxidized iron.

Reference to Munsell color charts aids consistent characterization of the weathered zone of the rocks as "high chroma," and the unweathered zone as "low chroma," many of the samples being of the hue, IOYR. Low chroma, as with soils, refers to chromas of 2 or lower.

The acquisition of several thousand feet of 4.6 cm solid test cores from forty-five boring sites in north central Preston County shows that the Lower Mahoning sandstone is a prominent stratum overlying the Upper Freeport coal in regions not readily exposed as old high walls or current mining operations. Chemical analyses for total sulfur in selected increments throughout this stratum, where one or more overlying sedimentary materials intervene between it and the land surface, show a composition similar to that displayed by unweathered Lower Mahoning obtained during blast hole drilling operations.

The samples from core S-2 were specifically chosen to include black partings, presumed to be small local coal lenses; the greater sulfur content of these sections is apparent. Data on changes in sulfur with depth in several predominantly shale overburdens indicate more total sulfur in the weathered zone of these shales than in that of the Lower Mahoning sandstone described above.

The Harlem coal seam (Site W) is relatively close to the land surface; this situation emphasizes that low sulfur material is not necessarily always present in near-surface strata. The material at Site T is also shale, but varies widely in total sulfur content. The lower four meters of material, lying just over the coal, is high in organic matter and of easily recognizable black color. The composition emphasizes the need to bury this kind of material to prevent pollution problems.

Petrographic Evidence of Pyrite

Petrographic observations of thin sections indicate that pyrite occurs throughout most rocks buried deeper than about 6 meters below the present land surfaces. Feldspars throughout the sections have altered to kaolinite and (or) dickite or they appear to be in process of such alteration. Persistence of kaolinite from the unoxidized to the oxidized rock and soil zones is apparent.

Pyrite, on the other hand, appears relatively stable at depths below about 9 meters, but is sparse or absent in the top 6 meters where segregated hydroxides are prominent. Several modes of occurrence of the pyrite are recognizable and have been distinguished.

Modes of Occurrence of Pyrite and Associated Minerals:

Study of thin sections of the acid sandstones indicates that several modes of occurrence of pyrite can be distinguished. These modes include:

Euhedral (both single crystals and clusters) pyrite along boundaries between individual sand grains (mostly quartz but some feldspar grains). Some replacement of quartz and feldspars is involved.

Euhedral crystals along hairline cracks within individual sand grains. In some cases, the crystals appear to be merely embedded within the grain with no crack evident, but it is assumed cracks were present and some were removed by sectioning. Replacement of quartz or feldspar by pyrite is indicated since crystals are larger than observable hairline cracks. Preferential replacement of feldspars compared to quartz is apparent in some cases.

Euhedral crystals disseminated among small (generally silt sized) grains. It seems likely that these crystals are secondary and not detrital although they have not clearly replaced silt grains. Probably they have grown in voids in the original sediment.

Crystals (or clusters) with mosaics of secondary kaolinite "books" usually bordering feldspar grains. The kaolinite appears to have formed from feldspars which enclosed the pyrite. More precisely, the pyrite formed before the kaolinite formed by diagenesis in a strongly reducing and slightly acid to alkaline environment. Persistence of the kaolinite in the weathered near-surface environment and in all horizons of the present soils indicates stability of the kaolinite through a wide range of geochemical conditions.

Crystals attached to the walls of voids within the rock but only partially filling the voids. Obviously, these crystals grew from material moving in through the void and (or) from material on the surface of the grain bordering the void (quartz or a feldspar).

Pyrite embedded within amorphous products of near-surface weathering. Usually the amorphous material is brown, yellow or black and may prevent positive identification of the pyrite. However, opacity to intense light and variable degrees of reflectance of indirect light suggest likely presence of some pyrite crystals. Sections across the strongly stained material sometimes reveal embedded pyrite crystals.

The pyrite occurrences identified in acid sandstones all appear to be secondary to sedimentation. The size of crystals and clusters observed is highly variable. Some reflecting surfaces that are too small for positive identification may be pyrite grains smaller than 2 microns in diameter. These small faces appear black. The total mass of such material may be small but its reactivity may be high. Crystals approximating 5 microns diameter are readily identified. The diameter of some individual crystals or clusters is as great as 100 microns or even larger, but 5 to 50 microns is a common range.

Thin-section observations suggest that accessibility of pyrite to the near-surface weathering agents is a variable influencing rates of acid formation as well as type of pyrite and grain size. It seems likely that within the zone of weathering, the degree of encasement of the pyrite within weathering products such as iron, aluminum and manganese hydroxides should have a marked influence on the rate of pyrite oxidation regardless of the chemical or biological steps involved. Appearances suggest, also, that crystals which have grown from hairline cracks within sand grains are less accessible to weathering agents than crystals on grain boundaries or lining larger voids.

Acid-Producing Potential of Pyritic Materials

The acidity that may develop from the oxidation of pyrite in overburden materials has been determined in two ways. Using the stoichiometry between pyrite and potential sulfuric acid it can be calculated that for material containing 0.1% sulfur, all as pyrite, complete oxidation will yield a quantity of sulfuric acid that will require 6,250 pounds of calcium carbonate to neutralize one thousand tons of material.

Analyses for total sulfur in old (5 to 30 year old) spoils indicate that considerable sulfur remains for at least this long. To be sure, it has not been demonstrated that the persistent sulfur content is entirely pyritic. In fact, with many kinds of old samples, much of the sulfur may be in organic or other forms. If so, this might influence the failure of the peroxide method to give consistent results with old spoils.

The potential acidity that a spoil material can develop has been demonstrated for some materials by direct chemical measurement. Treatment of pulverized (250μ) spoil materials with 30% H_2O_2 results in the oxidation of reduced sulfur to titratable sulfuric acid.

An equivalence between titratable acidity generated upon H_2O_2 treatment and total sulfur content has been shown in a comparative study involving selected sandstone and shale samples. Statistical treatment of the data shows the close relation between total sulfur and titratable potential acidity in 69 shale and sandstone samples. However, with 44 samples of old spoil, representing a wide variety of materials in terms of texture, organic content, and degree of weathering, no consistent relationship was obtained.

Data obtained from analyses of fresh sandstone and shale samples show that the hydrogen peroxide oxidation procedure can be effectively used to measure the amount of acid that certain rock specimens can be expected to generate upon prolonged exposure to weathering.

CHEMISTRY, MINERALOGY AND WEATHERING OF PROFILES

The rock chip samples taken from locations A, C, E, H and L were analyzed for pH, exchangeable bases, exchangeable aluminum, free aluminum, manganese and iron, total iron, total calcium, potassium, copper, manganese and zinc. These analyses were carried out for several reasons: to measure native acidity of rock materials; to determine the potential of this material to produce acidity; to establish areas (depth) rich in pyritic or other potentially toxic material; to evaluate nutrient contents which could become available to plants due to weathering of these rocks; and to compare properties of rock materials with that of the soils which are found above these unweathered rocks at the above-mentioned locations.

Samples were ground to pass a 60 mesh sieve. Rock chip and soil pH measurements were made with a glass electrode pH meter, using 2:1 rock chip:water ratio and 1:1 soil:water ratio with a 30-minute equilibration period. Ammonium acetate (pH 7) extractable Ca, Mg, and K; and KCl extractable Al were determined using a Perkin-Elmer 403 Atomic Absorption instrument.

Chemical Analysis of Rock Chip Samples

Rock chip samples, taken in 12.5 inch increments, were analyzed for distribution of exchangeable bases, Al and pH, at several locations. Rock chip samples started below the soil with sample increment No. 1, 3 to 5 feet below original land surface, and continued to near the coal seam.

pH: At Site A and Site L the pH of rock chip samples ranged from 5.3 to 5.8 within the weathered zone (samples 1 to 20), and in the unweathered zone, from 6.3 to 8.0 for the A series (samples 21 to 38) and 6.6 to 7.3 for the L series (21 to 48). No significant difference in pH with depth was observed at the other three locations (C, E and H). At these locations pH was in the acid range. High pH at A and L (unweathered zone) may have resulted from alkaline earth cations or carbonates deposited during sedimentation.

Exchangeable Bases: The top 20 feet of rock material at all five locations were low in exchangeable bases (Ca, Mg and K) and low in pH. Below this depth, exchangeable bases at locations A, L and H increased several fold. Increase in sulfur content was also observed at these locations below the 20 feet depth. Petrographic studies of this material showed the presence of pyrite.

Lack of bases in the weathered zone and accumulation of bases below twenty feet indicates that the high concentration of hydrogen ions in the upper horizons removed Ca, Mg and K from the upper weathered zone. High concentration of hydrogen ion and absence of pyrite in the upper layer was a result of weathering.

The presence of hydrogen ions around silicate minerals intensifies the weathering process. The decomposition of silicate minerals is accomplished through exchange reactions in which H ions replace metallic ions, which weakens the structure and accelerates additional weathering. There seems to be no doubt that hydrogen ions furnished by the oxidation of pyrite initiated disruption of crystal structures of minerals of argillaceous rock and this resulted in increasing the contents of exchangeable Al and free iron oxides in the upper horizons.

Elemental Analyses: Elemental analyses of rock chip samples were carried out to help determine the chemical and mineralogical characteristics of the rock. This type of information is of importance to the fundamental interpretations of the chemical processes of soil development and as a background to soil fertility and toxicity interpretations. Information gained from these analyses also contributes to formulation of optimum plans for revegetation and control of soil erosion and water pollution.

An important factor controlling the rate of breakdown of rocks and minerals and genesis of secondary products is the quantity and quality of water percolating through the weathering environment. Weathering reactions are accelerated by repeated flushings of rainwater which remove soluble constituents from the mineral surfaces.

Calcium and Potassium — In general, the upper twenty feet of rock chip samples at all locations contained only small amounts of Ca. Generally, Ca varied between 0.05 and 0.2%. The upper zones at all locations were also low in sulfur, pH and exchangeable Ca.

Below the weathered zone, total Ca in rock chip samples at locations A and L increases several fold. This was also reflected by high pH values and prominence of Ca among the exchangeable cations in lower zones at these locations, as well as the presence of free carbonates. Leaching of basic ions had not occurred in these regions. Accumulation of free carbonates probably occurred during the deposition of sedimentary material.

At the other three locations (C, E and H), Ca concentration of the lower zones did not vary significantly from the upper zones. The absence of free carbonates at these three locations indicates a difference in the conditions during deposition of the sedimentary material. From these low total Ca results one could easily visualize that exposing lower rock strata which are rich in pyrite at locations C, E and H will increase the difficulty of revegetation. These results also help in explaining the varying degrees of difficulty in establishing plant cover on apparently similar spoil materials.

Potassium is generally high at locations E and C; however, distribution of K shows no specific trend relative to changes in rock depth. Mica was distributed unevenly throughout the rock as indicated by X-ray diffraction patterns and by petrographic studies. Since no other K bearing minerals, e.g., feldspars, were observed in rock chip samples, it appears that K contents are contributed largely by the mica components of the rock material. The slight difference in the exchangeable K in weathered and unweathered zones can be explained on the basis that acid leaching of this rock material intensified weathering of mica in the weathered horizons, and this may have replenished exchangeable K in this zone.

Mine spoil material derived from both weathered and unweathered zones, Sites E and C, may be capable of supplying adequate amounts of K to plants so that additional K at the time of revegetation may not be needed.

Total and Free Fe, Total Mn and Cu, and Free Mn and Al — Colored compounds, especially iron oxides, are clearly visible and often are interpreted in the field, especially in distinguishing between highly weathered and nonweathered rocks and in determining soil drainage or wetness. Like Al and Mn, iron oxides are affected by the processes of weathering and have an importance in studies of soil genesis.

Iron and Mn are not only plant nutrients, but they also play a part in controlling the availability of phosphorus to plants under acid conditions. Determinations of Fe, Mn and Al in the rock strata and soil profile help, therefore, in describing the type, the distribution, the direction and the extent of weathering and may be used to interpret fertility problems related to phosphate availability and toxic levels of Fe, Mn and Al.

Total and Free Iron — Although distribution of free iron oxides is somewhat erratic with rock depth, the overall trend is a decrease in the unweathered rock. This is not an unusual property of rock material. With the exception of the E location, free iron oxide contents vary between 0.5 and 3.0%. This can be considered normal for this type of rock material in a weathered zone.

It is interesting that at location L, only a small fraction of the total ion oxide is present in free iron oxide form, while at other locations, between 50 and 85% of the total iron oxide is present in the free iron oxide form in the weathered zone. The increase in the free ion oxide/total iron ratio indicates more advanced weathering in the upper zones at location A, C, E and H. This also indicates that further exposure of this zone will not produce more reactive iron and as a result, maintenance of phosphorus availability may be eased.

Considering total levels of iron oxides at F, exposure of this material may result in increasing levels of free iron oxides, which could intensify phosphate fixation if proper liming is not carried out prior to applying phosphorus fertilizers.

Since the quantities of pyrite which were found at each location were too low to account for total amounts of iron oxides, some of the total iron must have come from another source. Mica which was common in these rocks (petrographic observations) is believed to have contributed to the remaining amount of total iron oxides.

Total and Free Mn — Free and total Mn concentrations in the rock were generally low, with the only major variability found in the upper weathered zone at L. It is interesting to note that in most cases, between 80 and 95% of the total Mn is present in free Mn form. This means that most of the total Mn is present in easily reducible form. At H, total Mn concentrations are as low as 10 ppm. Addition of lime to this spoil may lead to insufficient amounts of available Mn for some plants. On the other hand, manganese toxicity to plants may occur at L under very acid conditions.

Free Al — The high concentration of free Al, in addition to free Mn and Fe oxides in the upper 20 feet of rock strata, indicates intensive weathering. There is a relatively wide range in the amounts of free Al between locations studied; however, free Al contents are generally low below 20 feet at several locations and are clearly related to the exchangeable Al. Evidently free Al in the weathered zones of this rock material resulted from reactions of clay minerals with the acids produced by the oxidation of the sulfides. Free Al present at this stage may not pose difficult problems in the establishment of cover crops provided proper treatment with lime is carried out in advance of revegetation.

Total Cu — Concentration of Cu varied with location, between 4 and 26 ppm at location H and over 100 ppm at location C. In normal soils, over 50 ppm Cu is considered very high and below 10 ppm is considered low. At other locations, Cu was between 20 and 50 ppm. The total content of trace elements can be a good indication of their potential availability to plants.

Study of several West Virginia soils show Cu content between 10 and 20 ppm, which is acceptable for normal plant growth. Soils which will develop from weathering of rock at location C will certainly contain higher amounts of Cu.

Application of waste materials such as sewage sludges, in which Cu content may be around 1,000 ppm may make revegetation difficult on such mine spoils as C.

Plants grown on C rock in the greenhouse showed yellowing of leaves which may be an indication of Cu toxicity.

Chemical Characteristics of Several Soil Profiles

Soils at all locations, except E and C, were very low in exchangeable bases (Ca, Mg and K). Oxidation of the pyrite has resulted in leaching of parent rock with sulfuric acid before exposure to soil forming processes. As a result, these soils (except E and C, which were recently limed) are high in exchangeable Al. It is interesting that the soils are similar to the rock, between 70 and 100% of the total iron being present as free iron oxide in the upper horizons, while the lower horizons show a marked decrease in free iron oxide compared to total iron content.

Soils at locations E and L, which are high in total iron oxide, show a very small fraction of the total iron as free iron oxide. Low free iron oxide content of E and L soils suggest less intensive weathering of this soil. Mechanical analyses for these two soils, as discussed later, do not support this contention. Variation in the free iron oxide/total iron could be due to variation in the source of iron in the parent rock. Biotite minerals which carry iron are more resistant to weathering than pyrite, and may have influenced the free iron oxide concentration at the above-mentioned locations. Petrographic analyses of parent rock at E and L locations showed presence of mica.

Mechanical Analysis

Soils at five locations were subjected to mechanical analysis for total sand, total silt and total clay. Results at E and L are based on air dry, organic-matter-free basis; at locations A, C and H on the basis of air dry weight with organic matter included.

The particle size distribution of the soils at locations A, C and H was similar, with slight variability in the sand and clay contents of the H soil, which contained more sand and less clay. In these soils total clay contents ranged from 14 to 24%, being rather uniform with depth. These soils were well drained and consisted of coarse texture which reflected the influence of parent rock.

Soils developed at locations E and L showed greater variability in sand and clay fractions. Soil at E location contained less sand and more clay than the soil at location L. Both soils showed an increase in clay in the lower horizons, indicating illuviation of clay material. However, lack of prominent clay skins indicated that textural difference in parent rock may, in some cases, have contributed to greater percentages of clay in the lower horizons (B horizon).

There are some evidences, such as the decrease in sand in the B horizon of the E soil, which indicated a probable textural variation in the soil parent rock. This may explain differences in the development of different soils at locations E and L. From the results of mechanical analyses, one could visualize the influence of texture of parent rock on the development of these soils.

Mineralogical Characteristics

Clay minerals are of major importance in influencing the physical and chemical

Mine Spoil Potentials for Water Quality and Erosion

characteristics of soils and other earth material. They may play an important part in the mechanical stability of extreme slopes and consequently soil erosion or slippage. Since knowledge of the minerals in the clay fraction would contribute toward a fundamental understanding of the chemical and physical properties of mine spoil material and its reaction to weathering or practices, detailed mineralogical studies of rock and soils were carried out.

Mineralogical analyses were made on coarse clay (2 to 0.2 μ), medium clay (0.2 to 0.08 μ) and fine clay (0.08 μ) fractions separated from rock and soils. Magnesium saturated samples of each size fraction were glycerol saturated and parallel-oriented on glass slides for X-ray diffraction analysis. Diffraction patterns were obtained also on K saturated samples after the following treatments: (a) K saturation and air drying; (b) K saturation and heating at 110°C; (c) K saturation and heating at 300°C; and (d) K saturation and heating at 550°C.

X-ray patterns of oriented, Mg saturated, glycerol solvated clays separated from rock material from two locations (L and H) show strong peaks at 7 and 10 A indicating that kaolinite and mica are the dominent clay minerals in rock at both locations.

Differential thermal analysis indicated the clay fraction of the rock contained 30 to 40% kaolinite. Total K analyses of rock material showed 18 to 30% mica. Small peaks at about 14.4 A in the upper zones reflect small amounts of vermiculite at both locations. After K saturation the 14.4 A unit cells collapsed to 10 A, confirming presence of vermiculite. Samples taken from lower depths (unweathered zone, rich in pyrite) showed no 14 A peak.

Mica contents increase with increase in depth. These findings indicate that vermiculite formed during or after pyrite oxidation which may have removed K from mica. Fine clay (0.2μ) separated from rock showed a peak at 17.7 A indicating the presence of montmorillonite. This was observed in fine clay separated from both weathered and unweathered zones. Apparently during pyrite oxidation, some montmorillonite, which is not very stable under acid conditions, was not destroyed.

Mineralogy and Mottling of Soils

Distribution of mineral species in soil clays with depth was determined. X-ray diffraction analysis of the Mg-saturated samples from H location show vermiculite as the dominant clay mineral in all horizons and at all depths. Kaolinite ranked second in abundance in all horizons in this soil. Mica and montmorillonite clay mineral were absent in the upper horizons as no 10 A and 17.7 A peaks were observed. Increase in the intensities of 10 A and 17.7 A peaks with depth indicated presence of mica and montmorillonite at lower depths.

Soils taken from locations E and L showed that kaolinite (7 A peak) was the principal clay mineral with smaller amounts of vermiculite and mica (14 and 10 A peaks) in the upper horizons. In the lower horizons the amount of vermiculite decreased and mica increased. A small but prominent peak at 18 A was also present at lower depths. This indicates the presence of montmorillonite.

Mineral contents of all soil profiles are strongly influenced by the underlying parent rock, which contains kaolinite, mica and small amounts of vermiculite and montmorillonite. In general, kaolinite and mica are the major clay minerals, followed by vermiculite and montmorillonite in the lower soil horizons. In the upper horizons where intensive weathering has taken place, kaolinite and vermiculite are the dominant clay minerals with minor amounts of mica and traces of montmorillonite. It is of interest that in all those profiles, there is an alteration continuum.

In weathered rock and B horizons, there is expanding vermiculite which collapses on K saturation. Vermiculite in A horizons of all soils collapses only on heating to 300° or 550°C. The reason for the prominence of the vermiculite in the A horizon is not certain. Probably, the increase in the proportion of vermiculite to mica nearer the surface resulted from the removal of weathering products in solution and continuous replenishing of vermiculite by weathering of mica. This interpretation is supported by increase in mica with increasing depth of soil profile.

The reason for the presence of montmorillonite (least prominent clay mineral) under acid weathering conditions is not certain. It has been accepted that (2:1) layer silicate minerals such as montmorillonite are stable only in basic soils or under essentially saturated conditions, but where leaching is severe as in southeastern United Sates, kaolinite type minerals are expected. Considering the low pH values of these soils and weathered rocks, it is assumed that montmorillonite, occurring originally in parent rock below the water table, was encased and protected by stable crystalline or amorphous fines during pyrite oxidation and soil development.

In the course of morphological studies of soils at several locations, vertically elongated, low chroma (gray) tongues were observed in subsoils (B horizons) and in underlying sandstones at locations E and L. The physical appearance of this sandstone in the weathered zone was similar to that of the low chroma sandstone in the unweathered zone (pyrite rich) and could be mistaken for pyritic sandstone by mine operators. As a result, physical, chemical and mineralogical studies were carried out to determine the nature and source of this material.

Field observations showed that the low chroma surfaces of the sandstone may be continuous with vertically elongated low chroma zones in overlying subsoils. Chemical studies of this low chroma sandstone showed very low percentages of total sulfur (0.001%), total iron (1.10%) and free iron (0.50%). Similar total sulfur (0.001%) and more iron (total, 3.71%; free, 1.61%) were found in the high chroma interior. Mineralogical studies of low chroma and high chroma subsoil showed that low chroma material is devoid of mobile minerals (montmorillonite and amorphous), and rich in less mobile kaolinite and mica. On the other hand, high chroma material contains more amorphous fines as well as appreciable amounts of vermiculite and montmorillonite.

From these studies, it was concluded that low chroma material in the soil and weathered rock is mainly a result of localized leaching of colored compounds (oxides and hydroxides of iron). In the field, it should be remembered that low chroma sandstone or soil occurring naturally on outside surfaces of rock or soil fragments, is not likely to contain pyrite or toxic acid. It is the low chroma (gray) unweathered sandstone interiors which are pyritic and potentially toxic.

Applicability of Mineralogical Data: Results indicate that mica has weathered to vermiculite in soil profiles and weathered rock materials studied. Moreover, vermiculite clay carries high negative charge and as a result increases cation exchange capacity. This prevents rapid leaching of exchangeable elements (either toxic or favorable) to ground water. Clay minerals like vermiculite and montmorillonite would also serve as a "buffer" by their reactions with the acid produced by the oxidation of sulfides. The reaction of acid with clay minerals may increase Al activity in solution when Ca contents are low. This requires limestone applications for establishment of cover crops.

Since clay minerals play a prominent part in determining physical and chemical properties, there is an urgent need for quantitative clay mineral information in correlated mine spoil materials. Such data should afford a means of discriminating spoils with sufficient buffering and water retention capacity to inhibit leaching of toxins and to favor maintenance of desirable pH and nutrient levels for initial and for long-time growth of cover plants.

ROCK WEATHERING

Simulated Weathering Experiment

This part of the project involved the use of specially designed cells (modified plastic shoe boxes) to contain selected rock materials during simulated weathering. Two hundred grams of one of the rock types, low chroma sandstone and high chroma sandstone or low chroma shale were placed in a cell and exposed to a continuous flow of moist air. The cells were leached twice a week for the first eight washings and once a week for the next six washings using the following scheme: To each cell 100 ml of distilled and deionized H_2O was added in 50 ml increments.

The sample was allowed to soak for 10 minutes after each increment and then the cell was gently agitated to thoroughly wash the sample before the water was drained. After the sample was washed with both increments of water, the washings were centrifuged and the volume of each was recorded. Any residue from the sample left in the centrifuge tubes was returned to the proper cell. The data for the three sample types were the mean of three replicates; individual measurements were essentially the same.

Conductivity measurements were taken with a Wheatstone bridge and a pipette cell with a constant of one. Then the washings were split with half being saved to be analyzed for Al, Fe, Ca, Mg and K. The other half was analyzed for the potentially free acidity of the rock material.

The conductivity of the distilled and deionized water was checked each time the cells were leached and subtracted from the conductivity of the washings. The initial washings were highest in electrolytes for all samples and then the conductivity decreased to its lowest point on day 22. The low chroma sandstone and low chroma shale were quite close in all readings and on day 55 until the end of the experiment both samples showed a steady increase in electrolyte content of their washings. The high chroma sandstone sample was very low in all readings with its high point being on day 44.

The potentially free acidity that was produced by the low chroma sandstone and low chroma shale was low when compared to amounts produced from the same samples by the peroxide method. The amount of acidity produced by the high chroma sandstone was comparable for both methods. The low chroma sandstone had the largest and quickest release of acids even though the low chroma shale has the largest percentage of total sulfur. Both low chroma sandstone and low chroma shale showed a steady increase in acidity produced at the end of the experiment.

The total amount of Fe released was highest for the low chroma sandstone sample. Both the high chroma sandstone and low chroma shale released very little Fe, but in release of Al the high chroma sandstone sample is highest with the low chroma sandstone being the next highest. This could be the result of the high chroma sandstone sampled being weathered.

The release of basic cations was as expected with the low chroma shale being highest in amount of Ca and K. The low chroma sandstone released a large amount of both elements but the high chroma sandstone released only trace amounts; however, the low chroma sandstone was higher in release of Mg then the low chroma shale and the high chroma sandstone released the least.

The data indicate that the release of elements during weathering is slow. The low chroma sandstone had the fastest rate of weathering. This may seem contradictory to what has been previously stated but when comparing the amount released with the total amount of each element, it is quite evident that this is correct. The low chroma shale has a very slow release for all elements except Ca. The high chroma sandstone had the lowest total amount of all elements and the slowest release except for Al, which indicated a previous weathered condition.

General Observations

Observable disintegration of sandstones and shales in mine spoils indicates rock differences that are not fully understood. Some stones disintegrate quickly, whereas others that appear similar persist. Since secondary quartz growth is commonly evident, petrographically it is likely that some resistant Lower Mahoning sandstone is effectively cemented with secondary quartz although cementation is not sufficient for classification as a quartzite.

Most shales that disintegrate readily are fine textured (clays), whereas the most resistant shales (or siltstones) contain higher proportions of silt and may be partially cemented either with carbonates or quartz.

Freeze-thaw and wet-dry sequences or cycles and falling water drops have been used to separate sandstones and shales into groups differing in relative stability, but calibrations against field behavior in spoils have not yet been established.

EVIDENCES FROM OLD MINE SPOILS

Samples from old spoils that have remained barren for 5 to 10 years indicated acidity (pH 3.5) was the primary influence preventing revegetation. Nutrient levels ranged from very low to medium with highest quantities of nutrients in Bakerstown spoils.

Bulk densities, coarse fragments, textures and water retention capacities indicated ranges which should accommodate plant growth, although drouthiness would be expected with the most sandy and stony spoils, especially with shallow rooting species.

Twenty-five year old spoils near Morgantown (Canyon) from surface mining of Pittsburgh coal are satisfactorily covered with forage grasses and legumes and contain near normal percentages of soil organic matter, although analyses for total sulfur indicated 0.5 to 1.2%. This is enough sulfur to create excessive acidity and toxicity if present as reactive pyrite. However, considerable coal is present in the spoil, which probably means the sulfur remaining is in organic forms and is not converted to mineral acids until the coal is destroyed by oxidation.

Iron ore spoils of shaly lower Pennsylvanian, moderately acid materials, abandoned 70 to 130 years prior to sampling provided evidences that the natural soils were superior in bulk densities (lower), porosity (higher), soil structure development, infiltration, nitrogen or organic matter, surface texture (more loamy), and smoother land surfaces. On the other hand, mine spoils were superior in depth for plant rooting, total available water holding capacity, certain plant nutrients, and gentler slopes on benches. Forest site quality, pH and soil mineralogy were not greatly different between natural soils and mine spoils.

MICROBIOLOGICAL INTERACTIONS

Specially designed miniature lysimeters simulating mine soil land were tested for rating the activity of chemoautotrophic iron and sulfur oxidizing bacteria on pyrite or rock materials containing variable percentages of pyrite.

Malfunction of some deeply buried interception funnels indicated needed improvements of design, but partial data showed that burial of pyrite at a 3 inch depth in natural loamy soil of pH 6.7 reduces the rate of release of ferrous iron and sulfate compared to ½ inch depth of burial. Moreover, percolation downward through the 4 foot column of soil removed most of the acid and iron, presumably by reaction with exchangeable cations in the soils.

There is some evidence that alkaline earth carbonates may inhibit bacterial oxidation activity, possibly by preventing pH of 5.0 or less as needed to favor sulfur and iron oxidizing organisms. In some particular rock materials an insufficient quantity of accessible oxidizable substrate may have prevented microbial growth even through appreciable percentages of pyritic sulfur (0.20 to 0.75%) were present.

INTERACTIONS WITH PLANT COVERS

The effectiveness of vegetative cover in reducing water pollution from disturbed land areas has been well documented. Observations made during the course of this project have been primarily related to the thriftiness of plants in local spoil areas composed of specific types of rock materials.

Although there has been little time since the verification of a sulfur-free zone in coal overburden that would allow operators to selectively place this material in positions favorable to establishing vegetation, in several Upper Freeport spoils the sulfur-free rock has "accidentally" been placed on the spoil surface in localized spots. Likewise the acid-producing low chroma massive sandstone occurs in various size fragments on the surface of many spoils.

Overburden materials of some other coal seams, such as the Lower Kittanning and Bakerstown, contain bases, present as carbonates, frequently in sufficient quantity to effectively neutralize acid formed from oxidation of pyritic sulfur, which is present at levels comparable to those in Lower Mahoning sandstone or greater. The higher resulting pH of many of these spoils, along with higher quantities of plant nutrient elements derived from breakdown of shales, and improved water holding capacity obtained in the finer textured weathering products has resulted in highly acceptable revegetation programs on most spoils of these coal seams.

An example of complete failure of revegetation efforts on an older Bakerstown spoil near Albright, Preston County, can be attributed to the inopportune placement of black, highly pyritic shales from just above the coal seam on the spoil surface during the mining operation. The particularly toxic shale in this location attains a thickness of at least 5 meters, and although base status is comparable with overlying strata, the sulfur content is consistently between 1 and 6%, requring tremendous quantities of neutralizing material wherever oxidation attacks finely divided forms of the pyrite.

Spoils resulting from surface mining of Lower Kittanning coal in northeastern Preston County have been planted with forage mixtures of bird's-foot trefoil and grasses with great success. The overburden material between the coal seam and the original soil is shale containing, below the weathered zone, the equivalent of 6 tons (average) of calcium carbonate per thousand tons of material.

This same material averages 0.2% total sulfur, with the major portion being within 2 meters of the coal seam. Inasmuch as the basic carbonates solubilize rapidly in acid whenever formed from pyrite oxidation, the net pH of the spoil and drainage waters remains near neutrality.

Spoils composed of Lower Mahoning sandstone have been partially covered with thriving stands of bird's-foot trefoil and tall rescue where moderate amounts of lime and fertilizer have also been applied. Localized spoil areas consisting of the weathered (high chroma) sandstone, which has been shown to be free of pyrite, can be depended upon to serve as an acceptable growing medium if the plant nutrients removed during the weathering processes over geological time are replaced, and some lime is applied to counteract the natural acidity the leaching processess have produced. Areas of spoil dominated by the low chroma, pyritic sandstone are found to be barren of vegetative cover except where extremely heavy rates of lime have been applied.

Most of the spoils resulting from mining the Upper Freeport seam consist of a mixture of high chroma, weathered sandstone and low chroma, pyritic sandstone. Where adequate lime and fertilizer have been applied, and seeding time was at the optimum part of the growing season, excellent ground cover has been obtained with the grass/legume mixtures commonly used. Areas containing a large percentage of coarse rock fragments (nearly always of low chroma, and pyritic) and devoid of finer particles could be expected to remain barren.

Observation of several Upper Freeport spoils and a few formerly highly acidic Pittsburgh spoils suggests that rapid establishment of vegetative cover of any type tends to reduce the rate of generation of acid in the spoil. This would be expected, theoretically, because of increased carbon dioxide and reduced oxygen concentrations associated with respiring roots or decomposing organic matter.

Although with proper practices, forage grasses and legumes, seeded alone or in combination with trees, are providing quick cover needed to control erosion and sedimentation on many spoils, it is apparent in this region that native trees will soon convert most mine spoil lands to woodland unless tree seedlings are controlled.

In this connection, it appears that some landowners who prefer forages rather than woodland on gentle slopes have failed to realize that vigorous forage stands and growth require repeated applications of lime and fertilizers as well as proper grazing and clipping management. Moreover, wise use of mine spoil land is to be expected only when capabilities of particular spoils are well understood, the same as with other soils.

STUDIES OF EFFECTS
OF MINE DRAINAGE

The material in this chapter is excerpted from:

PB 210 709
PB 217 872

EFFECT OF STRIP MINING ON WATER QUALITY

Since 1958, the Ohio Agricultural Research and Development Center has been studying the physical and chemical changes occurring in weathering coal spoils. This work, using lysimeters, has indicated the pollution potential of 19 Ohio spoils, but results are based on small samples. Early in 1965, when strip mining of Middle Kittanning coal began on the watershed of Little Mill Creek, it was possible to extend and corroborate the lysimeter work with a field study.

Lysimeter Study

The lysimeter work at the OARDC has revealed a number of relationships between spoil pH and quality of water coming from Ohio's strip mined areas. At the start of the lysimeter study, 19 spoils were classified into one of four groups:

> Toxic — pH of surface material less than 4.0 on more than 75% of the area.
> Acid — pH of 4.0 to 6.9 on over half of the surface area.
> Calcareous — pH above 7.0 on over half of the surface area.
> Marginal — One-half to three-quarters of the surface area with a pH below 4.0, the remainder being calcareous, acid, or mixed.

On the basis of this classification, the lysimeter study included 3 acid, 1 marginal, 5 toxic, and 10 calcareous spoils, plus a sample of Wooster silt loam subsoil for comparison. The spoils were allowed to weather outdoors in plastic cylinders 1 foot in diameter and 4 feet deep, drained by gravity, with no surface runoff allowed.

Studies of Effects of Mine Drainage

To examine the combined effect of precipitation and duration of weathering, regression equations of the form

$$\mathrm{Log}_{10} Y = b_0 + b_1 X_1 + b_2 X_2$$

were fitted to the lysimeter data. Y was pounds of soluble salts, sulfate, calcium, magnesium, aluminum, iron, or manganese leached per acre per year. X_1 was years of weathering, X_2 was annual precipitation, and b_0, b_1 and b_2 were regression coefficients found by least squares analysis. A separate analysis was made for each spoil classification.

A majority of the regressions were significant at the 5% level of probability, or better. Chemical yield decreased with duration of weathering, and increased with increasing precipitation. There were large differences between the four types of spoil in both amounts of chemicals produced, and rate of change in chemical yield with time. For example, the three toxic spoils had the highest initial rate of sulfate production, but by the eighth year of weathering, the marginal and toxic spoils were yielding approximately equal amounts of sulfate in the lysimeter leachate.

Beaver Creek Study

The influence of strip mining on water quality was studied in Kentucky. Comparisons were made between chemical composition of water from mined and unmined parts of the Beaver Creek watershed. Prior to strip mining the study areas were similar in their chemical water quality. Chemical weathering of iron disulfide exposed by mining operations induced large changes in composition of streamflow. Unmined portions of the watershed yielded 26 and 28 tons of dissolved and suspended solids per square mile in 1958, while the watershed with 6.4% of its area disturbed by mining yielded 189 tons of dissolved solids and 1,930 tons of suspended solids per square mile.

Streamwater in the unmined part of the watershed contained less than 30 ppm of dissolved solids and had pH in the range 5.0 to 7.2. Between one-third and two-thirds of the dissolved solids came from precipitation, and the remainder from weathering processes. Ions of major importance in the water from the unmined area were calcium, magnesium, bicarbonate, and sulfate, while silica accounted for about 25% of the dissolved solids.

The mined portion of the watershed had similar chemical water quality prior to mining. Following mining, however, dissolved solid content ranged from 80 to 1,500 ppm, with a median value of 310 ppm. The dissolved solids included aluminum, manganese, iron, calcium, magnesium, and sulfate. Bicarbonate was absent, and pH ranged from 2.5 to 4.2.

The Mining Operation

The watershed of Little Mill Creek is underlain by sedimentary rocks to a depth of at least one mile. Middle Kittanning or No. 6 coal, part of the Allegheny series of the Pennsylvanian system, outcrops in a number of places along the divide between Mill Creek and Little Mill Creek. Near the confluence of Little Mill and Mill Creeks the outcrop is at about 1,040 feet, while in the study area at the upper or northern end of the stream, the coal outcrops at about 1,160 feet above sea level.

The overburden above the coal is Lower Freeport sandstone and shale, Upper Freeport sandstone, and a very small area of sandstone of Conemaugh age. Neither Lower Freeport limestone nor Upper Freeport limestone have been found where due to outcrop in Coshocton County. Thus there is little, if any, limestone in the overburden of the study area. Stripping of overburden began at the head of Little Mill Creek in late winter of 1965.

Sequence of Mining: A total of 31.29 acres were leveled as of March 30, 1970, north of State Route 643. The highwall had been completely backfilled and the area planted to pasture. Mining continues, however, on the south side of State Route 643. During the study, the watershed area affected by surface mining increased from 0 to 9%.

Watersheds: The validity of the comparison between the mined and unmined parts of Little Mill Creek depends on their similarity before mining. Adjacent watersheds may be compared on the basis of area-elevation-distribution graphs, showing the percent of the watersheds lying between various elevation limits. Watersheds 5 and 91 are quite similar in this respect (Figure 8.1). They are also quite similar in slope distribution and land use, although watershed 5 has more cropland and less meadow than watershed 91.

FIGURE 8.1: AREA-ELEVATION-DISTRIBUTION GRAPHS OF WATERSHEDS 5 AND 91

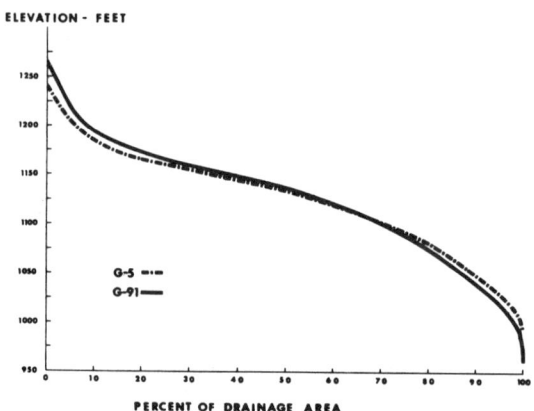

Source: PB 217 872

Materials and Methods

Streamflow Sampling: Four stream gauging stations, G-5, G-11, G-91 and G-92, on the Little Mill Creek watershed, were used as sampling points. G-5 is roughly 4,000 feet downstream from the mining area. G-91 gauges an adjacent watershed, quite comparable in size, geology, and land use, but without any mining activity. Many of the differences between water samples at G-5 and G-91 are considered due to the mining upstream from G-5.

G-92 receives runoff from both G-5 and G-91, while G-11 is on a downstream tributary unaffected by surface mining. Thus G-11 represents a second control for comparison with the mined area, while samples from G-92 contain water from G-5, G-91, and a third area of 278 acres.

Hand samples were taken each week from the weir notch in 1 quart polyethylene bottles, and stored in a refrigerator prior to analysis. Rate of streamflow at the time of sampling was noted. If there was no streamflow, no sample was taken.

An automatic sampler was installed at G-5 to provide representative samples of high streamflow. This sampler consisted of a Leupold-Stevens flowmeter coupled with a Nappe Corporation pumping system. All flows above 2.75 cubic feet per second were sampled at a rate proportional to flow.

Chemical Analysis: The concentration of sulfate, manganese, iron, aluminum, calcium and magnesium was determined in each sample. Manganese was measured by the persulfate method and sulfate by the barium sulfate turbidimetric method. Calcium, magnesium, iron and aluminum were determined on a direct-reading emission spectrograph. The pH of stored samples was measured with a glass electrode pH meter.

Statistical Methods: The concentrations of the various chemicals, determined on the weekly samples, were transformed to logarithms. Mean log concentrations and standard errors were computed for each sampling station. In a least squares analysis, the relationships between streamflow as the independent variable and log concentrations of potential pollutants as dependent variables, were computed. Correlations between log concentrations of the various potential pollutants were also determined.

Duration curves were prepared showing the proportion of water samples having pH, sulfate, manganese, calcium, magnesium, iron, aluminum, or iron plus manganese concentrations above various limits. Such curves are useful in predicting the frequency with which troublesome concentrations of potential pollutants occur. The average transport of the various chemicals, in pounds per week, was computed from the relationship:

$$\text{Weekly transport} = \frac{\text{ppm} \times 604{,}800 \text{ sec} \times 62.4 \text{ lb} \times \text{flow in cfs}}{\text{cu ft} \times 1{,}000{,}000 \times \text{week}}$$

From the concentrations and steamflow records obtained with the automatic sampler, transport of chemicals by high streamflows was estimated by the formula:

$$\text{Transport} = \frac{\text{flow} \times 62.4 \text{ lb} \times \text{ppm}}{\text{cu ft} \times 1{,}000{,}000}$$

Mine Pumpage and Drainage: At times, drainage water accumulated in the mine pit, interfering with operations. It was released into Little Mill Creek, either by gravity flow following cutting of a drainage way, or by pumping it up over the highwall and into the creek. The amount of this mine drainage could be measured at G-5, since it appeared as excess runoff not associated with precipitation.

Results and Discussion

Table 8.1 gives the mean concentrations in ppm, average pH and mean ± 2 x standard error for the potential pollutants at the four sampling locations. There were significant differences between sampling points for concentrations of all chemicals except aluminum. In Table 8.1 if (Mean ±2SE) of 2 sampling points does not overlap the chances are at least 99 out of 100 that the means are really different. Average concentrations of sulfate, manganese, and magnesium are quite clearly higher at G-5 and G-92, where streamflow included water from the mined area than at G-91 and G-11; pH is lower at G-5, but not at G-92. The mean sulfate concentration observed at G-5 was double that at G-11 and nearly three times the mean concentration at G-91.

This is doubtless due to the oxidation of pyritic materials exposed during coal mining. The higher concentration of manganese at G-5 results from the action of sulfuric acid, formed by pyrite oxidation, on manganese compounds in the sedimentary rocks.

TABLE 8.1: POTENTIAL POLLUTANTS

Potential Pollutant	G-5 Mean	G-5 Mean ± 2SE	G-11 Mean	G-11 Mean ± 2SE
$SO_4^=$	91.1	85.6 - 97.0	46.2	43.3 - 49.2
Mn	.46	.38 - .56	.018	.015 - .022
Ca	34.6	33.5 - 35.7	42.3	41.0 - 43.7
Mg	12.1	11.5 - 12.7	7.3	7.0 - 7.7
Fe	.10	.091 - .119	.08	.071 - .093
Al	.17	.15 - .20	.15	.13 - .18
pH	6.5	6.44 - 6.59	7.1	7.04 - 7.20
Fe + Mn	.82	.71 - .94	.10	.09 - .12

Potential Pollutant	G-91 Mean	G-91 Mean ± 2SE	G-92 Mean	G-92 Mean ± 2SE
$SO_4^=$	35.3	33.2 - 37.5	62.5	58.5 - 66.7
Mn	.019	.015 - .023	.062	.052 - .070
Ca	27.8	26.9 - 28.6	36.8	35.6 - 38.0
Mg	8.4	8.0 - 8.9	10.0	9.6 - 10.5
Fe	.11	.096 - .125	.08	.074 - .097
Al	.14	.12 - .17	.14	.12 - .17
pH	7.05	6.97 - 7.13	7.07	6.99 - 7.16
Fe + Mn	.14	.12 - .16	.20	.17 - .23

Note: Mean concentration is the antilog of the mean log concentration.

Source: PB 217 872

One might expect higher concentrations of iron and aluminum in water from the mined area. It is likely, however, that considerable iron and aluminum in solution in water near the strip mine has precipitated out before reaching G-5. Manganese remains in solution until most of the iron has precipitated.

Duration Curves: Duration curves were prepared, showing the percent of water samples having pH, sulfate, manganese, calcium, magnesium, iron, aluminum, or iron plus manganese concentrations above various limits. Used with a set of limiting concentrations for various beneficial uses, duration curves give a good idea of the seriousness of stream pollution. For example, the recommended limiting concentration of sulfate in water used in the dairy industry is 60 ppm. About 5% of the samples at G-11 and 20% of the samples at G-91 exceeded this concentration.

On the other hand, 40% of samples at G-92 and 70% of those at G-5 were above this limit. There were rather clear-cut differences between the mined and unmined portions of Little Mill Creek in frequency of higher concentrations of sulfate, manganese, magnesium and iron, and in frequency of lower pH levels.

Correlations: The extent of correlation between concentrations of potential pollutants was examined. Correlation coefficients between variables were often higher for the mined than for the unmined portions of the watershed. Logarithms of calcium and magnesium concentrations were quite well correlated. At G-5 and G-92, correlations between pH and log concentration of the other variables were always negative—high concentrations of hydrogen ion were accompanied by high concentrations of other solutes.

There were also a number of statistically significant correlations between concentrations of samples taken with the automatic sampler. Especially high positive correlations were found between sulfate and manganese, sulfate and iron, manganese and iron, and calcium and magnesium concentrations.

Mine Pumpage and Drainage: Measurable amounts of water from the mining operation were pumped or drained directly into Little Mill Creek on 186 days during the 236 week study period. The average amount released was 1,626 cubic feet or 12,163 gallons per drainage event. The daily maximum was 15,836 cubic feet. A total of 302,402 cubic feet of mine drainage, representing 0.5% of the flow at G-5 during the 4½ year study period, was observed. Mine drainage such as this is typically of low pH, and contains high concentrations of sulfate, iron, manganese, and aluminum.

Average Yield per Acre per Week: Average yield per acre per week of sulfate, manganese, magnesium, iron, and aluminum was higher from G-5 than from the other watersheds (Table 8.2). If the higher yields at G-5 are attributed to 31 disturbed acres of the watershed, then the annual per acre yield of the disturbed area can be computed. Such computations indicate that the actual mined area yielded an average of 28.2 lb of sulfate, 0.69 lb of manganese, and 2.4 lb of magnesium per acre per week. For comparison, average weekly per acre yields from similar spoils during the first year of the lysimeter study were 40, 0.04 and 4 lb of sulfate, manganese, and magnesium, respectively.

Relation Between Streamflow and Concentration of Potential Pollutants: Linear regressions of log concentration on log streamflow were computed. For the statistically significant regressions, the regression coefficient was negative with the single exception of the relationship of log sulfate concentration to log streamflow at G-91. As streamflow increased, concentration of potential pollutants decreased. The regressions for calcium and magnesium accounted for a considerable proportion of the variation in the dependent variables, with R^2 values up to 0.59. Regressions of manganese, iron and aluminum on streamflow were not statistically significant.

TABLE 8.2: YIELD OF POTENTIAL POLLUTANTS BY WATERSHED

Potential Pollutant	Average Weekly Yield Per Acre (Pounds)			
	G-5	G-11	G-91	G-92
Sulfate	3.894	2.180	1.395	2.422
Manganese	.062	.001	.001	.013
Calcium	1.005	1.655	.840	1.070
Magnesium	.521	.311	.284	.343
Iron	.061	.006	.008	.008
Aluminum	.115	.011	.009	.017

Source: PB 217 872

High Streamflows: During high streamflows, the mean concentration of potential pollutants was higher than the mean concentration of hand samples taken at G-5 (Table 8.3). This was especially true for sulfate, manganese, magnesium, iron and aluminum concentrations. High streamflows transported rather large amounts of chemicals, especially sulfate (Table 8.3). Regression analysis indicated a negative relationship between streamflow and concentration of the hand samples; as streamflow increased, concentration usually decreased. Why, then, were high concentrations observed during the high flows sampled with the automatic sampler? One explanation may be that heavy rains were effective in removing soluble salts from the mined area, whereas lighter rains were not.

TABLE 8.3: MEAN CONCENTRATIONS AND TOTAL TRANSPORT OF POTENTIAL POLLUTANTS AT G-5

	Total Transport (pounds)	Samples of High Flow Concentration (ppm)	
		Automatic Sampler	Hand Samples
$SO_4^=$	80,473	204.0	91.1
Mn	1,608	2.74	0.46
Ca	18,738	*42.3	34.6
Mg	12,231	31.1	12.1
Fe	354	1.04	0.10
Al	8,334	5.32	0.17
pH		4.43	6.5

Note: Mean concentration is antilog of mean log concentration, except for pH.

Source: PB 217 872

Summary and Conclusions

Surface mining of coal on the upper reaches of Little Mill Creek affected 31 acres or 9% of watershed 5 and was accompanied by increased concentrations of sulfate, manganese and magnesium, and hydrogen ion in water samples taken 4,000 feet downstream. Duration curves developed from weekly hand samples showed that stream concentrations of manganese near the mining area exceeded drinking water standards 80% of the time, while manganese concentrations in the part of the creek not affected by mining exceeded these standards only 13% of the time.

During high streamflows, sulfate and manganese, sulfate and iron, manganese and iron, and calcium and magnesium concentrations at the sampling point nearest the mine were highly correlated. Correlations of pH with sulfate, manganese, calcium, magnesium, and iron were all negative in samples affected by mining.

Statistically significant regression equations relating sulfate, calcium, and magnesium concentrations to streamflow at all sampling points showed that high concentrations were associated with low streamflows, and vice versa, with the single exception of sulfate concentration at G-92. Automatic samples of high streamflows, however, had generally higher concentrations of chemicals than the hand samples. Thus it appears that for the mined area of the watershed there was a threshold raindrop energy and flow rate above which the detachment force of raindrops and velocity of flowing water could move chemicals into and through the stream system from sources not available to lesser storms.

ACID MINE POLLUTION EFFECTS ON LAKE BIOLOGY

Description of Study Area

Six strip mine lakes in Indiana were studied in order to provide certain fundamental limnological information as a basis for sound management of such lakes. The aim was to select a reasonable number of lakes embracing the widest possible pH range.

An effort was made to select lakes that were as uniform as possible with regard to morphometry and other factors, differing mainly in their degree of recovery from acid pollution. The six lakes selected for study range in pH from 2.5 to 8.2. All are in basins formed as the final cuts of strip mining operations. The lakes have been designated by Roman numerals I through VI in order of ascending pH. Lakes II through V were under continuous surveillance from July, 1969, through December, 1970. Lakes I and VI were monitored throughout 1970. Following is a brief description of each of the lakes.

Lake I: This lake varies in apparent color from a rather turbid red-brown ("tomato soup") to a brilliant red-black. Lake I is, at three meters, the shallowest of the study lakes, and it also has the least area and volume (Table 8.4). Lake I was formed in 1940.

Lake II: This lake has the long narrow outline typical of strip mine lakes. The water is generally a very clear green due to low turbidity. The most striking thing about Lake II is the uniformity of chemical and physical conditions that it exhibits. This lake tends to be remarkably constant in pH, dissolved substances, etc., both throughout the water column and throughout the year. Thus the ranges of the various chemical and biological parameters are quite restricted for Lake II as compared to the others. Lake II was formed in 1960.

Lake III: This lake is located adjacent to Lake IV on one subunit of the Patoka Fish and Game Area of the Indiana Department of Natural Resources. Lakes I, V, and VI are on other subunits of this management area. Lake III has greater flow-through than the other study lakes. It has, apparently as a result of this, greater short-term changes in such parameters as pH, temperature, and turbidity. This effect is confined mainly to the surface layers because of a stratification situation. Lake III has greater variation in pH from inlet to outlet, from surface to bottom, and over the annual cycle than any of the other lakes. Lakes III and IV were completed in 1958.

Lake IV: This lake is directly connected to Lake III by a very short channel approximately 4 m wide and 20 cm deep. The direction of water flow is from IV to III at all times. This lake has the greatest maximum depth of the six. Lake IV has had a fish population for several years, and during times when water quality moderates, these fish may invade Lake III. During the winter of 1969-70 the stream that formerly flowed into the upper end of Lake III changed its course to the lower end of Lake IV. No water quality or biotal changes have been observed that can be attributed to this change in stream channel.

Lake V: This is the most irregular of the lakes in shape. There are five small islands in the main body of the lake and three long fingers pointing to the southwest. Lake V was fertilized during the summer months in an attempt to increase fish production. Lake V was formed in 1940.

Lake VI: This lake is one of the smaller of the study lakes and is roughly Y-shaped. Construction of an access road in the autumn of 1969 allowed this lake to be added to the series as a relatively high pH lake that was not fertilized or otherwise managed. Lake VI was formed in 1950.

Control Lake: This lake was selected as representative of nonstrip mine small lakes of the region. It is owned jointly by three rural families upon whose property it was constructed in 1963 for recreational purposes, primarily for fishing by the owners. Because of time limitations and relatively poor access this lake was visited less frequently than the others, but it has provided some comparative data on water chemistry and biota. Table 8.4 contains a summary of morphometric data for the six study lakes and the control lake.

TABLE 8.4: LAKE MORPHOMETRIC CHARACTERISTICS

Lake	Year Formed	Total Length (m)	Mean Width (m)	Maximum Depth (m)	Mean Depth (m)	Surface Area (10 m)	Volume (10 m)
I	1940	415	43	3.0	1.9	17.7	34.5
II	1960	1,300	73	6.0	3.8	94.5	355.3
III	1958	975	49	8.0	5.6	47.5	265.0
IV	1958	900	56	10.5	5.5	50.4	275.4
V	1940	960	31	5.5	3.4	29.5	100.6
VI	1950	510	41	7.0	4.4	20.8	90.8
Control	1963	850	95	7.0	3.0	80.8	242.3

Source: PB 210 709

Methods

Sampling Program: Schedule — Initially an effort was made to visit each lake at two-week intervals. This proved impractical, mainly because of the time required to process the resultant biological collections. Hence with the addition of Lakes I and VI to the series, the schedule was revised to one of three- to four-week sampling intervals.

Stations — During the first six months of the study (July to December 1969) horizontal variation in environmental parameters and plankton was monitored by studying two to four stations in each lake. The magnitude of horizontal variation encountered was quite small in all cases. It was, therefore, judged profitable to extend the overall range of the lake series by adding Lakes I and VI at the expense of the multiple stations. Throughout 1970 a single station in the deepest area of each lake was sampled (exclusive of benthos grab samples and various other kinds of biological collections).

Physical and Chemical Parameters: Water samples were collected with two types of nonmetallic water samplers, a 3-liter van Dohrn bottle and (very briefly during the winter of 1970) a 4-liter Kemmerer. The water samples were transported to the laboratory in polyethylene bottles. Water temperature and light penetration were measured in situ. Turbidity, pH, specific conductance, and dissolved oxygen were determined in the laboratory by conventional methods.

Ionic determinations were made by an outside consultant. Ions monitored were calcium, magnesium, sodium, potassium, total iron, dissolved iron, manganese, aluminum, chloride, sulfate, silica, nitrate, and total phosphate. In addition, total acidity, free mineral acidity, hardness, and dissolved solids were determined. Ions were monitored for the surface and bottom strata on a schedule of approximately two-month sampling intervals.

Biological Parameters: Plankton was collected in three different ways: (a) by means of a Wisconsin style tow net, (b) with a 10-liter Plexiglass plankton trap, and (c) by centrifugation of 1-liter water samples in a Foerst centrifuge. All plankton samples were preserved with acid Lugol's solution immediately upon collection, except when live collections were required to facilitate identification. Both trap and centrifuge samples were taken at 1-meter intervals throughout the water column. Depending on the density of organisms, either the entire sample or appropriate aliquots were counted microscopically.

Benthos was collected quantitatively with a Ponar Grab Sampler. Samples were taken from all areas of each lake, and initially, at all depths. Early experience indicated, however, that samples taken from the deepest parts of Lakes III, IV, and V never contained organisms. Therefore, later sampling was conducted mainly in the shallower areas of these lakes (where the bottom was within the mixolimnion). An effort was made to sample each 1-meter depth interval proportionately to its percentage of total bottom area.

Each collection consisted of ten samples (grabs), which was about the maximum that could be processed carefully within a reasonable period of time. The ponar sampled an area of 0.05 m^2 and collected approximately 10 kilograms (clay) or a bit less (leaves or loose detritus) of bottom materials per grab. The samples were placed individually in 20-liter plastic buckets for transport to the laboratory. Ordinarily, initial processing was completed within 24 hours of collection. Each sample was washed through a graded series of three large (35 x 48 cm) screens of mesh sizes: 121.0 mm^2, 1.0 mm^2, and 0.25 mm^2. Organisms were picked from the sorting screens and preserved in formalin. Subsequently, each species was counted and weighed (constant weight at 105°C). In addition, qualitative surveys of the bottom fauna were conducted at irregular intervals with dip nets and artificial substrate samplers of the condenser type.

Fish were collected with funnel traps constructed of hardware cloth and by means of seining in shallow water. Specimens kept for analyses were preserved in formalin. Vertebrates other than fish (frogs, snakes, muskrats, and beavers) were not studied specifically. Although the compilation of exhaustive species lists was not a major objective of this work, as many taxa as possible have been identified or at least recognized.

Morphometric Correction — The following method was adopted as the most straightforward approach to compensating for morphometric differences between the study lakes. For each category of organisms (zooplankton and benthos) dealt with, in each lake, the data were treated by 1-meter depth intervals. The mean standing crop for each depth interval was then multiplied by the percentage of the lake (percent area for benthos and percent volume for zooplankton) at that depth.

Results and Discussion

The primary aim of this section is to analyze those trends in the study lakes that seem to be associated with increasing pH. Such trends would be expected in both the biotic and abiotic components of the ecosystems. During the course of the study, however, it became apparent that a second pattern, in addition to increasing pH, was present in the six lakes. This was the persistent chemical stratification or meromixis of Lakes III, IV, V and VI. This pattern is reflected in the annual variation in temperature and conductivity and in the differences between the surface and bottom waters in many physical and chemical parameters.

Temperature and conductivity diagrams indicate that each of the meromictic lakes is behaving as two separate lakes: an upper one that overturns twice annually and a lower one that does not. This condition is substantiated by a complete lack of overlap in the observed ranges of conductivity and of calcium and other ions in the surface vs bottom waters of Lakes III through VI.

The influence of meromixis on the ecosystems of these lakes seems quite marked in Lakes III, IV, and V, but less so in Lake VI. It affects the distribution of heat, turbidity, and dissolved substances, and the penetration of light. These, in turn, have a profound influence on the biological communities of the lakes. The overall impact is such that meromixis is probably as responsible as pH for the differences observed between the six ecosystems.

In the following discussion the ecosystem of each lake in turn is briefly described. The emphasis in these descriptions has been placed on annual cycles. This is followed by a discussion of those trends in the ecosystem series that seem attributable to the pH spectrum or to the influence of meromixis.

Patterns Within the Lakes: Lake I — Lake I was quite acid throughout the period of study, varying in surface pH from 2.5 to 3.2 with 2.8 to 2.9 typical. The greatest vertical variation in pH observed in any of the lakes occurred in Lake I in March 1970 when a surface-to-bottom difference of two pH units (3.2 to 5.2) was noted. Thermal stratification persisted throughout 1970 except for periods of near homothermy in mid-August and from late September to mid-October. The integrity of stratification was maintained through the warming period, indicating that the entire water mass was heated by direct insolation. A maximum temperature difference of 10.2°C (surface-to-bottom) was observed in mid-May 1970.

Dissolved oxygen was generally low, extremely so during July and August when the concentration was less than 1.0 mg/l throughout the water column. This period of extremely low oxygen was associated with high water temperature and partial overturn—a combination that almost certainly produced a high chemical oxygen demand by upwelling of reduced solutes from the deeper anaerobic strata. The highest observed concentration of dissolved oxygen was 7.2 mg/l at a depth of 1.0 m under a 15 cm ice cover in late January 1970.

This was only about 55% saturation at the in situ temperature of 3.3°C. Despite relatively high turbidity, light was probably adequate for photosynthesis throughout this shallow lake at most times. During much of 1970 light reaching the bottom equaled or exceeded 1% of surface illumination.

In general, the dissolved substances as indicated by total dissolved solids and conductivity and many individual ions were high in Lake I. The surface water in particular had a much greater load of solutes than the other mine lakes. This was not universally true, however, since some ions (e.g., Na^+, K^+, and Cl^-) were higher in some of the other lakes. The deep water of Lakes III and IV exceeded those of Lake I in concentrations of the major cations and, in Lake III, in total solutes.

Although it might be argued that Lake I is meromictic, it exhibited virtual surface-to-bottom uniformity of water quality (pH 2.8 to 2.8, temperature 20.6° to 20.1°C, conductivity 4,350 to 4,350 μmhos) on September 29, 1970. This lack of vertical stratification and the presence of dissolved oxygen in relatively deep water (0.4 mg/l at 2.5 m) suggests that autumnal overturn was complete although of short duration, as indicated by the stratification of physical and chemical parameters on the preceding and following sampling occasions (September 8, October 14). On the latter date, the beginning of inverse thermal stratification was observed (surface 7.9°, bottom 9.5°C). Thus Lake I is probably best regarded as having a marked tendency toward meromixis that is imperfectly realized because of its relatively shallow unprotected basin.

The biological communities of Lake I were marked by very low faunal diversity. In the zooplankton, for example, only five taxa were recognized compared to 20 and 21 species, respectively, in Lakes V and VI. A total of 12 animal taxa were found in Lake I. Standing crop of zooplankton was low, Lake I ranking fifth and only slightly exceeding Lake III. In total benthic biomass, however, Lake I ranked third, with markedly greater standing crop biomass than III, IV, and V. In both zooplankton and benthos a single species dominated. These were the rotifer *Brachionus urceolaris* and the midge larva *Tendipes* sp. which made up, respectively, about 99% and 85% of total biomass.

Fish were not present in Lake I. Muskrats maintained a lodge in a stand of cattails (*Typha angustifolia*) at the northern end of the lake. Other aquatic vertebrates (amphibians, reptiles) were never observed, even though several species were common around the shores of Lake V a scant hundred meters distant.

Lake II — Lake II differed in several ways from the other mine lakes. It was marked by relatively great uniformity of physics and chemistry both over the annual cycle and throughout the water mass. Total pH variation during the 18-month study period was 3.0 to 3.4. A weak stratification of pH (never exceeding a surface-to-bottom difference of 0.3) sometimes developed during periods of thermal stratification. Although this lake differed little from Lake I in surface pH, it was quite different from Lake I in various limnological and biological features. Many of these differences were at least partly related to stratification differences.

The thermal regime of Lake II was marked by long periods of homothermy and relatively small temperature differences during stratification (the only observations of surface-to-bottom differences greater than 5.0°C were on May 3 and May 21, 1970). The maximum observed dissolved oxygen was 10.5 mg/l (at 1.0 under 21 cm ice in late January 1970), which, at 3.2°C, constituted approximately 80% saturation. In midsummer dissolved oxygen values were quite low throughout the water mass (3.5 mg/l or less in July and August 1969 and again in August 1970). Thus Lake II shared with Lake I the combination of low pH, high water temperatures, and low dissolved oxygen throughout the water mass in midsummer. This environmental combination is probably strongly limiting to many species of aquatic organisms. Lake II had greater transparency than the others. Light at the bottom (6.0 m) was never less than 1% of surface illumination and frequently exceeded 10%.

The surface and bottom ranges for total dissolved solids and for conductivity were more similar than in the other lakes. The surface concentrations of total dissolved solids and of most ions in Lake II ranked second to Lake I. In deep water, however, concentration values were exceeded by Lakes III, IV, and V and closely approached by Lake VI.

Although little different in pH, Lake II had almost twice as many animal taxa as Lake I (a 92% increase). This increase was most striking for the insects, with 11 taxa added and but one lost. It seems likely that this greater faunal diversity is related to the lower concentrations of solutes in Lake II.

Zooplankton standing crop was high, second only to Lake V. As in Lake I, most of this biomass (over 98%) was due to *Brachionus urceolaris*. Lake II had the greatest standing crop of benthos, nearly twice that of second-ranked Lake VI, of which about 84% was *Tendipes* sp. larvae. As in Lake I, over 95% of the total benthic standing crop consisted of herbivores. There were no fish in Lake II even though local fishermen reported having stocked it with their surplus catches from time to time. Because of its clear green water, Lake II was popular locally for swimming and water skiing until the area was closed for remining in January 1971.

Lake III — Lake III had the most complex pattern of pH variation of the six mine lakes studied. The water mass near the outlet was consistently 0.2 to 1.0 pH units lower than that near the inlet. This was the only regular horizontal difference discovered in any of the lakes. Lake III had a greater rate of flow-through than any other, and the higher downstream acidity was probably due to the leaching of acid-forming materials from a coal seam, mainly underwater, exposed in the old highwall. An acid heterograde pH curve with the minimum at intermediate depth was typical from July 1969 to August 1970. This type of pH curve is usually associated with meromixis. Lake III had a fairly well-marked trend toward higher pH during 1970, with no values less than 4.5 observed after April.

The pattern of temperature variation indicates the meromictic nature of Lake III very clearly. The high surface reading of 33.8°C in late May 1970 was the highest temperature measured in any of the lakes during the study. Temperature variation at the bottom encompassed the very small overall difference of 1.8°C, and the maxima and the minima lagged two to three months behind those at the surface. Dissolved oxygen was restricted to the uppermost 2 m during July and August 1969, leaving about two-thirds of the water mass anaerobic.

This was repeated in slightly less extreme form in the summer of 1970. Even during such extreme midsummer conditions, in contrast to Lakes I and II, oxygen at the surface rarely fell below 4.0 mg/l. The highest observed concentration was 9.9 mg/l under about 2 cm ice in late December 1970 (about 87% saturation at 8.4°C). The deepest observed penetration of dissolved oxygen was 6.0 m during the periods of partial overturn in spring and autumn.

Turbidity values were generally higher in Lake III than in the other lakes, especially in deep water. This was mainly due to the precipitation of dissolved substances (as hydroxides of iron, etc) at the interface between oxygenated and anaerobic strata. Usually, the turbidity maximum occurred at the level of this interface. During the partial overturns of spring and autumn, the surface water was colored a dark red-brown by the precipitates that resulted from an upwelling of anaerobic water with its heavy load of reduced solutes. The depth of the 1% level of light penetration typically approximated the depth of the oxygen interface.

Thus the depth of the 1% level was typically 2.0 to 4.0 m in summer and 4.5 to 6.0 m in winter. A marked decrease in turbidity in May 1970 was accompanied by increased light penetration to a 1% level of 7.0 m. This greater transparency was maintained, except for periods of turbidity from clay particles washed in by heavy rains throughout the remainder of 1970.

The contrast between the surface and bottom concentrations of solutes in Lake III was extreme. Its deep water had the highest conductivity and the greatest mean concentrations of total dissolved solids, calcium, magnesium, and total hardness of all the lakes studied. The surface water, however, had the lowest observed concentration of total solids (965 mg/l in March 1970) and ranked fifth in mean total dissolved solids, calcium, magnesium, sulfate, and total hardness (exceeding but slightly the surface values of Lake VI).

Faunal diversity was greater in Lake III than in Lake II with nine more taxa (about a 39% increase) recognized. In Lake II the zooplankton and benthos were quite unequal, contributing respectively 26% and 70% of the known taxa. In Lake III, however, the two groups were nearly equal with benthos constituting 47% and zooplankton 50%. This sharp reduction in relative diversity of the benthos probably resulted from the anaerobic bottom conditions and very steep-sided basin shape of Lake III. Some major groups, including Mollusca, Annelida and fish made their first appearance in Lake III.

The standing crop of both zooplankton and benthos was lowest in Lake III. Fish were not observed or trapped in that lake before June 1970. This may have been due to the moderation of pH and other environmental conditions as noted above. Although small (unidentified) fish were observed on two or three occasions, the only specimens taken in over 500 trap hours were seven green sunfish caught on October 30, 1970. It appears likely that colonization of Lake III by the established fish population of Lake IV will occur if the environment remains moderate.

Lake IV — Lake IV had, with the exception of a single low surface reading (4.5 under ice cover on January 26, 1970), a quite moderate range of pH variation. Apart from this unusual value, the lowest observed surface pH was 6.35. As in Lake III, the pH minimum (very rarely as low as 5.7) frequently occurred at intermediate depth.

The pattern of temperature variation was also quite similar to that in Lake III. Again there was a very slight overall annual variation at the bottom (1.9°C). Such uniformity of temperature in deep water is typical of meromictic lakes, in general, but as in the shallower lakes of this study, this may be modified by depth and transparency. Dissolved oxygen was generally greater than in the previous lakes.

Even in midsummer surface values were rarely less than about 7.0 mg/l, with oxygen penetration at least 5.0 m. The maximum observed penetration was 5.6 mg/l, at 8.0 m in December 1970. The overall maximum dissolved oxygen observed was 12.7 mg/l at 1.0 m under about 1 cm of ice in December 1969, which at 9.1°C was about 110% saturation. The highest surface concentration (11.3 mg/l in December 1970) was about 100% saturation. Lake IV was the first in the series in which 100% saturation at the surface was usual.

Turbidity was less and transparency slightly greater in Lake IV than in Lake III. Again the 1% level of light penetration was often associated with the oxygen interface.

As in Lake III, the ranges of such general indicators of dissolved substances as total dissolved solids, conductivity, and total hardness were nonoverlapping for the surface and bottom strata. For most of the major ions, Lake IV ranked fourth in surface concentration but third or even second in bottom concentration. It had the highest mean concentrations of dissolved iron, potassium, and chloride.

Forty-nine animal taxa were recognized in Lake IV, an increase of 17 species or 53% over Lake III. There was a striking increase in the number of vertebrate species, from one each in Lakes I, II, and III to a peak of seven in Lake IV. Four more plankton species, and seven more benthic species, were found in Lake IV than in Lake III.

Zooplankton biomass was greater in Lake IV than in Lakes I and III but very much less than in the other three lakes. Similarly, the biomass of benthic animals was much less than in Lakes I, II, V, and VI and only slightly greater than in Lake III. Only Lake VI exceeded Lake IV in fish biomass.

Lake V — Lake V had a range of pH variation nearly encompassing those of Lakes IV and VI. Thus a more distinct series might have been had by omitting Lake V. It has been retained in the series, however, because it is managed for sport fishing (i.e., fertilized in summer) by the Indiana Department of Natural Resources. The other lakes, in contrast, have not been disturbed since their formation except for the construction of launching ramps.

Only a moderate overall range of pH variation was observed in Lake V. The maximum vertical variation was found in late July when pH descreased from 8.2 at the surface to 6.3 at 2.0 m, then increased slightly to 6.6 at the bottom. As in Lakes III and IV, a pH minimum at intermediate depth was not unusual.

Although the overall deepwater temperature variation was greater than in Lakes III and IV, Lake V must also be considered meromictic. Some partial mixing obviously occurred at the times of overturn, but the data for dissolved oxygen and conductivity and for solutes indicate rather complete chemical separation of the surface and bottom water masses. Oxygen penetrated to the 4.0 m level in March 1970 but never reached the deepest (6.0 m) strata. The maximum surface value observed was 14.6 mg/l in June 1970 (about 190% saturation).

In September 1969 a maximum concentration of 25.7 mg/l was observed at 2.0 m, about 278% saturation. Dissolved oxygen in excess of 20 mg/l was observed at intermediate depths on several other occasions. Surface oxygen was never observed to be less than 7.0 mg/l. In contrast to the previous lakes, the highest dissolved oxygen concentrations in Lake V occurred during the warmest part of the year (April through October), probably because of massive algal blooms that were apparently a response to the fertilization of the lake.

Such blooms and the accompanying large populations of zooplankton were, in part, responsible for the high turbidity and restricted light penetration characteristic of Lake V. There was, also, a marked turbidity maximum at the lower limit of dissolved oxygen as in Lakes III and IV. It was usual for the 1% level of light penetration to occur within that boundary zone.

Again in Lake V, the general indicators of dissolved substances (total dissolved solids, conductivity, total hardness) had nonoverlapping ranges for the surface and bottom water masses. The dichotomy is less well-marked than in Lakes III and IV, primarily due to higher surface concentrations of major ions (Ca, Mg, SO_4) in Lake V.

Although the total number of animal taxa known in Lake V is the same as in IV, there are marked differences in fauna between the two. Of the 15 taxa present in Lake IV but not in Lake V, only one was replaced by a taxon found in one of the previous lakes in the series (an unidentified rotifer species also found in Lake III). Four species of fish were taken in Lake V compared to five each in Lakes IV and VI.

The standing crop of zooplankton in Lake V greatly exceeded that of any other lake. The weighted mean value of more than two million individuals per m^3 indicates a quite impressive level of production. In contrast to plankton, benthic biomass was relatively low in Lake V. Here, again, extensive areas of the bottom are anaerobic throughout most or all of the annual cycle. One curious feature of the benthos in Lake V is the apparent imbalance between the herbivores and the predators. Usually, the generalization is made that each trophic level is exceeded by the previous one upon which it feeds by a factor of about 10. This will be reflected in the biomass unless a marked difference exists in the rate of production.

Since there is no very good reason to think that the herbivore and predator benthos in these lakes differ greatly in their rates of productivity (typically one generation annually), the biomass of herbivores is expected to be roughly ten times that of predators. It can, of course, be many times greater. This expectation was reasonably well fulfilled in all the lakes except Lake V, where the predators were about five times greater in biomass than the herbivores. It seems likely that these predators, mainly immature Odanta, feed on the abundant small fish that feed in turn on the heavy crop of plankton and seek shelter in the same beds of dead leaves where the dragonfly naiads are most common. Many small (2 to 5 cm) specimens of *Lepomis* and *Micropterus* were taken in the benthic samples along with the dragonflies.

Fish biomass was quite low in Lake V. No very satisfactory explanation for this has been found. The plankton populations could certainly support very large numbers of small fish adequately. In fact, large samples of immature fish were seined easily in shallow water. It may be that the physical and chemical conditions, especially in summer when oxygen is restricted to the upper 2 m, limit the survival of more mature fish.

Lake VI — Lake VI had relatively little pH variation throughout the annual cycle. The surface water remained slightly alkaline, and there was no overlap in pH ranges for the surface and bottom water masses. Temperature variation resembled the pattern found in Lakes III through V, but was less extreme. A relatively wide temperature range (8.6°C) was measured in deep water. Dissolved oxygen varied moderately in the surface waters. The deepest stratum (below 6.0 m) was essentially anaerobic from late August to mid-October 1970, but had reasonable oxygen concentrations during most of the remainder of the year.

The maximum observed concentration was 16.8 mg/l at 5.0 m in September 1970 (about 205% saturation). This high concentration at intermediate depth was due to photosynthetic oxygen production by the dense growth of *Potamogeton* that carpeted Lake VI at all depths from about 1.0 to 6.0 m in summer and early autumn. The greatest observed open surface concentration was 11.8 mg/l in December 1970 constituting about 110% saturation. In general, Lake VI had lower turbidity and greater light penetration than any other except Lake II. Light intensity at the bottom frequently equalled or exceed 1% of surface illumination, except that in late summer and autumn the plant growth mentioned above shaded out the deepest meter or two.

Solute concentrations in Lake VI, especially in deep water, were noticeably less than in the other lakes. This was true for total dissolved solids, conductivity, and most ions. The observed ranges of surface and bottom concentrations, however, were mostly nonoverlapping, indicating rather rigid chemical stratification. Lake VI was meromictic during 1970, but the monimolimnion was essentially limited to the deepest meter. The stability of this stratification was probably much less than in Lakes III through V.

Animal diversity was greatest in Lake VI, although habitat diversity appeared to be no greater than, for example, in Lake IV. Four more taxa were found than in Lakes IV and V. The fauna of Lake VI was reasonably well balanced with 22 taxa of zooplankton, 25 of benthos, and 5 of fish. Beavers and muskrats, which occurred in Lakes IV and V, were never observed in Lake VI.

The standing crop of zooplankton in Lake VI ranked third, about 9% of that in Lake V. The weighted mean total of nearly 200,000 individuals per m^3 was much greater than those of Lakes I, III, and IV. In benthic biomass, Lake VI ranked second to Lake II. Although the standing crop was only about half as great, benthic production in Lake VI may have actually been even higher assuming fairly heavy predation by its relatively large fish populations. The biomass of fish in Lake VI was about three times that in second-ranked Lake IV. The average weight of individuals caught, however, was markedly less than in Lakes III, IV, and V.

Control Lake — The control lake had an observed pH range of 6.9 to 7.7, within the ranges of both Lakes V and VI. It must be noted, however, that this and the other data for the control are based on only two complete series of samples (June and December 1970). In June, the lake was thermally stratified and anaerobic below 4.0 m. In December, it was virtually homothermal (7.0°C at the surface gradually shading to 6.6 at the bottom) with 10.0 mg/l or more dissolved oxygen throughout. This lake probably follows the temperature regime typical of small ponds and lakes in southern Indiana. If so, it circulates twice annually in spring and autumn, with the possibility of prolonged circulation in years of mild winters with no ice cover.

Turbidity was higher in the control lake than in several of the strip mine lakes. This turbidity was partly due to very dense plankton concentrations on both sampling dates. The 1% level of light penetration was at 2.5 m in June and 4.0 in December.

The most striking limnological difference between the control and the strip mine lakes was, of course, its very such lower concentrations of dissolved substances. Even Lake VI exceeded the control by a factor greater than ten in total dissolved solids and conductivity. The control did have, however, higher concentrations of certain ions (notably Fe and Al) and SiO_2 than Lake VI.

Because of the difficulty of access and limitations of time, extensive quantitative sampling of the fauna was not undertaken. No absolute statements about either the diversity or standing crop of animals are possible for the control. A few general statements based on limited sampling and informal observation can be made. There is no reasonable doubt that the diversity of most ecological groups was greater than in any of the strip mine lakes.

A rapid survey of the plankton samples taken in June and December revealed no fewer than 25 taxa of zooplankton, more than were found by careful enumeration of 100 or more samples each (including qualitative samples) taken at all seasons in any mine lake. In the few quantitative samples taken, the number of individuals (total) was higher than in even those from Lake V. The benthic fauna appeared not to exceed Lake VI in biomass, but was probably more diverse. Fish were taken in the control lake by angling with rod and reel on several occasions. The several species introduced have become well established.

Patterns Between the Lakes: The lakes were originally arranged on the basis of increasing pH but that simple series was unexpectedly complicated by the meromictic nature of some of the lakes. Although the two factors are superimposed in nature, an attempt has been made to distinguish between the influences on the ecosystems of pH and of meromixis. The very interesting patterns of environmental stratification that often develop in a meromictic lake, illustrated by the following example drawn from Lake III, can have pronounced effects on the biocoenosis of the lake ecosystem.

The Influence of Meromixis — On the morning of July 15, 1970 the weather was pleasant (27°C, overcast, light variable breeze) at Lake III, and the green water was unusually transparent for that Lake (1% = 6.5 m). There was, in fact, no visible surface evidence of the remarkable complexity that existed in the subsurface environment. The metalimnion was shallow (1 to 3 m), and the curve of dissolved oxygen closely paralleled that of temperature, with anaerobic conditions below 3.0 m. The vertical distribution of pH was quite complex with minima at 2 to 3 m and 6 to 7 m and maxima at 0 to 1, 4 to 5 and 8 m. Turbidity increased abruptly from 6.6 mg/l at the surface to 19 mg/l at 3 m. Dissolved substances, as indicated by conductivity, increased stepwise with major steps (chemoclines) at 5 to 6 and 7 to 8 m. Thus even by these rather crude measures, environmental stratification was indeed complex.

The influence of this kind of stratification on the benthic fauna is evidenced by the results of bottom fauna sampling on the day in question. This particular example was chosen partly because the benthic grab samples included a greater range of depths than was usual in Lake III. At this time, only about 24% of the bottom area was within the aerobic zone (shallower than 3 m).

Two kinds of benthic animals were taken, both of them immature stages of Diptera. These two groups (the true midges of the family Chironomidae and the "biting midges" or "no-see-ums" of the family Heleidae) occurred in the benthos of all six lakes. Four of the six samples taken above the lower limit of dissolved oxygen (3.0 m) had at least one organism, while none of the four samples taken deeper had any macroscopic benthic animals. Densities of approximately 100 larvae per grab sample (as in samples 3 and 4) indicate a density in the lake bottom sediments on the order of 2,000 larvae per m^2. This was quite high for Lake III, but would be only a moderate density for Lake II.

Nine taxa of zooplankton organisms were taken on July 15, 1970. Most of these occurred at all depths, with greatest density at 2 m and least at 7 to 8 m. Total rotifers, for example, decreased from 10,000 individuals per m^3 at 2 m to 2,300 per m^3 at 8 m. In contrast, one rotifer (*Hexarthra* sp.) was most numerous at 8 m. Plankton animals are better able to tolerate anaerobic conditions than benthos, or so it seems from these results. The zooplankters in the deepest stratum can reach oxygen by swimming or floating vertically upward a distance of a few meters. Benthic animals, on the other hand, would have to crawl much longer distances horizontally (assuming they do not swim very readily). In fact, however, the chironomid and heleid larvae were facultatively pelagic in Lakes I, II, and III, occurring sporadically in plankton samples at various depths and times without apparent relationship to any of the environmental parameters measured. Nevertheless, these larvae were recovered from only about 9.5% of the bottom samples taken at anaerobic depths, always in relatively low densities. Thus meromixis appears to have a much more limiting effect of bottom fauna than on zooplankton.

Certain characteristics of strip mine lakes seemingly predispose them to behave in meromictic fashion. The water typically has a heavy load of dissolved substances. During summer thermal stratification, the concentration of solutes in the surface water can be decreased by dilution with rain water and the precipitation of reduced solutes by oxygen. At the same time, the dissolved substances in deep water can be increased by the addition of water with a very high concentration of solutes (leached from the cast overburden or from exposures of coal and shale in the highwall by surface and subsurface drainage) which flows downward to its density level.

Further, solubility is higher for some substances in deep water due to anaerobic conditions, and thus some precipitates (e.g., hydroxides of iron and manganese) can be redissolved when they settle below the lower limit of oxygen. If the density difference resulting from these processes becomes greater than that caused by cooling of the surface water in autumn, the chemical stratification will be perpetuated through the winter months. Inverse thermal stratification will exist until the following spring when partial overturn returns the lake to summer thermal stratification. The narrow deep basin shape (small surface area) of typical strip mine lakes and the sheltering spoil banks and highwalls combine to reduce the effect of wind mixing to a minimum. Thus even a relatively small density difference can be perpetuated.

The processes noted above (dilution, precipitation and resolution, leaching) are in operation throughout the year, continually increasing the density difference between surface and bottom strata. The stability of stratification is directly related to this difference and may be great as in Lake III, or relatively little as in Lake VI. In greater or lesser degree, however, it is clear that meromixis is not directly related to pH and therefore disturbs the simple pattern of increasing pH in the series. Hence the influence of meromixis must be taken into account in the analysis of trends that follows.

Environmental Patterns – pH: Although the original selection of the lakes and their arrangement into a series of increasing pH was based on relatively few samples, the series remained valid after 18 months of study. The most striking differences between the annual cycles of pH in the six lakes were in the vertical stratification and in the magnitude of variation with time. Vertical stratification of pH persisted throughout the period of study in the four meromictic lakes (III–VI). These vertical pH differences were greatest during summer thermal stratification, at which times Lakes I and II also exhibited pH stratification. Lakes III, IV, and V had the greatest overall surface pH variation and Lakes I and II, the greatest overall bottom variation. The magnitude of pH variation with time, like that of vertical stratification, appeared to be related more to meromixis than to relative placement on the pH scale.

Temperature: The most obvious difference between the patterns of temperature variation in the study lakes was that between Lake II, with its prolonged periods of homothermy and complete circulation, and the other five more or less permanently stratified lakes. Lake I, while thermally stratified throughout most of 1970, was essentially homothermous in mid-August and late September. This lake is probably best regarded as transitional between typical holomictic lakes (e.g., Lake II and Control) and the meromictic lakes. The overall temperature differences in the bottom strata provide a good indication of these differences in thermal patterns.

Certain anomalous thermal patterns usually associated with meromixis were observed in the study lakes. For example, dichothermous temperature distributions (the minimum temperature occurring at intermediate depth) occurred in Lakes I and II in March 1970, in Lake III from February through June 1970, in Lake IV from February to early August 1970, and in Lake VI during March and April 1970. Mesothermy (maximum temperature at intermediate depth) was observed in Lake I in August 1970, Lake IV in early November 1969 and 1970, and Lake V in April 1970. The complex temperature distribution in which one or more maxima and one or more minima occur at intermediate depth is known as poikilothermy and is very rare. Poikilothermous curves were observed in Lake III throughout November and early December 1969, and in Lake V in late October of 1969 and 1970.

Dissolved Oxygen: In the six lakes studied, both the level of dissolved oxygen and the percentage saturation increased with increasing pH. This relationship was modified by meromixis in that, with the exception of Lake VI, the deep waters of the chemically stratified lakes remained anaerobic throughout the period of study. Only Lakes II and VI had appreciable amounts of dissolved oxygen in deep water during a significant part of the annual cycle. The only lake that did not fit into the trend of increasing oxygen with increase in pH was Lake V which had unusually high levels of dissolved oxygen and also very high relative saturation. The fact that the peak occurred in Lake V (rather than Lake VI) probably resulted from the photosynthetic oxygen production by massive blooms of algae that occurred in Lake V in response to the fertilization program mentioned previously.

Thus the differences between the various patterns of dissolved oxygen variation observed in these lakes can be attributed, at least in part, to the different pH levels, to meromixis vs holomixis, and to differences in photosynthetic activity. pH probably acts indirectly, acidity increasing the total solutes (because many substances are more soluble at low pH) and leading, in turn, to increased oxygen demand by the reduced solutes.

Optical Properties: Turbidity and transparency did not appear to be related directly to pH, but were influenced by the stratification patterns. High turbidity was frequently observed in a stratum of water corresponding in depth to the lower limit of oxygen in the meromictic lakes (except VI). The two lakes with oxygen in deep water during the greater part of the annual cycle (Lakes II and VI) had strikingly lower turbidity, especially at the bottom. The control lake had quite high turbidity on the two days that it was measured. Transparency, as indicated by the 1% level, was approximately inversely related to turbidity. Lakes II and VI were the most transparent, Lakes III and V least transparent. Lake I had greater transparency (as indicated by the shallowest observed 1% level) than expected on the basis of turbidity.

Turbidity increased both above and below Lake II on the pH scale. However, this relationship was obscured by meromixis and the attendant precipitates. It seems reasonable that strip mine lakes of about pH 3.0 to 4.0 may be lowest in turbidity because below about pH 3 the solubility of iron increases abruptly (leading to turbidity from precipitates), and above that pH 4 organic turbidity of various sorts (plankton, increased decomposition, etc.) increases.

Dissolved Substances: In most natural freshwater lakes the major cations are Ca^{++}, Mg^{++}, Na^+, and K^+ and the major anions are HCO_3^-, $SO_4^=$, and $CO_3^=$ in order of decreasing concentration. The proportions are much less constant than is the case in oceanic seawater.

Perhaps the most straightforward measure of ionization in a liquid is electrical conductivity. The range of conductivity values reported for freshwater lakes is about 9 to 400 micromhos, but some saline lakes may exceed 60,000 micromhos, and certain industrial effluents are much higher. The surface conductivity values in the study lakes tended to stabilize at pH values higher than 3.5 although at different levels. These might be controlled by geological differences (i.e., differences in the composition of the coal or associated materials) or in part by differences in age. The latter seems unlikely, however, since the pH and conductivity were not well correlated with age.

In the strip mine lakes of this study, the ionic situation is complex. In typical freshwater lakes the dominant cations are considered to be Ca^{++}, Mg^{++}, Na^+, and K^+ in that order. In the acid mine lakes, iron, manganese, and aluminum tend to exceed sodium and potassium. Iron and manganese are also high in the monimolimnion of the meromictic lakes. In the surface waters of the less acid or slightly alkaline mine lakes, the usual situation (sodium and potassium in third and fourth rank) prevails. In normal lakes the dominant anions are HCO_3^-, $SO_4^=$, Cl^-, and $CO_3^=$ in decreasing order of abundance. The same ions, but in different order of abundance, occur in the mine lakes, with carbonate excluded at low pH values (less than 8.4).

In summary, certain trends in the ionic composition of the study lakes may be noted. pH influenced the concentration of several ions. Calcium, magnesium, iron, manganese, aluminum, zinc, sulfate, and chloride varied inversely with pH. Sodium and bicarbonate were directly related to pH. Ions that were markedly higher in the monimolimnion of meromictic lakes (probably due primarily to the redox situation) included calcium, magnesium, iron, manganese, sodium, sulfate, and bicarbonate. Some ions including calcium, magnesium, iron, manganese, sodium, zinc, and sulfate, were much higher than in natural freshwater lakes.

The magnitude of seasonal variation in concentrations for most of the ions measured was relatively small. The most prominent kind of trend found was an increase in concentration—especially in deep water—during maximum summer stratification. Such a pattern was observed for calcium, magnesium, iron, sodium, potassium, manganese, silica, and sulfate in at least some of the lakes, although not in Lakes I and II. No other marked seasonal trends were observed. The control lake had consistently lower values for nearly all ions than any of the mine lakes. This lake fell well within the range of water chemistry for natural freshwater lakes and it is probably representative of natural ponds and small lakes of the region.

Biological Patterns — Diversity: There was clearly an increase in overall faunal diversity with increasing pH in the lakes of this study. The rate of increase, however, was not uniform either for the total fauna or for the ecological subgroups. The overall rate of increase was greater at low pH (Lakes I through IV) than at higher pH (Lakes IV through VI). This was generally true of both zooplankton and benthos, and vertebrates were most diverse in Lake IV. The increase in benthic taxa between Lakes I and II was as great as the total additional increase from Lake II to Lake VI. This may reflect the influence of meromixis in Lakes III through VI on the conditions of life for the bottom fauna. Similarly, the greatest increase in number of zooplankton taxa occurred between Lakes II and III (nine taxa added of the total increase of 16 from Lake I to Lake VI).

It is of some interest to consider the relative importance of addition and substitution in the faunal changes that occurred through the lake series. The number of taxa carried over from one lake to the next at each step was typically about 75%, but only about half (56%) were carried over from Lake II to Lake III and nearly all (94%) from Lake III to Lake IV. The increase in faunal diversity may reasonably be regarded as primarily a result of addition of taxa. The total change from Lake I to Lake VI was from 12 to 53 taxa or an overall increase of 41 (342%). Eight of the 12 taxa found in Lake I were also present in Lake VI. These groups must be regarded as quite eurytopic with regard to pH, solutes, etc.

Standing Crop: Two general features of the overall mean biomass are of interest. First, the rank order of the lakes was different for each of the three faunal categories, no single lake dominating all others. In general Lakes II and VI were high, Lakes III and IV low, and Lakes I and V variable. Second, the magnitude of the differences between lakes was generally large, the highest exceeding the lowest by factors of approximately 8, 20, and 285 (fish, benthos, and zooplankton).

The total percentage standing crop can be calculated for purposes of comparison between lakes. The apparent fallacy that results from combining data based on numbers and weight—"adding apples and oranges"—can be avoided simply by regarding the total percentages as "fruit." On this basis, it appears that there was a trend toward increasing biomass with increasing pH. This was not so straightforward, however, as was the case with diversity. Lake II was much higher and Lakes III and IV lower than would have been expected on the basis of pH alone. The differences between lakes in the patterns on circulation are probably responsible, at least in part, for this lack of agreement between pH and biomass.

The depth distribution of zooplankton was less influenced by differences in stratification patterns than that of benthos. With the single exception of Lake I, the zooplankton was distributed remarkably uniformly throughout the water column. In Lake I the zooplankton was concentrated near the bottom for unknown reasons. Benthic biomass, on the contrary, had three distinct patterns of distribution with depth. The biomass tended to be greatest in deep water in Lake II, in the shallow regions of Lakes I, III, V, and VI, and at the surface and intermediate depths in Lake IV. More than half the total benthic biomass in Lake II was concentrated in the two deepest 1 m depth intervals, while Lakes I and VI had relatively little, and Lakes III,

Studies of Effects of Mine Drainage

IV, and V no benthos in their deepest areas. The benthic maximum between 5 and 6 m in Lake IV may have been partially due to the presence of a moderate shelf between 5 and 8 m. However, it is probably correct to view the lower limit of dissolved oxygen during maximum summer stratification as very important in restricting benthos to the shallower parts of these lakes. Thus it is not surprising to find virtually no benthic production in the deepest areas of Lakes III, IV, and V. Less readily accounted for is the low biomass below 4 m in Lake VI where dissolved oxygen was always adequate to 6 m and usually to 7 m.

The lower boundary of dissolved oxygen was also apparently important in the distribution of the planktonic ciliate *Euplotes*. Although most of the zooplankton organisms were distributed more or less uniformly as was the total, *Euplotes* tended to be concentrated at, or just below, the lower limit of oxygen where it presumably fed on the bacteria associated with that chemical boundary zone.

Zooplankton abundance was generally greatest in autumn and in spring or early summer, least in midwinter and midsummer. There were, however, exceptions such as midsummer peaks of abundance in Lakes III and V. The zooplankton biomass in Lake V fluctuated greatly during the summer of 1970, perhaps in response to changing phytoplankton abundance resulting from the artificial fertilization program. No other lake had such pronounced short-term oscillations in biomass.

Benthic biomass in all of the lakes increased through the autumn to a peak in November-December, then decreased abruptly to a low level in March. In Lakes II, III, and V the benthos remained low during the summer, but in Lakes IV and VI a secondary peak occurred in May-June. The rapid decrease in the benthic biomass in spring was associated with the emergence of insects whose aquatic larvae constituted the bulk of the benthos in all of the lakes except Lake VI (where gastropods made up over 50%).

The meromictic lakes did not have any characteristic pattern of seasonal changes in biomass distinct from that of the nonmeromictic lakes. The seasonal patterns do not appear to differ from those commonly observed in other kinds of lakes and ponds.

Some Hypotheses of Ecological Succession

This inquiry began with the idea that a series of coal strip mine lakes at different stages of recovery from acid pollution could be regarded as a series of stages in the primary succession of a single such lake. It was further suggested that such a set of lakes could be used as an analogue to test hypotheses about ecological succession in general. It has been suggested that ecological succession has the following attributes: (1) It is an orderly process of directional, and therefore predictable, community changes; (2) It results from modification of the environment by the community; (3) It culminates in the establishment of the maximum possible ecosystem stability.

Certain kinds of changes are regarded as generally characteristic of the process of ecological succession. These are here set forth in their simplest form: (1) The kinds of organisms present change continuously throughout succession;

(2) The number of taxa present increases initially, then stabilizes or declines; (3) Total biomass increases; (4) Net production decreases as a result of an increase in community respiration; and (5) Complexity of organization increases. All these changes are in the direction of increased homeostasis. This pattern occurs only in the context of a relatively stable environment. If the environment is unstable, an ecosystem will be selected that is composed of species with high reproductive potentials and broad environmental tolerances.

Before considering the data in this context, one must recall that at least three factors besides pH almost certainly have a strong influence on the strip mine lake ecosystems. The first is morphometry. The very restricted area of the shallow shore zone in most strip mine lakes may be an important factor limiting production of benthic organisms and organic production in general. Prospective spawning sites are, also, thus limited.

The second factor, meromixis, is related to morphometry in two ways. The narrow deep basin shape undoubtedly predisposes these lakes to incomplete turnover. In addition, such narrow deep lakes necessarily have a relatively small proportion of their volume in the upper circulating mixolimnion.

The third factor is the relatively low habitat diversity present in the mine lakes. Substrates present are limited almost entirely to clay, fallen leaves, and in some cases tree branches. In the two lakes where they occur in significant volume (IV and VI), rooted aquatic plants provide a substrate for some organisms.

All three factors probably depress both production and diversity in the study lakes. However, the first affects all the study lakes except I, and the third affects all six. Only meromixis has a pronounced differential occurrence—affecting primarily Lakes III, IV, and V (since the monimolimnion of Lake VI is limited to the 6 to 7 m stratum and dissolved oxygen is usually available at all depths). The rate of ecological succession is highly variable, depending on regional climate and geomorphology, kind of ecosystem, stage of the evolutionary process, and many other factors. Generally change is slow at first, then accelerated and finally slowed again in the late stages.

If the six strip mine lakes do in fact represent a series of stages in the sucessional development of an ecosystem, then some or all of the trends listed above should be present in the series. In this section the data are considered in relation to these hypothetical patterns of directional change.

The composition of the biocoenosis changes continuously throughout succession. This was certainly true of the fauna. Considerable change occurred at each step in the series including that from Lake IV to Lake V at which 14 taxa were replaced with no net change in number of taxa. Generally, the overall change in composition of the fauna was greater in the earlier stages of the series.

The number of taxa increases, especially in the early stages of succession. This, too, was generally true for the fauna. The data show a progressive increase in number of animal taxa except in Lake V. The rate of increase was greater in the early stages (i.e., acid lakes) than in the later ones.

Biomass increases during the course of succession. This was true only in a very general sense. A review of the data shows that faunal biomass was not well correlated with pH. Other factors, especially those associated with meromixis, were also important.

Net productivity decreases during the course of successions. An increasingly greater fraction of the gross productivity is used within the ecosystem as it matures leading to longer food chains and a greater standing crop of secondary, tertiary, and higher order consumers. The general increase in the proportion of predators among the benthos suggests an increase in net productivity, as does the increase in fish which are the top carnivores in those lakes where they occur. In Lake V, the increase in net productivity was apparently translated into a relatively very high biomass of benthic predators at the expense of the fish population (or perhaps a more correct view would be that other factors limited the fish and thus allowed invertebrate predators to increase).

Ecosystem organizational complexity increases during succession. This hypothesis implies such factors as increase in diversity of pigments and other biochemicals, increase in ecosystem compartments and alternative pathways, greater habitat diversity, and increase in the total amount of information in the system. One historical approach to assessing this aspect of ecosystems has been the study of the relationship between the number of taxa (biotic diversity) and the size of the area studied (mainly plant studies) or the number of individuals (mainly animal studies).

A much more comprehensive approach to the study of ecosystem complexity is that of systems ecology. In general, a systems analysis includes the identification and quantification of compartments (variables of habitat, resources, and biocoenosis) and their interconnecting pathways of material and energy transfer. An increase in the number of compartments implies more pathways, greater redundancy, increased feedback, and hence increased homeostasis and stability. In the lakes of this study, the only really significant differences in compartments appeared to be those of the biocoenosis.

That is, the environmental (habitat and resource) variables did not differ appreciably. From this it follows that the observed increase in biological compartments at various levels (i.e., the addition of major taxa such as rooted plants and fish, or of minor taxa such as additional species of rotifers or insect larvae) implies a corresponding increase in organizational complexity.

One line of evidence that bears upon this question is that of seasonal changes in biomass. If it is true that homeostasis, and hence stability, increases with increasing pH in the six lakes, then this should be expressed in terms of a sequential damping of the seasonal oscillations. An examination of the curves for zooplankton suggests that the more mature systems (Lakes IV through VI) do, in fact, have seasonal variations of somewhat less dramatic magnitude than the less mature Lakes I and II. The fertilization of Lake V in summer apparently resulted in dramatic oscillations between May and September 1970. The bottom fauna cycles are expected to be more oscillatory because of the emergence of aquatic insects at certain seasons. It appears, however, that Lake VI had much higher minima than the others. Thus, it would seem that the hypothesis of increasing organizational stability is supported, in a somewhat limited way, by the data.

Alternative Patterns of Stripland Utilization

General Utilization: The general question of the best use for abandoned coal strip mine areas has been much discussed. Of the many possible uses for old coal strip mine areas, the following would seem to be among the most practical.

Wildlife Habitat: Some areas of old strip mine land in the Patoka Fish and Game Area have been designated as wildlife areas. Even with rather extensive planting of food plots, the density of most species is extremely low. Perhaps improvements in reclamation procedures leading to increased plant growth, and hence in improved cover and food supply, would increase wildlife populations.

Farmlands: In many areas, including some of southwestern Indiana, the land stripped for coal was never very good farmland. Thus even if sources of acid pollution were covered and the topsoil restored to the surface, it would be suitable for little more than rangeland or pasture. In some areas old striplands have been restored to satisfactory crop production. Sludge from Chicago's treatment plants has been used to fertilize old striplands in Fulton County, Illinois, with striking results.

Residential: Abandoned strip mine lands can provide space for human habitation, either in the form of single-family dwellings or as multiple-dwelling complexes. This use is compatible with the above uses. While there are some very attractive farm homes, summer cabins, and permanent residences on abandoned striplands in Indiana, much greater utilization would be possible without overcrowding or despoliation of the landscape.

Recreational Areas: Some very attractive recreational sites have been developed on old strip mine areas. Such areas may include golf courses, camping and picnicking facilities, water sports areas, etc. Again, the potential for this sort of development on abandoned striplands has been little realized.

Sport Fishing: Sport fishing in the final cut lakes is quite compatible with all of these uses. Even the nonacid strip mine lakes in the study area do not usually provide very good fishing success. Some results of this study having implications for the formulation of management techniques that could improve fish production in such lakes are here briefly discussed.

Meromixis in these lakes has several implications for their successful management. Therefore, a regular schedule of limnological sampling should be undertaken on any lakes to be manipulated for sport fishing. Such a program should, at a minimum, include depth series for temperature, dissolved oxygen, pH, conductivity, and transparency. These data would indicate the presence or absence of meromixis and the gross effects of any management procedures undertaken. The two main direct effects of meromixis are: (1) Solutes become trapped in the monimolimnion. This might be desirable in the case of high concentrations of inorganic ions, but unfortunate in that of essential nutrients. (2) Dissolved oxygen is usually not present in the monimolimnion at any time of the annual cycle. Further, during summer thermal stratification the oxygenated epilimnion may be shallower than if the lake were holomictic. This confines many groups of animals to the surface strata where other factors (e.g., temperature) may be suboptimal.

Under some circumstances it would probably be possible to convert meromictic strip mine lakes to holomixis by pumping deep water up to the surface, most effectively at the time of partial overturn. If all sources of excess solutes could be stopped, a permanent change should be possible.

Basin shape is quite important to the production of benthic animals, which are, in turn, an important food source of some medium- and large-size game fishes. In the strip mine lakes, the steep-sided form of the basin restricts the shallow area of greatest benthic production. This effect can be compounded by meromixis. The concentration of bottom fauna in the shallows was especially marked in Lakes III and V, but only in Lake II was the distribution very uniform.

Thus one obvious way to improve bottom fauna production in such lakes would be to increase the extent of the shallow littoral zone. The strip mine lake most productive of fish in the Patoka Fish and Game Area is one in which the littoral was extended in the mid-1960's by construction of a low earth-fill dam. It is apparently the only strip mine lake in the region that supports fish populations approaching those of typical local nonstrip mine ponds. Whenever feasible, the use of this technique should materially improve fish production in strip mine lakes.

Fertilization, if practiced, would require careful monitoring and regulation. The resultant high algal production, such as observed in Lake V, can lead to heavy mortality among fishes and other animals because of (1) oxygen depletion during the dark period (or even on cloudy days) resulting from algal decomposition and the respiration of both algae and the dense zooplankton populations that accompany such blooms, and (2) toxic substances released by decomposing algal masses. Excessive algal production can also reinforce meromixis by adding a biogenic component to the crenogenic element already present.

One intentional effect of lake fertilization is the "shading out" of rooted aquatic plants. While excessive growth of the macrophytes may be undesirable, their complete and lasting eradication (as has been accomplished in Lake V) seriously reduces habitat diversity, which is already quite low in strip mine lakes. The rooted plants provide food as well as shelter for benthic animals, and are important in recycling nutrients from the bottom sediments.

RECOVERY OF ACID STRIP MINE LAKES

The material in this chapter is excerpted from:

PB 219 264

The literature abounds with references confirming the formation of acid mine drainage from the oxidation of pyrites and marcasites exposed during coal strip mining. The accumulation of these oxidation products along with their by-products formed from chemical interaction with the accompanying overburden collect in natural basins and constitute acid strip mine lakes.

Campbell and Lind (1) have noted that strip mine lakes recover naturally with time. Through a series of successional stages the acid and sulfate concentrations are gradually reduced, with concomitant reduction in the concentrations of the dominant cations; Ca, Mg, Al, and Fe. Such changes must precede the establishment of bicarbonate alkalinity, occurring in a later stage.

Decker (2) found that the addition of organic matter in the form of primary sewage sludge greatly accelerated the recovery process. In a matter of days the acidity, sulfate, and iron were drastically reduced appearing to follow the pattern outlined by Campbell and Lind for natural recovery. From such results, it was postulated that the factor limiting the rate of natural recovery of acid strip mine lakes was the accumulation of allochthonous organic materials.

These materials provide the organic carbon source needed by the sulfate-reducing bacteria whose metabolic processes derive energy from the reduction of sulfate. Sulfate-reducing bacteria release sulfide ions as a waste product. Decker further concluded that upon release sulfide ions combined immediately with two hydrogen ions which under low pH conditions were given off to the atmosphere as H_2S gas constituting the physical transport mechanism responsible for acid reduction in the natural and accelerated recovery of strip mine waters.

From the addition of a great variety of organic materials Ogg (3) concluded that the sulfate-reducing bacteria can utilize any biodegradable organic carbon sufficiently high organic carbon loading rates are required to achieve anaerobiosis.

Each of the above researchers observed a general decrease in each metal concentration as the recovery process progressed, however little more than empirical recordings were made of the phenomenon. Thus, it was the aim of the study described in PB 219 264 to determine the role of metals from the time of acid formation on the pyrite crystal to the time of lake recovery.

METHODS AND MATERIALS

This study was divided into three investigations. Investigation I was undertaken to determine if any iron-clay-aluminum interaction could be noted in naturally recovering strip mine lakes. Each of twelve lakes, located about ten miles northeast of Columbia, Missouri, were analyzed for pH, acidity, sulfate, iron and aluminum.

In Investigation II an apparatus, schematically outlined in Figure 9.1, was designed to maximize the clay mine water interaction by exposing a large clay surface area to a small unit volume of mine water. A flow rate of one drop of mine water every ten seconds to both the empty control tube and the tube containing clay was regulated by a series of Tygon tubes, Y-sections, and screw-pinch clamps.

The Pyrex tube containing clay was one inch in diameter, three feet long, and contained 130 grams of clay evenly distributed throughout the tube. The Pyrex control tube was of the same dimensions, but contained no clay. Its inclusion in the apparatus design allowed evaluation of any reactions between the mine water, Pyrex tube, and the atmosphere. As the mine water flowed out of the tubes it was collected and immediately analyzed for pH and acidity. An aliquot of each sample was acidified with a few drops of concentrated nitric acid and stored in glass vials prior to analysis for aluminum and iron by atomic absorption.

FIGURE 9.1: CLAY BUFFER APPARATUS: INVESTIGATION II

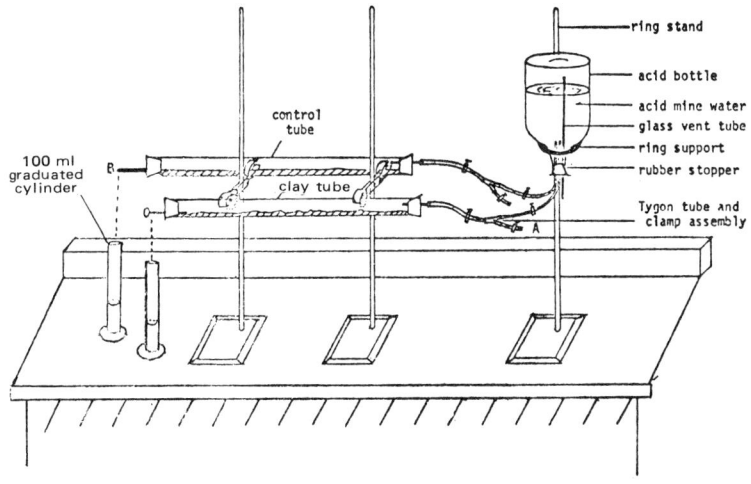

Source: PB 219 264

Each clay mine water reaction was continued for 12 hours with samples being taken approximately every half hour. Samples taken at Point A (Figure 9.1) were designated initial samples, those at Point B as control samples, and those at Point C as clay samples.

The acid mine water used to challenge the clay minerals was collected as drainage from a washed out tipple pond, located 11 miles northeast of Columbia, Missouri. Chemical characterization of mine water was always undertaken prior to use. Clay minerals used in this investigation included kaolinite, gibbsite, montmorillonite, and aluminum oxide.

In order to determine the role of iron and aluminum during the accelerated recovery process a segment of Decker's research was duplicated during Investigation III. One liter of fresh precharacterized primary sewage sludge was added to nineteen liters of precharacterized acid mine water in a 20 liter Pyrex carboy.

Anaerobic conditions, necessary for efficient sulfate reduction were ensured by purging the carboy with nitrogen for 12 hours after the addition of the sludge. Another 12 hour period permitted attainment of equilibria after which sampling of the system began on a regular basis, about every two days. Acidity, pH, and phosphorus were analyzed immediately, while sample aliquots were again acidified and stored for aluminum and iron determinations.

The method used for the analysis of total acidity was an ambient temperature titration with standard sodium hydroxide to a final pH of 8.30. Both pH and acidity were analyzed under nitrogen using a Fisher Accumet 320 pH meter. Iron and aluminum were analyzed on a Perkin-Elmer 403 Atomic Absorption Spectrophotometer. Sulfate determinations were performed on a Coleman:Hatachi 124 Spectrophotometer using a 1.0 centimeter cell width.

Total phosphorus and orthophosphorus were analyzed at a wavelength of 740 nm passing through a 1.0 centimeter cell on a Beckman DB Spectrophotometer. Sewage sludge required grinding in a glass tissue grinder prior to dilution.

Total organic carbon was determined on a Model 315 Beckman Infrared Carbonaceous Analyzer. Prior to the appropriate dilution the raw sewage sludge was finely ground in a glass tissue grinder, while mine water samples could be directly injected. Fresh primary sewage sludge was obtained from the city of Columbia, Missouri sewage treatment works on August 7, 1972.

RESULTS AND DISCUSSION

In the presence of air and water pyrite oxidizes to soluble ferrous iron and sulfuric acid. Various catabolic bacterial mechanisms appear to be associated with this reaction. Stumm (4) and Singer (5) have studied the oxidation of pyrite in great detail. They propose the following reaction scheme:

(1) $FeS_2(s) + 7/2\, O_2 + H_2O \rightleftharpoons Fe^{++} + 2SO_4^= + 2H^+$

(2) $Fe^{++} + 1/4\, O_2 + H^+ \rightleftharpoons Fe^{+++} + 1/2\, H_2O$

(3) $Fe^{+++} + 3H_2O \rightleftharpoons Fe(OH)_3(s) + 3H^+$

The overall reaction is:

(4) $FeS_2(s) + 15/4\ O_2 + 7/2\ H_2O \rightleftharpoons Fe(OH)_3(s) + 2SO_4^= + 4H^+$

The first step is the physical oxidation of the pyrite by atmospheric oxygen to ferrous iron, reaction (1). This reaction also produces two equivalents of hydrogen and sulfate ions. As soon as the ferrous iron is formed it is oxidized to the ferric form, utilizing one equivalent of acidity, reaction (2). Hydrolysis immediately results in the precipitation of ferric hydroxide and the release of three equivalents of acidity, reaction (3). The net result is the formation of 4.0 equivalents of acidity per mol pyrite oxidized, or 2.0 acidity equivalents per mol sulfate produced.

The ferric iron formed in reaction (2) may interact with pyrite forming more sulfuric acid, reaction (5).

(5) $FeS_2(s) + 14Fe^{+++} + 8H_2O \rightleftharpoons 15Fe^{++} + 2SO_4^= + 16H^+$

Sampling deep coal mine drainage Singer verified his proposed theoretical reaction sequence for the oxidation of pyrite by finding two equivalents of acid per mol sulfate. For further proof he sampled strip mine lakes and drainage from deep mines that had flowed over surface clays. There he found only 0.9 to 1.3 equivalents of acid per mol of sulfate. He also observed an acid reduction associated with the interaction of acid mine water with alumina (Al_2O_3). Although he presented no data he concluded that the acidity reduction was the direct effect of a clay buffer.

Campbell and Lind recorded 1.53 to 0.25 equivalents of acidity per mol sulfate present in three naturally recovering strip mine lakes (Table 9.1). This acidity to sulfate ratio indicated an acidity loss ranging from 0.47 to 1.75 equivalents when compared to the theoretical 2.0 value. Also observed in their acid lakes were unusually high concentrations of aluminum, calcium, magnesium, and manganese.

These field studies suggested an acidity reduction associated with the interaction of clay minerals and acid mine water. With Singer's postulate about the buffering capacity of clays it became apparent that Campbell's high aluminum values could be associated with this clay-acid neutralization.

TABLE 9.1: NATURALLY RECOVERING ACID STRIP MINE LAKES FIELD

Lake	pH	Acidity (mg $CaCO_3$/l)	Iron (mg/l)	Aluminum (mg/l)	Sulfate (mg/l)	$H^+/SO_4^=$ (meq/mmole)
A_1	2.9	2660	182	212	3340	1.53
A_2	3.2	615	15.6	129	1260	0.94
A_3	3.6	35	1.1	2.6	271	0.25

Source: PB 219 264

Investigation I

The field survey conducted in Investigation I recorded an iron, aluminum, and sulfate decrease as the pH increased throughout the recovery process (Table 9.2). These data agree very well with that recorded by Campbell and Lind. The data sets presented in Tables 9.1 and 9.2 have been combined in Figures 9.2 and 9.3. In Figure 9.3 the sulfate concentration increased as a function of acidity, however, the theoretical 2.0 equivalents of acid per mol sulfate were never observed. Those lakes with extremely low acidity to sulfate ratios may have never had a great deal of acid production in their water shed.

TABLE 9.2: NATURALLY RECOVERING ACID STRIP MINE LAKES FIELD DATA

Lake	pH	Acidity (mg $CaCO_3$/l)	Iron (mg/l)	Aluminum (mg/l)	Sulfate (mg/l)	$H^+/SO_4^=$ (meq/mmole)
L-1	2.70	641	27.4	73.1	1700	0.61
L-2	2.51	2232	201	2.8	3080	1.39
L-3	2.25	9810	3990	2.1	15000	1.26
L-4	3.30	51.5	3.1	2.6	356	0.38
L-5	2.65	480	24.3	45.4	1570	0.59
L-6	3.10	238	9.7	24.0	1580	0.29
L-7	4.30	24.5	0.25	1.5	792	0.61
L-8	7.09	12.5	0.13	0.30	372	0.64
L-9	7.37	8.53	0.16	0.22	1100	0.02
L-10	7.31	8.51	0.05	0.20	1150	0.01
L-11	7.40	8.52	0.13	0.21	394	0.42
L-12	4.46	52.6	0.17	3.55	1250	0.80

Source: PB 219 264

FIGURE 9.2: IRON AND ALUMINUM CONCENTRATIONS VERSUS pH

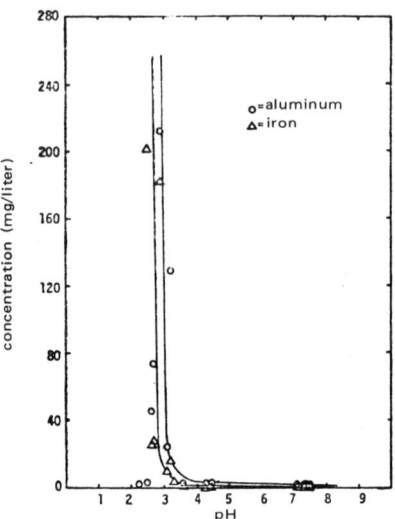

Source: PB 219 264

FIGURE 9.3: HYDROGEN ION TO SULFATE RATIO VERSUS pH

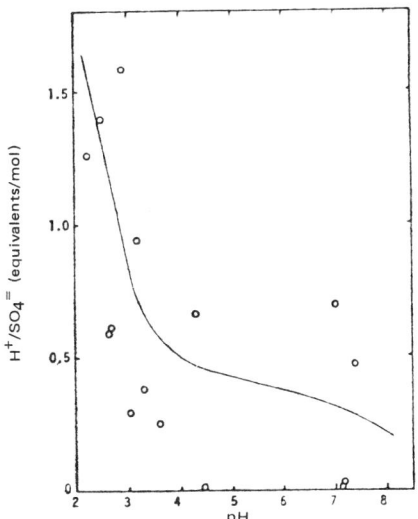

FIGURE 9.4: LOG IRON AND ALUMINUM CONCENTRATIONS VERSUS pH

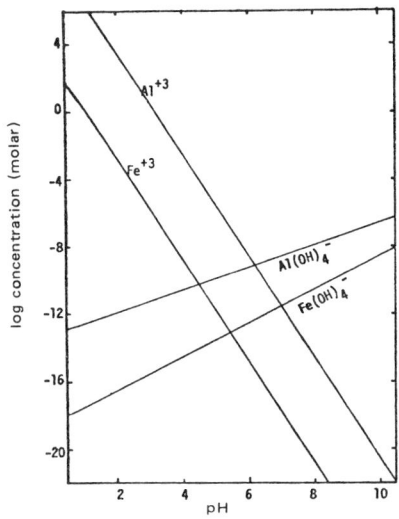

Source: PB 219 264

From Figure 9.4 it can be seen that ferric hydroxide is four orders of magnitude less soluble than aluminum hydroxide. Thus, it would be expected that iron would precipitate before aluminum, as observed from Figure 9.2, in the naturally recovering lakes. Figure 9.4 represents ideal laboratory equilibria K_{sp} data taken at 25°C at zero ionic concentration. This is not the case in the natural system and although the same solubility trends are observed between iron and aluminum hydroxides the higher pH where precipitation occurs naturally is probably due to metal and ligand complexing agents.

The representation of the fate of all iron to be ferric hydroxide and all aluminum to be aluminum hydroxide drastically oversimplifies the natural system. Oxides and oxyhydroxides of both Fe(II) and Fe(III) along with possible polynucleic aluminum complexes must play a definite role in this extremely complex chemical system.

Clay minerals play a critical role in the formation of acid strip mine lakes. In the presence of sulfuric acid and water kaolinite clay dissolves according to reaction (6).

(6) $\quad 6H^+ + Al_2Si_2O_5(OH)_4(s) \rightleftharpoons 2Al^{+++} + 2H_4SiO_4 + H_2O$

Six equivalents of acidity, generated from the oxidation of pyrite, are exchanged for two aluminum (III) species, silicic acid, and water. Because the oxidation products of pyrite are in intimate contact with the clay minerals in the spoil bank reactions (4) and (6) can be combined.

(7) $\quad FeS_2(s) + Al_2Si_2O_5(OH)_4(s) + 2H^+ + 15/4\, O_2 + 5/2\, H_2O \rightleftharpoons$
$\qquad\qquad 2Al^{+++} + Fe(OH)_3(s) + 2H_4SiO_4 + 2SO_4^=$

The actual contribution of silicic acid to this system has not been ascertained, however, it will be assumed that the deprotonation associated with dissociation of this acid will have a small effect on the net lake acidity. With this assumption a pH decrease would be expected from reaction (6), but a total titratable acidity change would not, since both $2Al^{+++}$ and $6H^+$ require six equivalents of hydroxide ions for neutralization. Some of the dissolved aluminum will hydrolyze either on the spoil bank or through the slow sequential hydrolysis and precipitation as gibbsite or bayerite.

In either case the acid associated with the aluminum will be felt during the recovery process, acting as another buffer in the mine water. The net sulfate concentration, as seen by reaction (7), is not affected and may be considered an indicator of net acid production in the watershed if natural sources of sulfate, such as gypsum ($CaSO_4$) are meagerly dispersed in the overburden.

Even though some of the clay dissolves and is responsible for the high aluminum concentration in the strip mine lake water, this is only a minor contribution to the net effect produced by the clay minerals. As the pyrite is oxidized the release of sulfuric acid greatly accelerates the natural weathering of these clay minerals.

It is probable that the majority of acid lost to the clays, though not directly associated with the complete dissociation of the clay minerals, is sorbed onto the many interlayer negative crystalline sites characteristic of all clay minerals. Such a sorption combined with ion exchange will not be associated with the release of any ionic species, but will result in the natural succession of feldspar bearing minerals to kaolinite, to montmorillonite, and finally to gibbsite. Such a process is characteristic of the geologic weathering of rocks and minerals and is believed to be responsible for the acidity loss associated with the interaction of acid mine water and clay minerals.

Investigation II

Investigation II was designed to determine the "buffering capacity" associated with various clay minerals and aluminum oxide. The designed apparatus (Figure 9.1) maximized the clay particle surface area per volume of mine water and permitted observation of the relative buffering capacity. Table 9.3 indicates the chemical characteristics of the mine water for each determination.

TABLE 9.3: CHARACTERIZATION OF ACID MINE WATER

Clay Mineral Used in Analysis	pH	Acidity (mg $CaCO_3$/l)	Iron (mg/l)	Aluminum (mg/l)
Gibbsite (Bauxite, Ark.)	2.51	6085	3180	117.8
Kaolinite	2.48	7034	2900	186.9
Montmorillonite	2.59	6623	3280	122.8
Aluminum Oxide	2.66	6083	3500	134.8

Source: PB 219 264

Montmorillonite, gibbsite, kaolinite, and aluminum oxide all produced the same trends by decreasing the acidity and increasing the pH (Figures 9.5 and 9.6). Montmorillonite, gibbsite, and aluminum oxide reduced both the iron and aluminum (Figures 9.7 and 9.8). The well defined crystalline nature, large particle size, and low cation exchange capacity of kaolinite are believed responsible for the aluminum increase and lack of iron decrease seen in Figures 9.7 and 9.8.

The relative amounts of pH increase, acidity, iron, and aluminum decrease are believed to be the result of the chemical nature of each clay mineral as indicated by their relative ion exchange capacities; montmorillonite 80 to 150 meq per 100 grams clay, illite 60 to 40 meq per 100 grams, and kaolinite 1 to 15 meq per 100 grams. Gibbsite and aluminum oxide are not considered clay minerals, but were found to have strong acid sorbing qualities. According to the recorded data aluminum oxide would have an ion exchange capacity equal to or greater than montmorillonite, while gibbsite would be ranked between montmorillonite and kaolinite in relative ion exchange capabilities.

From the above investigation the "buffer capacity" associated with the reduction of acid in mine water is most definitely a function of the clay with which the acid water comes into contact.

FIGURE 9.5: CHANGE IN ΔpH WITH RESPECT TO TIME

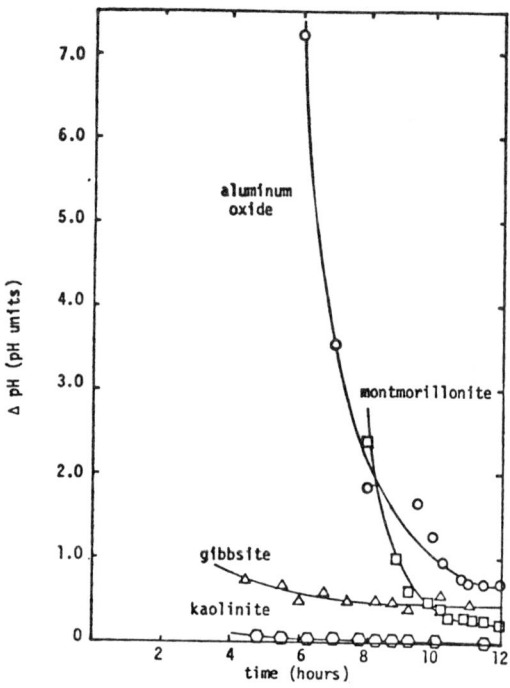

FIGURE 9.6: CHANGE IN Δ ACIDITY WITH RESPECT TO TIME

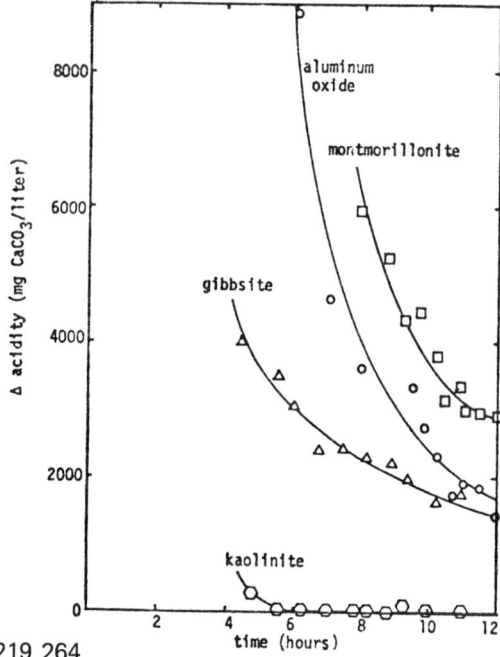

Source: PB 219 264

FIGURE 9.7: CHANGE IN Δ IRON WITH RESPECT TO TIME

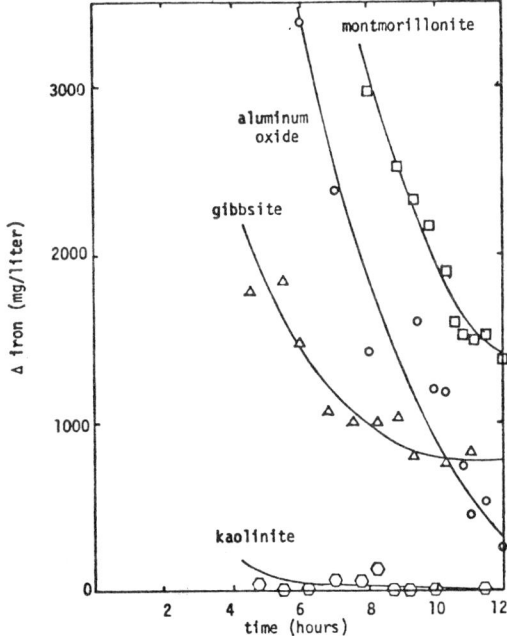

FIGURE 9.8: CHANGE IN Δ ALUMINUM WITH RESPECT TO TIME

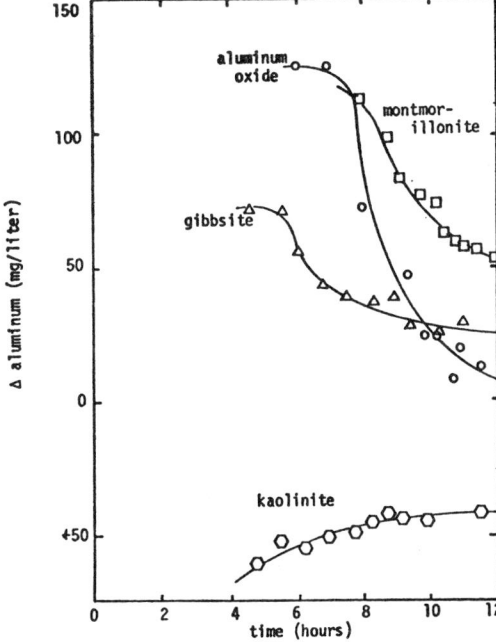

Source: PB 219 264

Each clay mineral differs in its ability to sorb ions due to the different surface charge density, number of exchange sites per area, and specific surface area, all of which play a role in determining the chemical properties of each particular clay. Such chemical properties are the result of a complex interaction determined by concentration, exchange positions, ionic nature, composition of solution, and temperature.

With such interrelated complex properties associated with clay minerals, iron and aluminum removal mechanisms are probably associated with cation exchange, precipitation as hydroxides and oxides, and possible entrapment in the gel that tends to form when water and clay combine. Complexes with other cations, silica, and/or ligands may also play a role. The primary removal mechanism of metals or acidity has not been elucidated in this study. However, it is believed that the data obtained from these laboratory studies were the result of the same mechanisms that function as "clay buffers" on the natural clay spoil banks.

Singer has mentioned the catalytic properties of clay particles in the oxidation of ferrous to ferric iron. Such a catalysis may replace or combine with aeration as the main iron oxidizing agent in the natural "clay-buffering" process. However, this is most likely only a transitional reaction leading to the final metal and acid removal initiated by the clay minerals.

Besides being lost to the clay, whether by sorption or dissolution, some of the acid is also lost in the dissolution of alkali metal compounds commonly associated with strip mine overburden.

(8) $CaCO_3(s) + 2H^+ \rightleftharpoons Ca^{++} + CO_2(g) + H_2O$

(9) $CaCO_3(s) + H^+ \rightleftharpoons Ca^{++} + HCO_3^-$

(10) $MgCa(CO_3)_2 + 4H^+ \rightleftharpoons Mg^{++} + Ca^{++} + 2CO_2(g) + 2H_2O$

(11) $MnCO_3 + 2H^+ \rightleftharpoons Mn^{++} + CO_2(g) + H_2O$

As these compounds are dissolved the water chemistry of the acid mine water changes. Ions such as Ca^{++}, Mg^{++}, Mn^{++}, Na^+, and K^+, replace the hydrogen ions. Such cations are not titratable in the pH range used in a normal acidity analysis and the result is an increase in pH and decrease in acidity. High Ca^{++}, Mn^{++}, and Mg^{++} concentrations, characteristic of acid strip mine lakes, have been verified by Campbell and Lind along with the presence of K^+ and Na^+.

Titration of 50 ml of mine water from the same lake Decker and Ogg used in their studies appears in Figure 9.9. The slopes of the line in the pH range 3.5 to 4.5 and 7.4 to 8.2 are indicative of two apparent buffers. The buffer intensity curve, shown in Figure 9.10, was calculated from Deckers Phase III measured sulfate concentrations and the assumption that natural mine water was in equilibrium with the carbon dioxide in the air (16 micromoles CO_2/liter).

Figure 9.10 shows that in naturally recovering acid strip mine lakes the sulfuric acid and atmospheric CO_2 buffers play a role in the total buffering of the water, but are not responsible for the majority of the buffer intensity in acid mine water. Figure 9.9 vividly shows the large discrepancy between the buffer intensity of these two buffer systems. The titration curve calculated from the buffer intensity curve closely approximates the titration curve of a 50 ml solution of

FIGURE 9.9: TITRATION CURVES

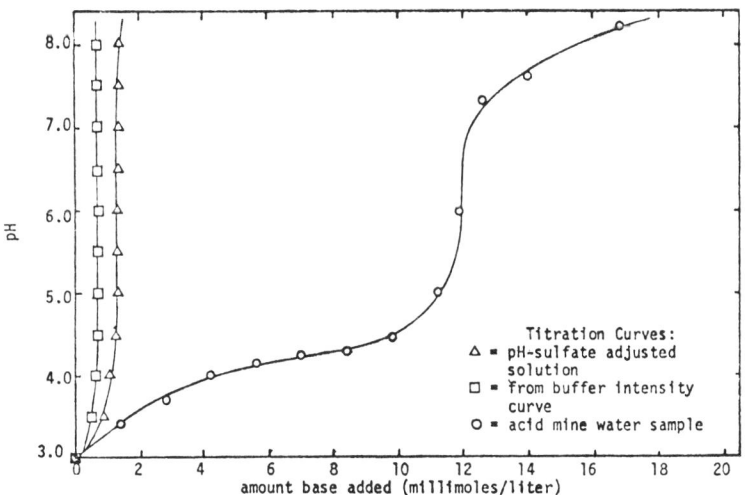

FIGURE 9.10: BUFFER INTENSITY CURVE FOR NATURAL ACID MINE WATER

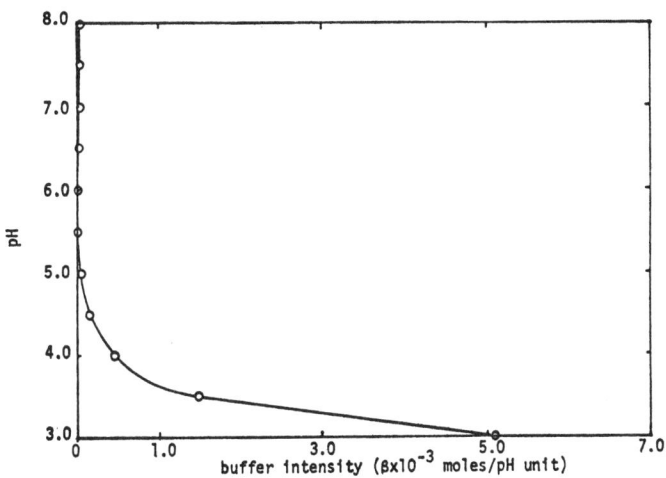

Source: PB 219 264

distilled water adjusted to a pH of 3.0 with concentrated sulfuric acid and titrated with a standard base (NaOH, 0.106 N). The dilute sulfuric acid titration curve accurately verifies the validity of the assumptions and the conversion of buffer intensity to millimoles base per liter.

The major buffer in the pH range 3.5 to 4.5 as indicated by the titration curve in Figure 9.9 is believed to be the high concentration of aluminum and iron. Iron would play only a minor role in the pH range of 3.2 to 3.4 because of its presence in such low concentrations resulting from a low K_{sp}. Aluminum, however, is very soluble, compared to iron, in the pH range of 3.5 to 4.5 where it may have been hydrated once, releasing between one-third to one-half of its total potential acidity. But, aluminum can still act as a strong buffer as the remaining one-half to two-thirds of its acidity is gradually released to the system through hydrolysis as the pH rises.

Investigation III

In Investigation III Decker's Phase III was duplicated in order to determine the role of metals during the accelerated recovery process. In this 88 day accelerated recovery, reducing conditions were artificially produced by a twelve hour nitrogen purge beginning immediately after the sewage sludge addition. This eliminated that amount of lag time required for the establishment of anaerobic conditions by methane forming bacteria.

The sewage sludge used had a pH of 5.80 and contained 337 mg/l iron, 280 mg/l aluminum, 835 mg/l phosphorus and 11,160 mg/l carbon. The mine water used had a pH of 2.78, an acidity as $CaCO_3$ of 656 mg/l, 26 mg/l iron, 81 mg/l aluminum, 0.05 mg/l phosphorus and 2.60 mg/l carbon. Acidity and pH (Figure 9.11) followed the same pattern as Decker recorded during his study.

The aluminum concentration (Figure 9.12) began decreasing immediately and was essentially completely removed by the time of complete recovery. Phosphorus was observed to be continually released from the sludge until the concentration increased to 1.17 mg/l after fifteen days. After this time it decreased rapidly until complete removal at 56 days. The sporadic iron values early in the study, like the increasing phosphorus concentrations, indicated the slow attainment of equilibrium between the sludge and the acid mine water. By the end of this investigation the iron concentration had been reduced to 1.0 mg/l.

During this recovery process the incremental pH changes caused the K_{sp} values of the various ionic species to be exceeded. In this study aluminum was the first observed species to precipitate. In Figure 9.13, the plot of Investigation III aluminum concentrations versus pH, closely follows the equilibrium data for aluminum hydroxide obtained from a K_{sp} value of $10^{-32.6}$. This strongly suggests that the aluminum is precipitating as a hydroxide. The plot of phosphorus decrease closely paralleled the decreasing aluminum curve (Figure 9.12). This might indicate that most of the phosphorus is precipitating from solution as aluminum phosphate.

Iron was not thought to play a part in the precipitation of phosphorus because the iron concentration remained constant as the aluminum and phosphorus curves decreased. Bacterial growth requirements also may be responsible for a portion of the net phosphorus removal. Iron precipitated later in the recovery process

FIGURE 9.11: pH AND ACIDITY VERSUS TIME

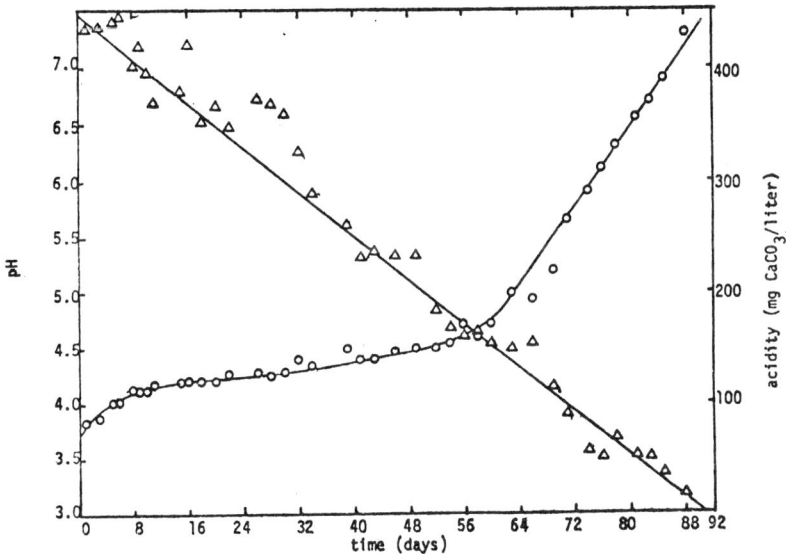

FIGURE 9.12: ALUMINUM, IRON, AND PHOSPHORUS CONCENTRATIONS VERSUS TIME

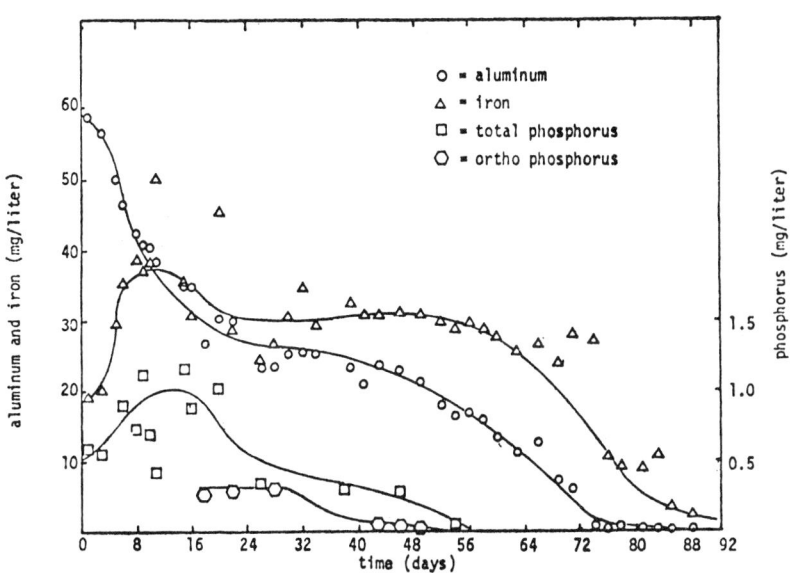

Source: PB 219 264

FIGURE 9.13: ALUMINUM CONCENTRATIONS VERSUS pH

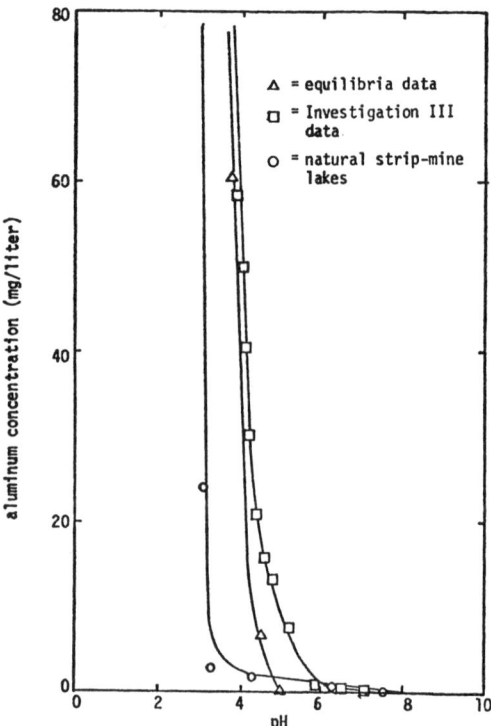

Source: PB 219 264

than did aluminum and phosphorus. The form of this precipitate was probably an iron sulfide. Seventy-two days into the recovery process the entire carboy turned black. A drastic iron decrease was associated with the formation of this black finely divided precipitate.

The mechanism responsible for the acidity loss was the reduction of sulfate by sulfate reducing bacteria followed by the release of hydrogen sulfide gas to the atmosphere. As the acidity was lost by physical transport to the atmosphere the pH very slowly rose until after the fifty-sixth day when it increased in direct proportion to the acidity decrease. This mechanism may be ideally represented by the following reactions.

(12) $2HSO_4^- \rightleftharpoons 2SO_4^= + 2H^+$

(13) $SO_4^= + 2C(org.) \xrightarrow{Bacteria} S^= + 2CO_2(g)$

(14) $2H^+ + S^= \rightleftharpoons H_2S(g)$

The overall reaction is:

(15) $2HSO_4^- + 2C(org.) \rightleftharpoons SO_4^= + H_2S(g) + 2CO_2(g)$

The incremental pH change, due to the bacterial reduction of sulfate (Figure 9.11), can be compared to an acid-base titration curve in which the acid, instead of being base neutralized, is reduced by combination with sulfide ions and released to the atmosphere as H_2S gas. Figure 9.14 displays the converted pH curve from Figure 9.11 in units of millimoles of standard base. It has been assumed in the construction of Figure 9.14 that the bacterial reduction of sulfate was quantitatively associated with any acidity loss.

Under these accelerated conditions carbon dioxide and hydrogen sulfide gases are produced in vast quantities. Figures 9.15 and 9.16 present data from Decker's Phase III study and indicate the relative amounts of CO_2 and H_2S present as the pH increased during the recovery process.

In Figure 9.16 it was assumed that the system remained saturated with CO_2 gas after a concentration of 1.35 meq carbon/liter (16.2 mg inorganic carbon/liter) had accumulated at a pH of 4.3. This is indicated by the straight line on the graph. Any increase in inorganic carbon above a pH of 4.3 was believed to be the result of the dissociation of carbonic acid to bicarbonate.

Knowing the equilibrium constants for sulfuric acid, carbonic acid, and aqueous dihydrogen sulfide a buffer intensity curve (Figure 9.17) based on sulfate, sulfide and inorganic carbon concentrations was constructed for the accelerated recovery process. The contribution of each ionic buffering species to the net buffer intensity at any pH is graphically portrayed by the ionization fraction plot appearing in Figure 9.18.

FIGURE 9.14: TITRATION CURVE FOR ACCELERATED RECOVERING ACID STRIP MINE LAKE WATER

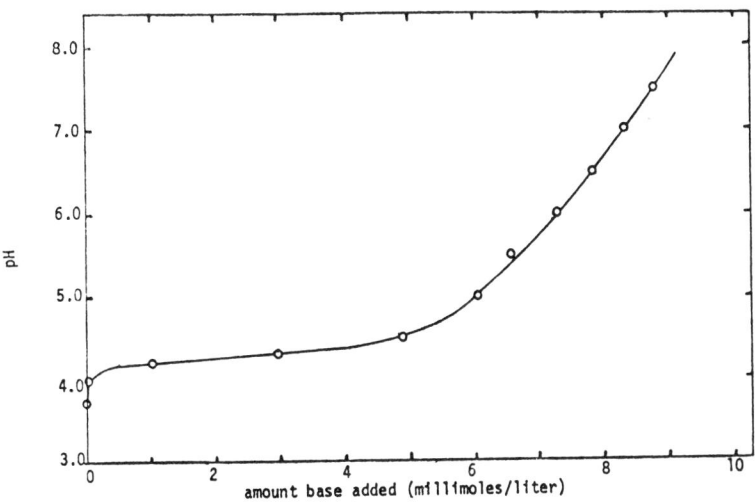

Source: PB 219 264

FIGURE 9.15: SULFIDE CONCENTRATION AS A FUNCTION OF pH

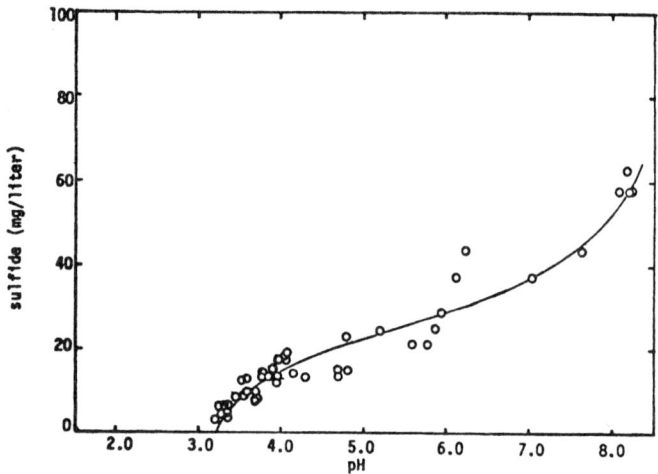

FIGURE 9.16: INORGANIC CARBON CONCENTRATION AS A FUNCTION OF pH

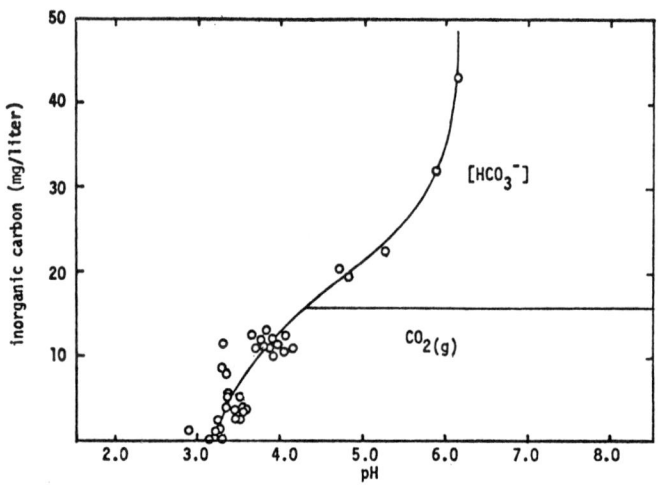

Source: PB 219 264

Recovery of Acid Strip Mine Lakes 217

FIGURE 9.17: BUFFER INTENSITY CURVE FOR ACCELERATED ACID MINE WATER

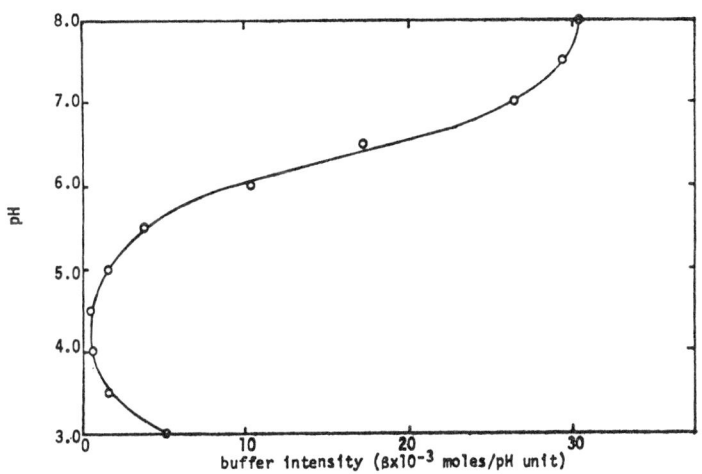

FIGURE 9.18: IONIZATION FRACTION PLOT

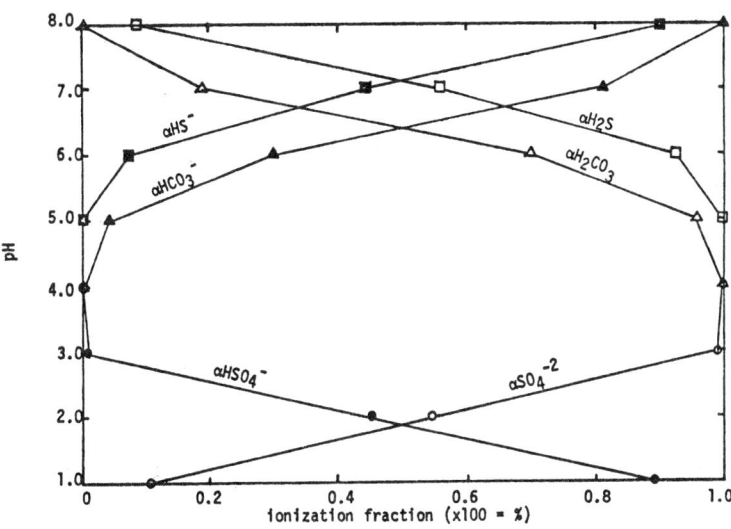

Source: PB 219 264

FIGURE 9.19: TITRATION CURVES COMPARISON

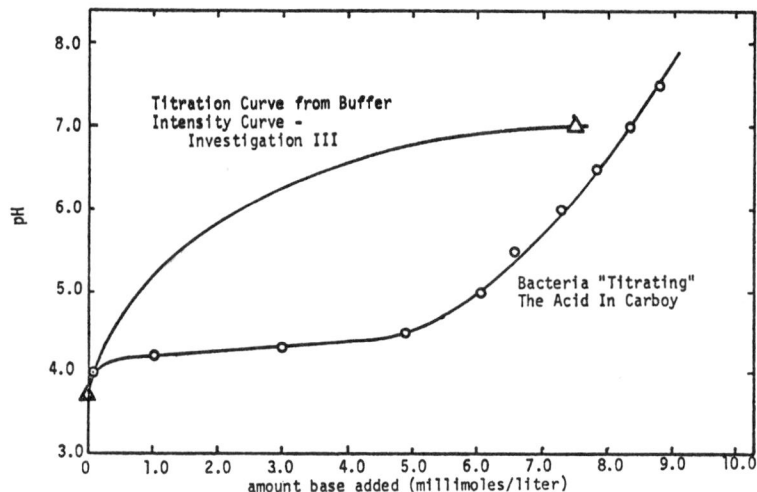

Source: PB 219 264

Figure 9.19 compares the curve shown in Figure 9.14 with a titration curve calculated from Figure 9.17. Again a discrepancy between the curve calculated from Figure 9.17 and the observed results recorded as pH change in Figure 9.14 is obvious. Even the great buffering influence that such high concentrations of CO_2 and H_2S gases exert on the system cannot be responsible for the total buffering effect. Aluminum and other metals, acting as buffers, are believed to be responsible for the increased buffer intensity.

The difference between the natural and accelerated systems may be easily observed by comparing the buffer intensity plots in Figures 9.10 and 9.17, or the titration curves calculated from buffer intensity curves plotted in Figures 9.9 and 9.19. The water associated with the accelerated recovery process has a much greater buffer capacity than that of naturally recovering lakes. This is due to the different amounts of alkalinity formed in each system. The amount of alkalinity in either system is in direct proportion to the amount of $CO_{2\,free}$ present in the water.

In the accelerated system the alkalinity at pH 7.0 was measured to be 7.5 meq per liter or about 1.5 millimols $CO_{2\,free}$ per liter. Every milliequivalent of alkalinity present represents another milliequivalent of acid lost and is equal to the corresponding reduction of one milliequivalent of sulfate (Reaction 16).

$$(16) \quad SO_4^= + 2C(org.) + 2H_2O \rightleftharpoons H_2S(g) + 2HCO_3^-$$

The high $CO_{2\,free}$ concentration, responsible for high alkalinity, was allowed to build up in the accelerated system since mixing and exposure to the atmosphere were minimal. A rapid constant rate of CO_2 production from bacterial respiration coincides with this high alkalinity and was only possible by providing a plentiful organic carbon source and anaerobic conditions. The result of such a high alkalinity was an increase in the amount and rate of acid loss in the accelerated system due to the preferential buffering of bicarbonate ions. The rate at which the bacteria are able to reduce sulfate is very slow under low pH conditions. This can be seen from Figure 9.20 in which the rate of sulfate reduction is expressed as a function of pH from Decker's Phase III research. Decker's sulfate determinations showed that to raise the pH from 3.2 to 4.2 took 35 days, whereas, from 4.2 to 5.2 required 10 days, and from 5.2 to 6.2 required only 6 days.

In the pH range of 4.5 to 5.0 the bacteria are no longer limited by harsh acidic conditions and sulfate reduction occurs at a constant maximum rate. As the growth rate increases, the concentrations of CO_2 and H_2S gases along with pH also increase rapidly. Beyond a pH of 6.4, which is the pK_1 for carbonic acid, dissociation favors rapid formation of bicarbonate and hydrogen ions.

(17) $CO_2 + H_2O \rightleftharpoons H_2CO_3 \rightleftharpoons H^+ + HCO_3^-$

This buffering system releases hydrogen ions and tends to maintain the pH near 6.4. Since the pK_1 of the H_2S buffer system is 7.0, almost all the sulfide release by the bacteria will be immediately exported as $H_2S(g)$ from the system. This preferential buffering of the bicarbonate can be seen in Figure 9.21. As the pH is further increased the bicarbonate buffer will be reduced and $H_2S(aq)$ will begin to dissociate. At this pH the system loses efficiency in the export of acid, for now a much greater $H_2S(g)$ concentration must build up in order to export the same amount of $H_2S(g)$ as was released at lower pH values.

(18) $H_2S(aq) \rightleftharpoons 2H^+ + HS^-$

FIGURE 9.20: CHANGE IN Δ SULFATE WITH RESPECT TO pH

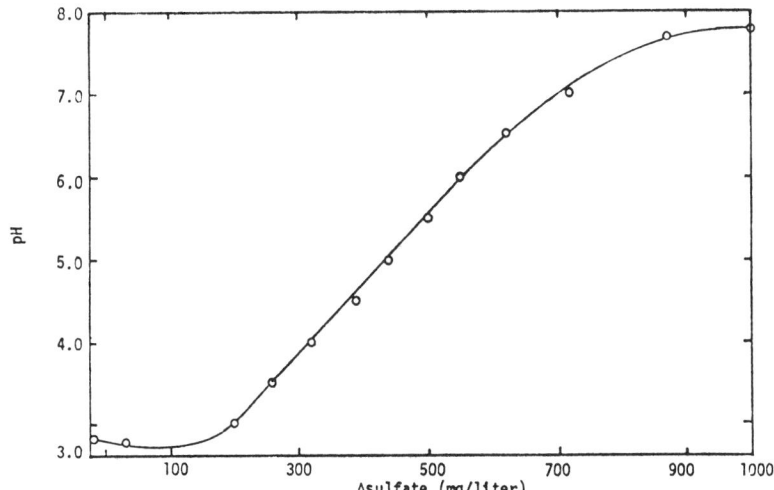

Source: PB 219 264

FIGURE 9.21: PREFERENTIAL BICARBONATE BUFFER IN NATURAL ACID STRIP MINE LAKE WATER

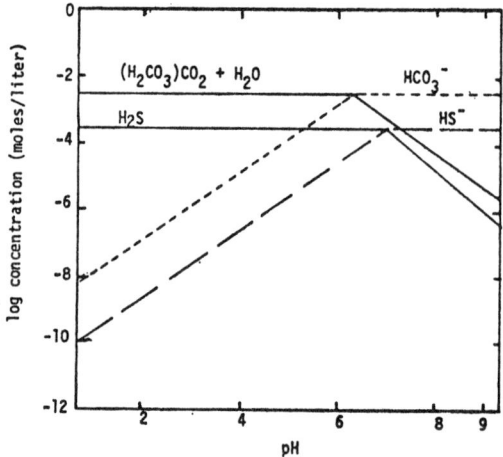

Source: PB 219 264

Natural acid strip mine lakes mechanistically recover in the same manner as outlined in the accelerated recovery process, however, there are a few basic differences. In natural acid lakes anaerobic conditions are usually seasonally sporadic and localized to the less mixed deeper arms of the lake. Also, since the pH is typically between 2.5 and 4.5, the bacteria have difficulty maintaining themselves. The very slow build up of allochthonous organic material, compared to the plentiful supply in the accelerated system, also contributes to the much longer recovery time of natural mine waters.

Combined with constant wind action slow bacterial growth rates do not permit supersaturation of CO_2 in the water. The water can only maintain an atmospheric concentration equal to about 16 micromols $CO_{2\,free}$/liter. Thus, the preferential bicarbonate buffer, facilitating maximum acidity loss will play only a minor role. Because of the slow release and accumulation of CO_2 and H_2S gases the metals will continue to buffer the system for longer periods of time in the natural recovery process than in the accelerated system.

In both systems the metals, primarily the aluminum, act as buffers in the pH range from 2.5 to 4.5. In natural recovery the effect of these metal buffers contributes markedly to the length of time required for recovery. When considering the chemistry of either system the dissolved metals contribute considerably to the net buffer intensity of the particular system. Some of this buffer capacity attributed to the metals, however, may be partially due to the presence of organic acids or silicates.

Essentially, the recovery of acid strip mine lakes is the slow "titration" of the acidity by sulfate-reducing bacteria. The buffers responsible for the characteristic buffer titration curves include HSO_4^-, $H_2CO_3(aq)$, $H_2S(aq)$, and ionic metal "buffers". Each is present in different amounts depending upon the system considered and the initial amount of acid present combined with hydrogeologic and environmental factors characteristic of the particular system. Thus, it would appear from the above research that the evaluation of the present state of recovery of any existing acid strip mine lake could be determined from a single acid titration curve.

The process involved in acid strip mine lake recovery is the biologically instigated neutralization of buffers. Essential to this process are the necessary anaerobic conditions and organic carbon sources needed for efficient bacterial sulfate and acid reduction. Such a process may best be summarized by the following reactions, where X = concentration of HSO_4^-, Y = concentration of metal buffering ions (M), and n = degree of metal hydrolysis.

Sulfuric Acid Buffer:

(19) $2XHSO_4^- \rightleftharpoons 2XSO_4^= + 2XH^+$

Metal Buffer:

(20) $YM(OH)_n^{+(3-n)} + Y(3-n)H_2O \rightleftharpoons YM(OH)_3(s) + Y(3-n)H^+$

Sulfate Reduction:

(21) $[X+\frac{Y(3-n)}{2}]SO_4^= + [2X+Y(3-n)]C(org.) \rightleftharpoons [X+\frac{Y(3-n)}{2}]S^= + [2X+Y(3-n)]CO_2$

Alkalinity Formation:

(22) $XSO_4^= + 2XC(org.) + 2XH_2O \rightleftharpoons XH_2S + 2XHCO_3^-$

Acidity Reduction:

(23) $[2X+Y(3-n)]H^+ + [X+\frac{Y(3-n)}{2}]S^= \rightleftharpoons [X+\frac{Y(3-n)}{2}]H_2S$

Overall Reaction:

(24) $2XHSO_4^- + [\frac{Y(3-n)}{2}]SO_4^= + YM(OH)_n^{+(3-n)} + [4X+Y(3-n)]C(org.) + [2X+Y(3-n)]H_2O \rightleftharpoons YM(OH)_3(s) + [2X+Y(3-n)]CO_2 + 2XHCO_3^- + [2X+\frac{Y(3-n)}{2}]H_2S$

RECOVERY OF ACID STRIP MINE LAKES—A SYNOPSIS

In the presence of air and water iron pyrite oxidizes to sulfuric acid and ferric hydroxide. The majority of the hydrogen ions associated with the sulfuric acid never reach the acid strip mine lake because as they flow over the overburden they are involved in a series of reactions that are responsible for the weathering and dissociation of rocks, clays, and minerals. The majority of the sulfate ions, on the other hand, do reach the strip mine lake and their concentration in the lake tends to indicate the amount of acid production in the particular watershed. The ferric hydroxide is also washed into the lake with the sulfate ions and settles to the bottom, however, a certain amount redissolves in the lake according to pH and K_{sp} limitations. Iron, sulfate, and hydrogen ions along with a host of acid dissociated ionic species, including aluminum, manganese, calcium, and magnesium, and allochthonous organic materials are constantly being washed into the acid strip mine lake.

It is these ions and organic matter that characterize the chemistry of these lakes. A small amount of buffer in the acid mine water is from the dissociation of HSO_4^-. Carbon dioxide and hydrogen sulfide gases also contribute considerably to the buffering of the water. However, it is the high concentration of such metals as aluminum and iron that make the greatest contribution to the net buffer capacity of the water. These metal buffers are responsible for the long natural recovery times associated with all acid strip mine lakes. The amount of such buffers depends upon the amount and type of clays and minerals dissolved on the spoil banks. Depending on the clay type more or less aluminum may be dissolved and allowed to flow into the lake. The greater the aluminum concentration, the greater the buffer capacity of the water, and the longer the pH of the water remains at low values.

The longer the lake remains at low pH values the more difficult it is for the sulfate-reducing bacteria to maintain growth and reproduce. However, the bacterial growth rate is also interrelated with the length of time needed to build up sufficient bacterial organic carbon sources and favorable geologic and limnological surroundings to establish anaerobic conditions in the lake.

Once these conditions are met the bacteria slowly "titrate" the acidity in the lake through the release of hydrogen sulfide to the atmosphere until all the acidic buffers tending to maintain the pH below 4.5 are exhausted. When this occurs lake recovery proceeds at a very rapid rate. It is due to this fact that few naturally recovering acid strip mine lakes are ever found in the pH range from 4.5 to 7.0. CO_2 and H_2S are now rapidly produced and exported to the atmosphere constituting a permanent acidity loss. Responding directly to these slow incremental pH increases the various metals and other ionic species precipitate from solution.

As more H_2S gas escapes the pH gradually becomes high enough for appreciable dissociation of carbonic acid resulting from bacterial CO_2 production. This results in the establishment of a bicarbonate alkalinity system. At this point phytoplankton and zooplankton begin to appear, establishing the necessary base of the food chain that will shortly support a viable aquatic community. Such a recovery process is believed to occur in natural recovering strip mine lakes. When recovered the water chemistry of these lakes is much like that of an early eutrophic lake except that the sulfate to bicarbonate ratio is high.

REFERENCES

(1) Campbell, R.S. and Lind, O.T., "Water Quality and Aging of Strip Mine Lakes", *Journal Water Pollution Control Federation,* Vol. 41, No. 11, 1969.
(2) Decker, C.S., "Accelerated Recovery of Acid Strip Mine Lakes", M.S. Thesis, University of Missouri, 1971.
(3) Ogg, C.W., "Organic Wastes of Acid Strip Mine Lake Recovery", M.S. Thesis, University of Missouri, 1972.

(4) Stumm, W., "Oxygenation of Ferrous Iron: Properties of Aqueous Iron as Related to Mine Drainage Pollution". In *Symposium on Acid Mine Drainage.* Mellon Institute, Pittsburgh, Pa., 51 (1965).
(5) Singer, P.C., *Oxygenation of Ferrous Iron,* Federal Water Quality Administration Report No. 14010 - 06/69, (1970).

AREA RECLAMATION PROJECTS

The material in this chapter is excerpted from
PB 221 337
PB 225 165
PB 226 905

As environmental concerns became more prevalent across the country, and as it became more apparent that technology could be applied to the diverse problems of reclamation if enough good information were available concerned individuals recognized the need to generate this information. It became quickly apparent that a realistic appraisal of strip mine problems, costs, and time factors could only be made by conducting actual demonstration projects.

GRUNDY COUNTY, ILLINOIS DEMONSTRATION PROJECT

The Development Plan

This project was designed by the Department of Conservation in Illinois and conducted and reported by James R. Johnson for the Illinois Institute of Environmental Quality. The plan as originally envisioned by the Department of Conservation was to reconstitute the unused portion of Goose Lake Prairie State Park as an active recreational element in an otherwise passive recreational area. This would provide an area for the needed parking, picnic, and other active recreational forms in the park without disturbing the refurbished prairie areas. Development of an "esker" land form in the center of the site with lakes skirting the form were part of the improvement concept.

The strip mine land reclamation project consisted of approximately 50 affected acres, of which more than 14.5 acres were covered with acid water ponds. Mining operations in this facility ceased in the early 1940s and the area left was subjected to severe erosion, acid-producing runoff, natural vegetative recovery, and aesthetic injustice.

Although the soils within the mine area were not thoroughly analyzed, certain physical properties were evident from the spoil banks and the adjacent lands. The spoil banks were generally inversions of the original soil profiles of the area where the more fertile soils were buried under forms of clay, shale and sandstone. Better than 75% of the surface areas within the project were not covered with vegetation of any species, leaving most of the spoils subjected to various forms of erosion. Where weathering had produced a moderately viable soil condition, in the valleys of the spoils or near the base of the spoils, some forms of vegetation were apparent.

Large sandstone blocks were disposed throughout the spoil areas, generally near the upper portions of the spoil banks. The weathering of these sandstone deposits produces a material, when intermixed with organic soils, that can add to the friability and drainage characteristics of any soil. They also add to the siltation problems of the water bodies present in the area. The adjacent "undisturbed" areas were basically light to medium-colored soils able to support vegetation.

Water in the area was highly acid and without corrective measures completely unsuitable for flora or fauna habitats. The five impoundments were variable in depth and nothing more than the low collection spots for all the internal drainage of the area. The northernmost pond area had a spillway type of ditching but apparently was functional only during periods of extremely high rainfall. Surface elevations of the impoundments varied from an elevation of 511.0 to 515.5 feet, or a difference of 4.5 feet between areas. There was no pattern of elevation changes other than the spoil banks around each area which would indicate that there was possibly no subterranean movement of water. The pH range of the water ranged from 3.3 to 3.6 with total and free acidity ranging from 631 to 1,250 mg/l.

The plan, as shown in Figure 10.1, involved the design of the esker form, and the alteration of the lake features by the redistribution of the spoil banks located on the site. The specifications were prepared to utilize existing earth construction equipment in an effort to relate cost of development with present technology. A portion of spoil banks was covered with regenerative species of trees and shrubs. In order to "reshape" the land, these banks would have been cleared of all vegetative material. Surface drainage in the area would be handled within the property confines.

Not avoiding the paramount purpose of the demonstration project, the basic development plan was to contain as many of the items normally expected to be encountered in other sites where reclamation is required. Some of the details of the plan are related to preservation of minor establishments or colonies of native prairie species. The vegetative reestablishment will be commensurate with the species located in the prairie and compatible to the species preserved on the site. Two Eurasian species will be used to provide temporary erosion control until the prairie grasses become dominant. Any natural regenerate species introduced in the area would be allowed to remain in order to establish a niche and maintain the homeostasis of the site.

The Construction Document was developed quickly in an effort to collect preliminary cost data prior to the end of the year, 1972. Following the preparation and submittal of the environmental statement, approval, and concurrence

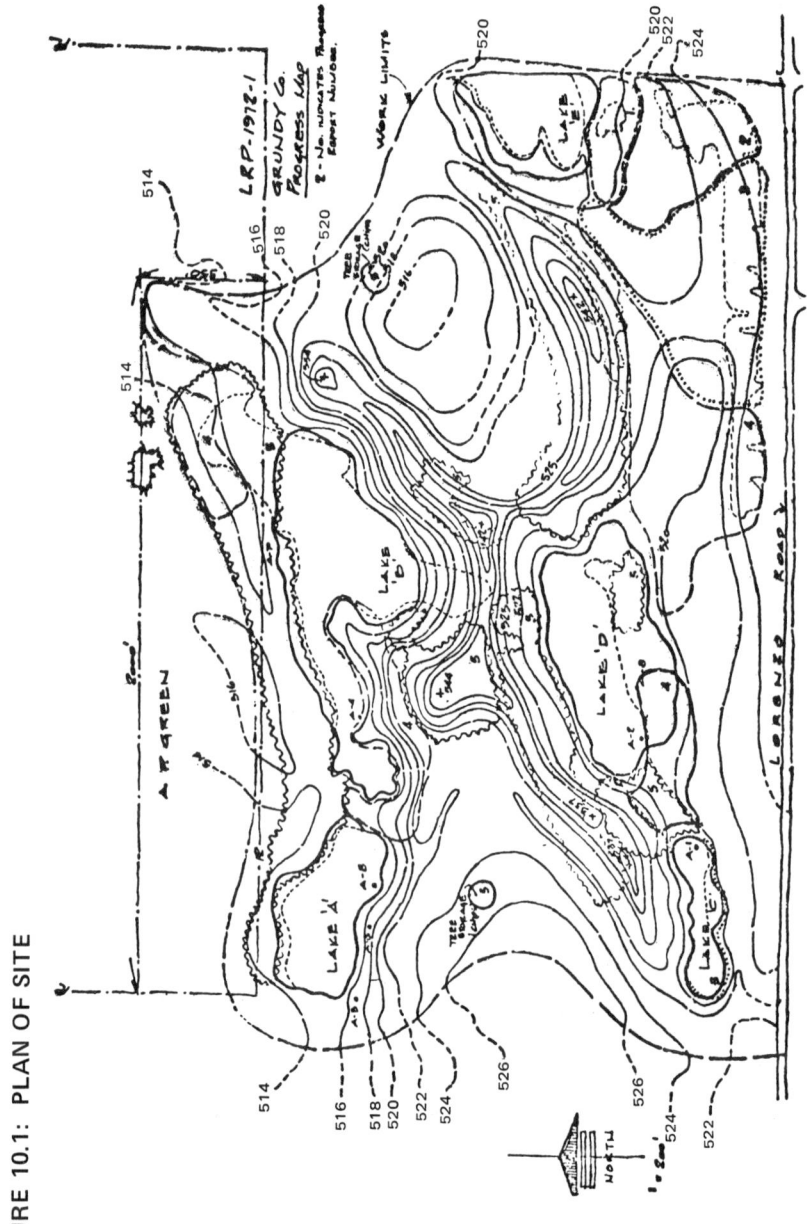

FIGURE 10.1: PLAN OF SITE

Source: PB 226 905

of all land agreements, and preparation of necessary design details, the Construction Document was prepared to culminate into a final working contract including all necessary plans, special provisions, and cost estimates.

In order to follow the normal procedures for public contracts, the proposal was submitted to the Department of Transportation to be let in accordance with all policies and procedures of that department, for a September 15, 1972, public letting. The basic purpose for this procedure was to provide a public awareness of the intended project and determine whether or not it would adhere to all State regulations regarding such contracts. In addition, it was advantageous that the contract work would be performed by those people normally engaged in heavy highway construction.

The contract was prepared in accordance with the design concepts of the Department of Conservation and included seven basic construction items: tree removal, spoil bank relocation, water feature alteration, treatment of acid water, topsoil placement, seeding, and tree transplanting. In order to gain a more complete dossier of the project, the breakdown in this phase would enable the project to provide initial cost breakdown for future reference, and assist in estimating for future bidding purposes. Field records provided the cost data even though the project was let as a lump-sum bid. The following is a brief description of each of the aforementioned items included in the demonstration operations:

(1) Tree removal dealt with the removal of existing deciduous regenerative material existing on the spoil bank. It not only included the complete removal, but also a prescribed method of disposal.

(2) Spoil bank relocation constituted approximately 85% of the major work to be done on the site. It involved the physical redistribution at prescribed locations and grade, by whatever necessary equipment would be required. This item was also provided for the opportunity to experiment with other forms of heavy equipment that would not normally be used under these circumstances.

(3) Water feature alteration dealt primarily with the alteration of the ground surface in relation to the existing bodies of water. It involved the actual reconstruction of certain or designated shore line areas, compatible with the original design concept. In addition, for shore lines that were not physically altered, it was intended to correct underwater slopes for safety purposes.

(4) Acid water treatment was designed primarily to investigate and observe several methods of neutralizing acid conditions in the existing lake areas. It should be noted that treatment of acid water was not necessarily limited to physical application of neutralizing agents to the water. This would also imply that modification of acid production areas would be required on adjacent slopes.

(5) Topsoil placement was intended to physically cover all disturbed areas with various thicknesses of topsoil and eliminate or reduce the evolutionary process of topsoil establishment. The source for the topsoil would come directly from the site by natural selection of heavily graded areas where organic soils become evident.

(6) Seeding consists of normal vegetative refurbishment on the entire

development area. It was intended to be completed in two phases. The first phase was the introduction of various species for erosion control purposes only, and the second was the inclusion of compatible turfs complementary to the existing prairie. This item also included any necessary soil amenities required to assist and sustain vigorous vegetative growth.

(7) Tree transplanting involved the physical transplanting of 4" x 6" caliper trees by mechanical methods. The trees were to be selected from similar existing eco-areas, in adjacent or nearby timber stands along the Mazon River.

The contract was proposed for a September 15, 1972 letting, as normally held by the Department of Transportation. There were four bidders on the proposed section and with the exception of one, all came from the general area of the demonstration site. The fourth bidder came from the extreme southern portion of Illinois, but was familiar with normal strip mining activities. A recommendation was made to award the contract to the lowest bidder; and on September 24, 1972, a contract was entered into by the Illinois Institute for Environmental Quality and two construction companies working jointly to perform the work as prescribed in the contract for the amount of $223,000.00. Actual construction was to start as soon as possible after the award date. Following all necessary bonding and exhibits requirements, construction began on October 11, 1972.

The Project Document

The basic components of this document contained the normal information required for documentation of a construction report. This included the summation of inspection reports (if required), scale verification, daily inspection reports, and a summary of quantities used. However, in order to gain a more complete dossier of the project, the addition of other items was necessary.

As part of the "standard" element of the document, rates for labor and equipment were inserted. Equipment capabilities of work (yield) were also included and documented according to source. Summary sheets for each work item as outlined in the contract were included to recapitulate daily work operations and their appropriate direct costs. The observations and sketches made in the field during construction were also to be included. To culminate each week's work on the project, there was to be a weekly progress report, a brief synopsis of progress, problems, instructions to the contractor, physical plan changes, etc.

Development of Cost

Since one of the more critical aspects of the project is the development of empirical cost-data for reclamation activities by contract, an explanation of the process of developing these costs is necessary. The costs were direct costs incurred by the project that were definable within the state's jurisdictional limits; that is, the costs of equipment actually operating, the operators' or laborers' rates, and the engineering time and expense. Overhead costs, fuel and maintenance for equipment, and profit were costs to the state by the contractor and could not be readily obtained. However some valid assumptions can be made about these costs, notably, a comparison between costs projected by the state and the actual monthly billing by the contractor based on percent of work completed at the time of the billing. This should reflect a difference of determination of cost; therefore, the overhead factor.

Area Reclamation Projects

To further describe cost, it was necessary to determine exactly what costs are to be examined, the cost of "what," so to speak. For the purposes of this project, this would imply the cost/yd^3 of excavation and filling (spoil bank relocation and disposal), cost/gal of pumping, cost per ton of neutralizing agent for acid water, cost per acre for tree removal, cost for vegetation establishment, and cost/yd^3 for topsoil placement and finishing. All of the above items are indicative of unit-price-bid items normally used in work of this type.

The proposal document gave no summation of specific quantities as pay or unit items. In acquiring cost, assumptions of quantities have to be made as related to each specific work item. Quantities of materials affected were determined by the use of manufacturer's specifications for the work capacity of each piece of equipment; e.g., as an International TD 25C can move approximatley 80% of 450 yd^3 of 3,000 lb material/hr, the 80% is a quantitative optimum time of operation per hour. This project reduced that set volume by the variables established in the strip mine area, as shown in the project document. Knowing the capability of the TD 25C, a formula can be developed to reflect cost/yd^3 for this project. For example:

(1) crawler cost/day + operators' cost (hourly rate + fringe benefits)/day = operating cost/day

(2) (yd^3/hr)(80%)(K) = (yd^3/hr)(operating time) = yd^3/day

$$\frac{\text{operating cost}}{\text{yd}^3/\text{day}} = \text{cost/yd}^3$$

This same formula was used for each crawler type. The yd^3/day cost was based on the number of machines, their types, and their operating costs; and an average of the yd^3 cost for all machines for that day. Similar procedures were to be used for each piece of equipment in accordance to:

(1) the task assigned (work item)
(2) the equipment capacity
(3) the operating cost.

It should be pointed out that the contractor agreed to provide cost data not normally disclosed, in an effort to provide a more complete and encompassing cost of the project. These figures would show incidental expenses charged to the project, project trailer cost, fuel, maintenance, company overhead, and profit. From the culmination of cost data the following should be available:

(1) An established unit-price per work item or pay item.
(2) Empirical data for future project estimation.
(3) Through projection, an approximate cost for programmed development of all reclamation activities in the State of Illinois.

Other Items of Interest

Besides the paramount business of cost-data accumulation, the reclamation project should provide material to support length of operating time (efficiency), observations of unusual site characteristics or alterations, and some yield data. In the monthly summation of work, substantiated statements were required to evaluate the working day or time factor especially in estimating future work.

This also included yield factors or components. These "yield" characteristics are supplementary, and complementary to manufacturers' specs for equipment. This information is definitely needed in order to make reasonable estimates of time, material, and equipment for each project. Although "all strip mines are different," there are still some conceptual items that are standard and basic to the reclamation activity. The yield data collected on this project can be the basis of all reclamation sites with the variables of each site built into the broader spectrum of reclamation knowledge and research. One item that might be unique to this project with regard to yield is tree removal. It was intended to determine the tons of chips available per acre for this clearing operation.

The Reclamation Project

Before any work was started on the project, a photographic record was made to record the site in its existing condition. On October 11, 1972, the contractor moved in the initial equipment to begin the construction of the project. This consisted of crawler tractors and two 6" diesel pumps. It was immediately apparent that a separate cost report would have to be prepared for the pumping operation. Although originally estimated as a part of spoil-bank relocation, pumping operations should, in the future, be a separate work-item, and also be considered as a separate pay-item.

As a part of the organizational process by the contractor, concerted efforts were made to perform two or more of the jobs concurrently. For example, the pumping of water out of the southeast corner was completed more efficiently because the spoils were being pushed into the final-cut area as the water was being withdrawn, displacing the water and keeping the water level high for efficient pumping.

This operation was started on October 11, 1972, with the two 6" pumps. Pumping was required to drain water areas that had to be filled with spoil and still keep the water on site. Two small levees were built on site to retard water movement out of Lake **B** into Lake **D** (Figure 10.1). Only one operator was required for this job.

Preliminary work prior to pumping consisted of the installation of a temporary 18"-CMP at the haul road area, and the construction of a small diversion ditch that would handle the pumped water. At the completion of the job, the CMP was removed and the ditch reshaped to blend with the original and proposed grade. This area fell within the proposed parking lot area desired by the Department of Conservation, and the location of the ditch was within their overall design concepts.

All pumping required for the initial phase of the construction project was completed by October 30, 1972. At the end of that period, nearly 900,000 gallons of water were pumped in 12 working days, or 75,000 per day. Cost for the entire pumping operation was $2,223.56, and when applied to actual number of gallons relocated, the cost/gallon for pumping is established at $0.0025.

It was observed during the pumping operation there was a relative change in the level of Lakes **C-D** and **B**. The water was raised in each of those lakes approximately 3 to 4 feet, and the man-made temporary levees retained the water within the confines of the mine property. There was some fluctuation in the pH

Area Reclamation Projects

levels during the pumping, resulting from mixing activities in the three lakes. The pH of Lakes **C-D** and **B** rose from 3.6 to more than 5.0 and remained at this level until heavy runoff from spoil relocation during the late winter months drained into these lakes.

At the completion of the pumping operations most of the final-cut area had been filled with spoil from adjacent land. A small hole was left at the extreme west end of the final cut to hold the silt and mud, and to contain superficial runoff during the remainder of the project.

Approximately 18 acres of the mine site was covered with regenerative-plant materials. Costs for tree removal, however, were based on 15 acres of actual removal done in accordance with the specifications prescribed in the contract.

The original intent of the contractor was to subcontract the tree removal work. The subcontractor brought the necessary equipment to begin the project, but found that the standard timber-cutting machinery could not cope with the existing site conditions.

The first attempt to cut the material was done with a scissors clamp and pincer knives mounted on a four-wheel drive "Bobcat" tractor. Tracks were added to the wheels for better traction on the one-to-one slopes. This piece of equipment was designed to cut material on reasonably level ground where the plant material did not exceed 10" to 12" in diameter. In addition to cutting the Bobcat clamped the tree for carrying to a stockpile or chipping area.

The problems with this machine were related to the steepness of slope while it was cutting and holding the trees. The tops of the trees and the angle of the spoils would change the center of gravity of the machine, causing it to slide down the slope or to become top-heavy and actually tip. Attempts were made to remove the plants by working both from the top and/or the bottom of the spoils. Even with the use of anchor machines, the cutting machine could not handle the removal operation.

The contractor released the subcontractor from the cutting responsibility, but held him to the required chipping process to be completed at a later date. No costs were developed for the mechanical process, due to the lack of completed work that would establish a cost basis.

It was decided that tree removal would be done with manpower and small 18" draw-bar chain saws. This method of tree removal was complemented with towing equipment to move the downed material to a holding area for chipping. Costs for this project were based on this method of tree removal.

It should be noted that, generally, vegetation would not be buried on site, but the reduction of the original 18 acres to the final 15 acres of tree removal was due to the removal and partial destruction of trees in the newly constructed haul-road areas and the spoil openings.

On November 8, 1972, laborers were brought in to start the sawing process. By November 21, 1972, after seven working days, all trees had been sawed down and the small crawler tractors were brought in to begin the stacking process. All stacking of plant material in the holding areas was completed on Dec. 13, or in 10 working days. By the end of November, however, most of the spoil areas originally covered by trees were cleared and ready for the continuing spoil-relocation process.

Cost for tree removal was determined from the amount of equipment and man-hours utilized per day and compared to a unit of measure accomplished in that day. Areas were determined from planimeter measurement rather than field measurement. The sawing costs for the entire project were determined to be $86.324 per acre. Yet the total cost of the tree removal was $550.00 per acre, excluding chipping.

Yield of chipped material was based on the weight measured in trucks, but for estimating purposes, yield was based on plant density and caliper. On this site, density of plants ranged from 150 to 175 plants per acre with an average diameter of 9". This population should yield 6 to 8 tons of chips per acre, produced at a cost of $15.00 to $18.00 per ton. With the chipping costs of the tree removal included in the price per acre, the final unit costs for tree removal was $716.00 per acre.

Spoil Relocation

The primary goal for the demonstration project was to determine the actual cost of relocating, removing, or disposing of all unsuitable spoil banks that did not conform to the intended grade and line established by the original plan. On Wednesday afternoon, October 11, two D-8 Caterpillar tractors were moved into the area to begin the "opening-up" of the reclamation site. As stated earlier, several operations were going on simultaneously. The most critical of these was the pumping of water and the physical movement of spoil banks into various areas throughout the site.

The Caterpillars were not necessarily doing spoil relocation for the first two days, but spent most of the time preparing exploratory areas for cross-section observations of the spoils and making roads to provide access to the internal areas. The observations which provided the preliminary aspect of how spoils were placed during the mining operation were basic indicators of how topsoil or organic soils could be isolated and reserved for surface use later in the project. Some of the spoils were made of consolidated materials, while others were completely composed of unconsolidated materials.

The pattern of these spoils within the site, even though appearing rather ambiguous, indicated where the shovel had been moved in the stripping operation, especially where the organic materials were placed in separate piles. This is quite evident in the final-cut area along the southeast corner of the project. The rock or consolidated material was placed in the random or conglomerate banks, as would be expected. The fact that the soil is separate in certain places was encouraging, as it would be easier to salvage for the placement on the relocated banks and easier to prepare for seeding. This would also reduce the cubic yards of borrowed topsoil required to cover the disturbed spoils.

In addition, some of the barren spoils on the north side of the property reflected a complete inversion of the original land profile—which means that some organic materials will be available there also, but will be under approximately 25 to 45 feet of spoil bank.

Figure 10.2 gives an example of the cross section of the spoil banks as they appeared after penetration by the dozers during the "opening-up" operation. The left side of the sketch indicates the type of spoil bank that was found on the north side of the property where no vegetation occurred, and an apparent inversion of all overburden materials was accomplished during the mining operations. The right side of the illustration indicates that there was apparently more care taken, either by accident or otherwise, in the placement of the spoils by the layering of various types of organic material. This is also indicated by the types of vegetation and the location of the vegetation on that spoil bank. This illustration is representative of all or most of the spoil banks throughout the reclamation site.

For the purpose of obtaining legitimate costs of moving the spoil banks in accordance with the plan, several things had to be accomplished before work commenced. The cost per hour was developed, prior to actual working conditions, for both the equipment and manpower utilized during the normal spoil bank relocation operation. From these figures, an hourly rate could be determined, and also (using manufacturers' recommendations on the amount of material that can be moved by each piece of equipment) the cost/yd^3 could be derived. For example, during the month of October, from the 11th to the 31st, approximately 92,000 yd^3 of spoil were moved. This was done at the cost of over $19,000.00, or at the cost of approximately $0.21/$yd^3$.

In addition, spoil-relocation costs were further broken down by machine type. There were three separate mechanical operations performed during the spoil relocation procedures. These were the normal dozer operations, a 1.25 yd^3 drag-line, and the use of Hancock or self-loading scrapers. Manpower and equipment costs were maintained for each of the operations, and the price/yd^3 of spoil relocation was determined as a separate item.

The final conglomerate of the three basic mechanical operations were then put together to come up with the unit cost for moving spoil banks at the rate/yd^3. Average cost/yd^3 for moving spoil banks by crawler-type tractor was $0.2071. The drag-line was averaging better than $0.53/$yd^3$. The self-loading scrapers averaged from $0.29 to a high of $0.34 a yd^3. These costs were based on the theoretical maximum operation time for each piece of equipment, the operators' costs, and the fuel cost in a normal eight-hour day. Actual cost/yd^3 of spoils was based on infield daily operating costs as they were maintained on the Inspector's Daily Report.

One of the more important operations of spoil-bank relocation was the continuing observations made of equipment, especially that equipment actually involved in the major work. From these operations it could be determined that standard heavy earth-moving equipment could be used, either as produced by the manufacturer or with slight modifications, and satisfactorily meet the needs for spoil-bank relocation.

The concern for equipment types in actual usage in spoil-bank relocation dealt with items such as large rocks or boulders that had to be moved, how they might affect self-loading scrapers, the maintenance of the drag-line in moving this kind of material, the replacement of teeth on scrapers and drag-lines, and any damage done to dozers while moving this type of material. Throughout the earth-moving operations, however, it was noted that the standard earth-moving

FIGURE 10.2: TYPICAL SPOIL BANKS

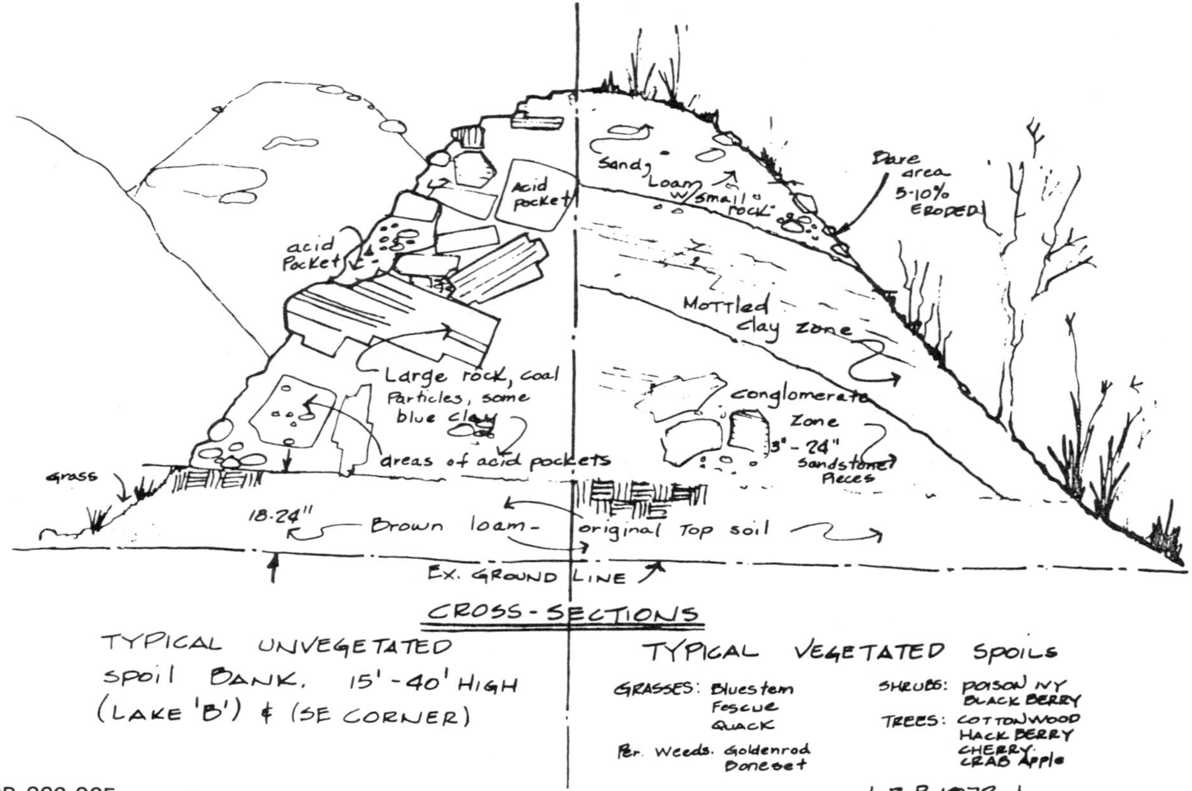

Source: PB 226 905

equipment normally used on highways could satisfactorily and efficiently move the spoil banks without any measurable degree of difficulty. It should be pointed out that teeth replacement on scrapers and drag-lines did occur more frequently due to the stones present in the spoil banks.

It should also be noted that dozers should work in pairs for two reasons: one, in case one machine is disabled for some reason, the second machine is near by to provide assistance; secondly, it was found that if the dozers worked in tandem with one another, the over-spill of material off the edges of the blades was reduced by 50 to 75%. On an experimental basis, four dozers were put side-by-side and as much as 125 yd^3 of spoil were moved at one time. This was done in several situations where a direct shove was all that was required to move the spoil from its original position to the desired location.

General Work Procedures

The contractor decided to work the spoil relocation in several phases throughout the site as a simultaneous operation. In addition, at a later date in the contract these dozers, working in these various operations, would also act as complementary equipment to both the drag-line and the self-loading scrapers. The contractor initiated the contract by working in the final-cut area on the southeast corner at the same time the drag-line was working the shoreline slopes around the small lake on the southwest corner. Upon moving in of the full complement of machinery required for the job, the contractor elected to work in three separate areas simultaneously.

The spoils on the A.P. Green property, north of the reclamation site, were worked into Lake B. In the southwest corner of the project, the contractor moved the spoils resulting from shoreline alteration for Lake C. These two operations were carried out simultaneously, with the work being done in the southeast corner of the project or the final-cut area. Work that proceeded on this basis normally produced between 10 and 12 thousand cubic yards of spoil moved per day.

The exterior of the site was the area in which the contractor generally did most of the work for the first three to four weeks. By proceeding in this manner, he provided time for the tree-removal operation to clear the areas in the center of the project. When the trees had been removed from the spoils in the center of the project, the contractor was then ready to move equipment into the center and begin the work on the esker form. The normal progression of work occurred in this fashion throughout the life of the contract during the spoil bank relocation phase. This work was completed on February 9, 1973, and the cost accrued for this phase of the project totalled $146,733.50. This involved the movement of 600,000 yd^3 of spoil bank. The total cost/yd^3 for this phase of the project, using this equipment, was $0.2424.

As a point of interest, it is well to note the difference in the volumes of dirt that were moved through the end of November. By that time, of the approximately 140,000 yd^3 of spoil bank that had been relocated somewhere in the site, 121,000 yd^3 of this was material moved by crawlers or dozers. The remaining 20,000 yd^3 was done by the self-loading scrapers. Day-by-day breakdown of each of these items was maintained in the Inspector's Daily Report of the project document.

It is interesting to note that all of the spoil on the site prior to the reclamation project is still on the site following all the earth-moving operations. The distribution of those spoils is indicative of what can be done when foresight or design is given to arrangement. Some of the spoils were used to fill existing voids on the surface, others were used to construct and establish land bridges to separate water features, and still other spoils were moved to provide sufficient material to establish a predetermined land form. On this particular site, no spoil material had to be hauled away. When spoil was used for land bridges or pond filling, however, certain aspects of soil structure had to be evaluated before complete filling could be accomplished.

The efforts to "bridge" the pond area with 8' to 10' of spoil material in the final cut area proved to be ineffective. The silt and the sand deposits in solution in the bottom of the pond had to be pushed forward by advancing the dry spoil at a predetermined rate. In other words, the spoil could not be pushed into the pond area from all sides. It had to be worked from one end and advanced to the opposite end. The land bridge between Lakes C and D was also made from spoil, and when completed, was approximately 175' wide with two to one slopes on each side and 275' long. Figure 10.3 is a sketch of how the spoil banks were moved to form the land bridge.

The filling of Lake B from the spoils on the A.P. Green property was accomplished in a similar manner. Figure 10.4 shows a sketch of the approximate height and width of the spoils that lay on the A.P. Green property that were used to partially fill the north end of Lake B.

Water Feature Alteration

Since the earth moving that was connected with the water feature alteration was of a different nature than the spoil bank relocation work, it was felt that it should be accomplished under a different contract. Basically, it involved a dragline that dealt with the relocation of very small quantities of spoil material in very tight isolated areas. It also dealt with the design aspect of the project by altering the shore lines of each of the lakes that were to remain on the site after the completion of the project.

Figure 10.5 shows the shoreline treatment that was attempted in this particular aspect of the work. The purpose for the variation of depth and slope was primarily for safety, assuming that the area was to be used by young people, and secondly, that because of the rock hardpan in some unmined areas, there would be undue wear on the equipment. After getting this portion of the project under way, it was learned that many of the areas adjacent to the ponds had been left unmined. Consequently, it was possible to pick up additional topsoil to be stockpiled for later use on the project. From an environmental standpoint, the shallow depth of the water along the edge of the ponds was compatible with the ecological systems in the adjacent prairie area. Therefore, there was little or no trouble in duplicating both the flora and the fauna that existed in that area.

This particular operation caused approximately 2,500 yd^3 of spoil material to be relocated per week. Some of the low volumes are attributed to the fact that some time was spent moving the equipment from one pond area to another on the site. As a result, the cost/yd^3 for moving material of this nature was a good deal higher than it was for dozer work.

FIGURE 10.3: CROSS SECTION OF LAND BRIDGE BETWEEN LAKES C AND D

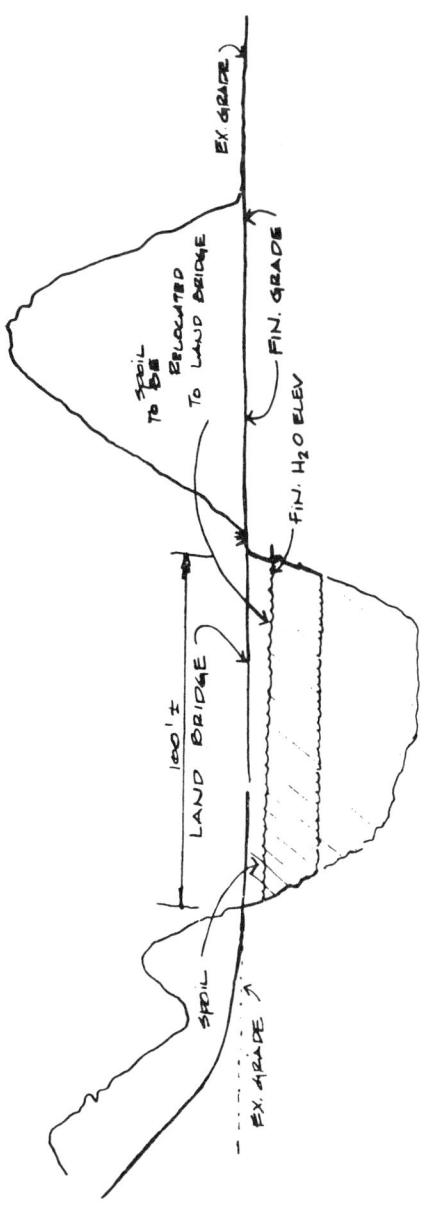

Source: PB 226 905

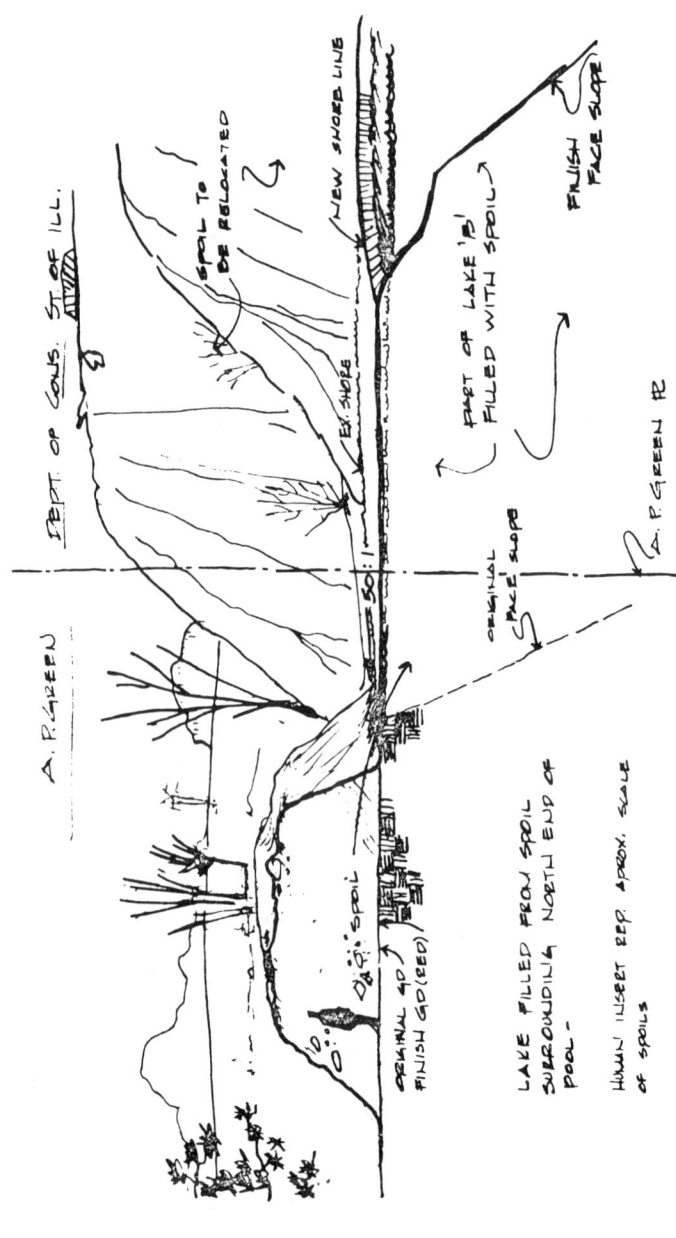

FIGURE 10.4: ELEVATION OF LAKE B NORTH END ON A.P. GREEN EASEMENT

Source: PB 226 905

Area Reclamation Projects 239

FIGURE 10.5: SHORELINE ALTERATION

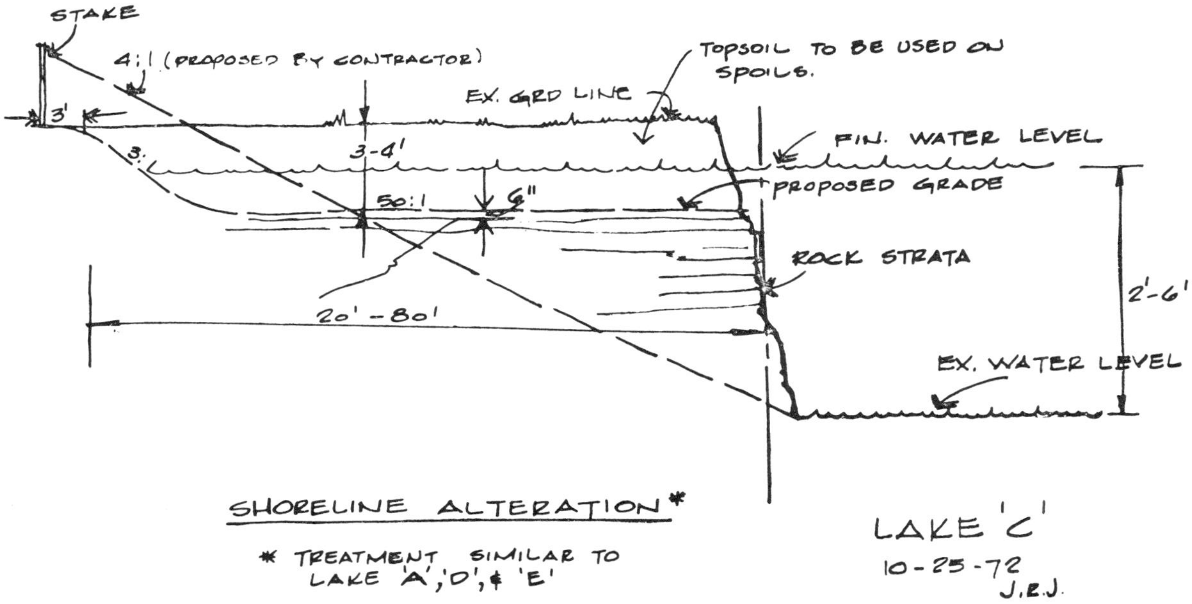

Source: PB 226 905

The time span for this particular operation was nearly identical to that used by the dozers working in the field, which started in the middle of October and ran to the end of February, 1973. Cost projected for this particular operation was in excess of $27,000.00. This was used to move over 41,000 yd^3 of material at a cost of $0.6684/yd^3.

Topsoil Placement

At the start of the project, it was anticipated that all of the disturbed areas would have to be covered with an organic material that would support vegetative growth. This topsoil was to come from off-site areas. It would have been maintained as a cost unit for borrow and for topsoil placement. Efforts were made during the preliminary stages of the project to negotiate for sources and cost of off-site topsoil. During the interim, it was agreed that Commonwealth Edison Company would supply approximately 200,000 yd^3 of topsoil from their new construction site immediately west of the strip mine area. The Department of Conservation also had a small project directly behind the Rangers' Headquarters, approximately one mile northeast of the strip mine area, and an estimated seven to eight thousand cubic yards of topsoil were to be left over as waste from the construction of a parking lot and small reflecting pond. The combination of the new sources would more than adequately supply the needs for topsoil anticipated for this project.

As the project progressed through the spoil relocation process, the observations made in the field proved that topsoil was available from on-site locations. During the early stages, though, it was impossible to determine the exact amount of yardage available on-site. Such items as the water-feature alterations with the dragline and preplanned spoil relocation disclosed several thousands yards of organic material that could be utilized as topsoil. The actual usage of this material for topsoil purposes would depend upon the nutrient value and the pH of those soils.

As the project progressed it became more evident that all the topsoil necessary on the project would be available from on-site. This was due to two factors: one, the preservation of organic soil areas that were found during the spoil relocation time; and, two, the amount of topsoil taken from undisturbed areas during the water feature alteration phase. Another benefit, during the relocation of spoil banks, was the mixing or dilution of undesirable materials with the organic soils that could not be salvaged. Many of the areas then did not require topsoil as a dressing. The areas then that did need topsoil dressing would be covered with from one to two feet of material that had been held in reserve during the various construction phases.

During the middle of February the contractor began to spread the organic soils left on the site on designated areas within the site. The equipment was the same as that used for spoil bank relocation, including the crawler tractors and the self-loading scrapers. The exact volume of material moved per day was considerably less than that moved during the spoil relocation, even though the same equipment was used, but this is primarily due to the fact that 75 to 80% of the work completed was done with the self-loading scrapers with some long-haul distances, and the actual cost incurred for material moved per cubic yard was more expensive when done by scraper than by dozer.

The completion of the topsoil placement was done by March 13, 1973, and at the end of that time the contractor had moved and spread in excess of 76,000 cubic yards of topsoil for a cost of $20,700.00, or $0.2733/yd^3. It is well to note that all topsoil came from on site and that no borrow was required. However, in other areas of the state, borrow will probably be required and therefore the cost of purchasing and hauling borrow is necessary for the development of a good unit price structure.

Soil tests were taken of the topsoil to check pH, and to check for the elements of nitrogen, phosphorus, and potassium, as well as traces of iron and sulfur. From this, requirements for seed bed preparation could be determined prior to seeding. Following the preliminary soil tests, a determination was made of the amount of agricultural limestone needed to supply the soil with the proper elements to raise the pH to a satisfactory level for good vegetative growth. It was determined that a total amount of three tons of lime per acre would be sufficient to raise the pH to approximately 5.5 to 5.8. This pH factor is compatible with existing prairie soils surrounding the site.

On May 16, 1973, the liming operations started with the application of approximately 30 tons of lime distributed at the prescribed rate. Soil preparation followed the liming operation to obtain a better mixture of the limestone into the soil. Liming and application of fertilizer was followed during the next few days by additional soil bed preparation prior to seeding. It should be noted that the seeding requirements as put into effect on this particular project deviated from the original plan.

The original plan for seeding provides a cool season turf for a short time or interim period of establishment, with final culmination in the uses of prairie turfs. The deviation occurred due to the time of seeding, and because the temporary erosion control measures originally anticipated were not required, nor was the cool season turf. By the time seeding took place in 1973, all that was required on the section was the use of the warm season or prairie grasses.

Fertilizer on the section was also changed from the original intended plan by a reduction of nitrogen and an increase in the phosphorus and potash requirements to amend the soil properly. Although the primary concern for vegetative growth was the establishment of a grass crop, the nutrient levels of the soils and the requirements to supplement those soils were based on agronomic research, i.e., normal agricultural crops require a certain number of food units or food elements per acre and the attempt was made to alter the fertilizer to be used on the job to bring the existing levels of the three basic nutrients up to the minimum requirements for agricultural or cropland areas.

The reduction of nitrogen was based primarily on the use of certain legumes in the grass seed mixture. It was felt that a natural source of nitrogen provided by the legumes would be more beneficial to the entire area than the application of a synthetic nitrogen that might ultimately leach from the soil prior to its consumption by the plants.

The legumes introduced into the crop were white or ladino clover and vernal alfalfa. The contractor elected to add one additional cereal grain crop to the already required cereal grain crop of oats. His addition of farm rye to the overall seed mixture gave an additional assurance of a quick germination crop that would provide cover, and eventually mulch, for the new prairie seedlings that would germinate at a later date.

Total acre equivalent weights of materials added per acre were as follows: in seed or grass crops there was a total of 62 lb of seed per acre at a prescribed mixture. Fertilizer was applied at the rate of 495 lb/acre, with a 9-12-32 ratio. Agricultural limestone was applied at the rate of 3 to 4 tons/acre.

It was determined by observation that the original requirement of application of two tons of straw mulch per acre would not be needed. Therefore, it was not used upon any portion of the job. The primary reason for this was the amount of moisture available in the soil and the amount of rainfall that the seed areas had received since the time of seeding.

Erosion occurring on the side slopes of the esker formation was minimal; therefore the straw mulch would not have been necessary for controlling erosion. Without the straw, the demonstration would provide a good basis for future determination of soil and seeding requirements. Mulch might be used on certain areas after individual identification of these requirements are made on each future site as they are developed. It must be kept in mind that the straw is primarily a moisture-retention element and not entirely designed as an absolute erosion-control material.

Equipment used on the job deviated from the normal type of seeding or construction equipment. The use of the hydro-mulcher meant little traverse by equipment over the area while seeding and no disturbance of a major portion of the entire area after it had been seeded. The amount of small stones and rock resulting from the mining operation also precluded the use of standard seeding equipment. These would have made an uneven seeding bed, would have provided an uneven seeding pattern, and would have been unnecessarily hard on the equipment used.

Soil preparation was done by a drag bar with 100 parallel teeth used for raking and scarifying. A normal disk or disk harrow would not have been effective on these particular slopes unless an outside source of topsoil was applied at any prescribed depth to counteract the influence of small rocks and other foreign material. It should be noted, however, that these rocks and chunks of other consoliated material will have little or no effect on the ability of the vegetative cover to grow on this area.

The cost for seeding of this reclaimed area was in excess of $5,000.00 for the 35 acres. This amounted to a cost of approximately $150.00 per acre for the seed, equipment, and manpower required. Because of the gentle inclines of the newly formed slopes, the contractor did not have to provide special equipment to apply fertilizer and limestone, and was therefore able to utilize existing agricultural equipment at a cost of approximately $1.00/acre. The success of this operation depends upon the percentage of germination, and the viability of the plant material once it begins to grow. Any bare areas that may result from incompletely neutralized acid pockets that might remain, or from undue damage to the soil by natural elements, must be repaired and reseeded to insure 100% coverage of vegetative material over the entire area.

Acid Water Treatment

Observations of the demonstration tract once again proved to be a valuable tool for the determination of the amount of work remaining on the site, as well as what can be anticipated for future work. Originally, it was intended that all pond areas would require some type of an acid treatment or neutralization of the water left on the site after construction. However, during the course of construction it was found that Lake A (see Table 10.1) had a test report of 5.9 for pH during December, 1972, compared to a 3.6 pH in the original sample taken in June, 1972. The major reason for this change in pH was not due to a neutralization within the water, but to the removal of the acid sources from the shoreline areas around that lake.

A test taken again in May and June 1973 showed the same overall effect in the remaining bodies of water and proved that there was only one lake, Lake B, that still required some type of neutralization. The original test for this lake in June of 1972 showed a 3.4 pH, but by December of 1972 the pH dropped to 3.1. Tests of the water taken again in May and in June showed pH ranging from 3.5 to 4.0. Lake C, D, and E reflected a pH reading of 6.0 to 7.0, in June 1973, while in December they too were in the 3.3 and 3.2 pH ranges.

From this it was determined that between the natural rainfall and the removal of acid pockets from the spoils originally in the soil that the lakes began to neutralize themselves, and the maintenance of these pH factors at the higher level would be sustained by the establishment of vegetation along the shorelines, eliminating or reducing to a minimum the amount of erosion carrying acid materials into the water.

The system for treating the acid lakes was varied several times during the course of the construction operation. This was done primarily to facilitate the use of various pieces of equipment that would be available during the time of acid water treatment. Due to the expense of the equipment available at this time to treat these waters and the types and amounts of material that were required to complete the operation, it was decided that the basic form of lime would be changed from a calcium carbonate to a hydrated lime and applied by some slurry method.

Acid water treatment procedures are based on the addition of lime in a slurry into the water body and the addition of circulation within the water body to provide a continuous distribution of lime throughout the lake. In order to provide the same techniques at the least amount of expense for the purpose of the demonstration, it was decided that the normal hydromulcher used for seeding operations would be used to apply the hydrated lime in a slurry form, and that circulation of the body of water internally would be done by the use of various pumps.

On June 12, the contractor had made arrangements to have on site the hydrated lime and the equipment to proceed with the acid water treatment. At that time there were four tests of the water taken at various points along the lake to show the pH range. These tests reflected a pH range from 3.5 to 4.0, as they had in May. Three pumps were installed at strategic locations. To determine the ability of the pumps to circulate water within controlled areas, each was equipped with a discharge hose, varying from 150' to 250' in length and a 50'

TABLE 10.1: RESULTS OF WATER TESTS

Test Required	1*	Filled 2	A	3**	C	D	4	B	5	Well (on site)
pH	3.6	–	5.9	3.6	3.3	3.2	3.3	3.1	3.4	8.1
Iron	20	–	0.2	31.5	21	2.7	59.5	22	52.5	61
Sulfate	840	–	700	940	2400	2900	500	2600	980	100
Total Acid	1020	–		586			1100		536	–
Total Acid (N)			0.1		0.21	0.20		0.22		–
Free Acid	180	–		78			150		95	
Free Acid (N)					.004	.004		.005		–

Figures given in mg/liter unless otherwise noted.
Tests taken: Lakes 1 through 5 June, 1972
Lakes A through E Dec, 1972

*Lake 1 was pumped into Lakes "B," "C," & "D" during pumping operations and was filled with relocated spoil.
**Lake 3 was divided by a land bridge formed with relocated spoil to form Lakes "C" & "D."

Source: PB 226 905

intake hose. The lime was pumped into the hydroseeder and, with a pump drawing water from the lake into the hydromulcher, was then discharged out onto the lake in a spray pattern. Two results of the use of hydrated lime, as determined by tests taken during and after the operation, were immediate; it achieved suspension and provided a reaction to the acid effect of the water. An accumulation of hydrated lime on the bottom of the lake also was evident from mud samples taken.

Figure 10.6 is a sketch of the location of the pumps and the hydrated lime source (the hydromulcher) and the common carrier hauling the material to the site. The fan-shaped areas show the approximate area of treatment as the lime was discharged by the hydromulcher. For the purposes of the demonstration, it was also necessary to determine the effectiveness of the pumps in establishing a circulatory pattern within the body of water. The location of these pumps by the sketch and the arrangement of the fan-shaped application areas show that some areas did not receive direct contact with the hydrated lime, but the pH checks of the water on the 21st of June showed that the pumps were effective in recirculating the water and bringing the pH to a level satisfactory for animal and plant life.

The amount of hydrated lime required for the treatment of Lake B was determined by the amount of acid present in 14,000,000 gallons of water. This basic determination showed that 60 tons of hydrated lime would be necessary to bring the pH of the water from its existing range to a living range of 6.5 to 7.0 or near neutral condition. It was determined, however, that after the first day of application of hydrated lime that 60 tons of lime would not be necessary to do the same job, and that only 40 tons would be required in order to provide proper coverage around the lake. The total amount of material then applied to Lake B was approximately 40 tons.

The cost for the treatment of this lake was based on the equipment used, the man hours required, and the cost of material plus demurrage spread over two days of actual application. Total application time was slightly over 3½ hours each day for two days, but the three pumps situated around the lake were run for 16 hours. This was to encourage additional circulation after application and to avoid any unnecessary concentration of $Ca(OH)_2$ at the source points. Total cost of the project was approximately $1,900.00 to treat the 14,000,000 gallons of water. The cost per ton applied was $46.75 or $0.0234/lb of hydrated lime.

After the final application of the material on the 13th of June, 1973, monitoring tests were run on June 21 that showed the pH was still at the range of 6.5 to 7.0. Tests taken in areas that were not in the application zones also reflected a pH of 6.5 to 7.0. The final assumption might be that the treatment of acid water in this particular pond by this method was successful and that the encouragement of growth around the shoreline could eliminate, or reduce to a minimum, the amount of erosion that would carry acid materials back into the pond.

A mud sample from the base of the pond taken on the 21st of June also showed that there was a great deal of lime still active at the base, and that it was still working as a neutralization factor when a sample of soil taken by the pond edge showed an acid pocket with a pH between 3.5 and 4.0. Acid produced from the moisture hitting the acid pocket and its runoff into the pond was still being effectively neutralized by the lime active in the pond itself.

FIGURE 10.6: LOCATION OF PUMPS AND HYDROMULCHER

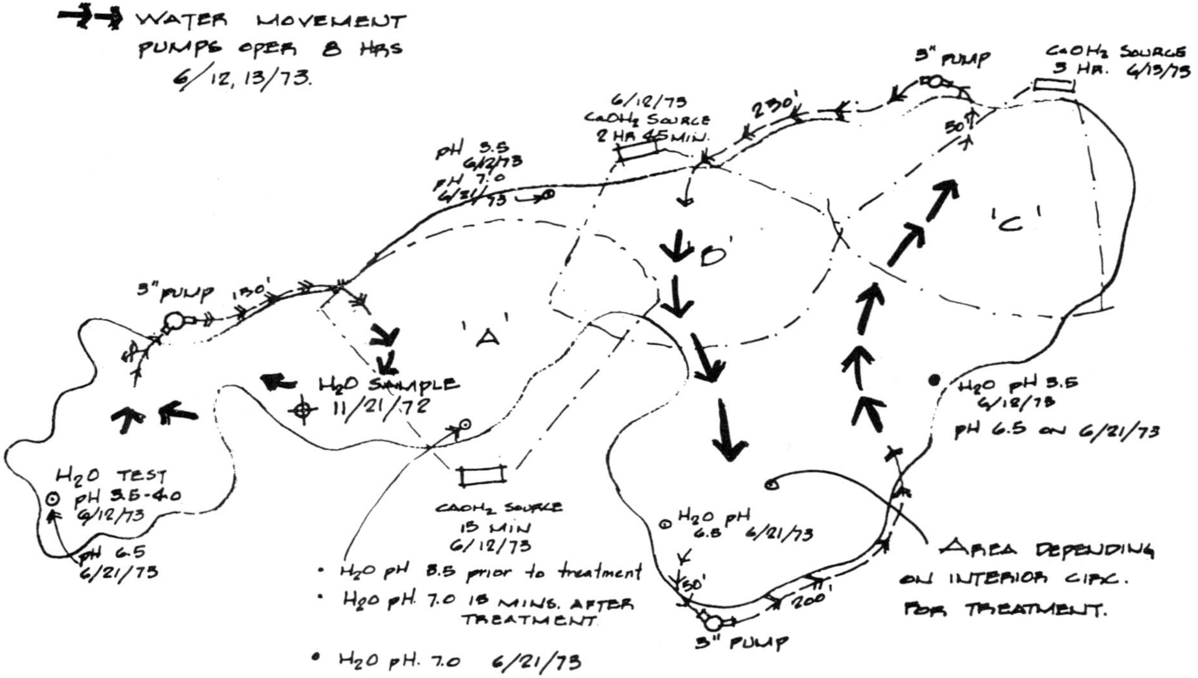

Source: PB 226 905

Tree Transplanting

Although tree transplanting was assumed to be one of the original elements of the contract, it was later determined that this was not a part of normal reclamation activities and therefore should be taken from or deducted from the contract price as awarded in September 1972. In addition, it was considered that tree transplanting, unless it was a reforestation project as a part of reclamation activities, was an aesthetic element of the project for the benefit of aesthetics alone, and that the Department of Conservation should be responsible for the planting of any additional vegetation following total reclamation of the project. Therefore, an agreed upon price of $7,000.00 was deducted from the original $223,000.00 contract.

Conclusions

It was the hope at the inception of this work that it would provide some of the necessary cost data, the efficiency factors, and even the feasibility of a statewide strip mine-reclamation program. The objective was to be documented by the physical evidence derived from performing one of these demonstration acts. Prior to that date, which was in early 1972, the only cost figures available to the State of Illinois were those performed or resulting from performances by other states and by existing strip mine operations in Illinois. These were rather ambiguous and lacked the necessary information that applies to an Illinois area mine situation.

From the cost incurred in this particular project, it is evident that the scope of cost is as variable as each site is variable through the state, and that each acre cost would have as much differential as do the more than 100,000 existing acres. As this document shows the actual cost incurred was in excess of $212,000.00. The contractor received $216,000.00 for the work he performed. Although the actual cost of $212,000.00 does not reflect actual overhead and profit, it does reflect an increase in the basic prices of equipment charges that would entitle the contractor to his normal margin of profit.

Table 10.2 is a summation of the development cost for the project and shows the periods of the preceding year in which these various work items were accomplished. It also reflects an approximation of the expenditures for each of those items during the month that the work was performed. Table 10.3, which is a summation of unit cost by work item, shows the unit of measure, the total number of units, the total cost for each of those work items, and a cost per unit for this project. It should be noted that these costs per unit include only the equipment, material, operating costs, and labor. Profit and overhead are not a part of that actual unit price.

Through the course of the project, an effort was made to procure the contractor's ideas and his concepts on equipment efficiency, plan preparation, and document preparation—plans that would allow him to come up with closer estimates of material involved—and, the opportunity for work-staging prior to actual letting of the contract. Due to the uniqueness of some of the work items, the contractor indicated that additional explanation should be made in the proposal to allow the contractor to be more informed of the work he is expected to complete.

TABLE 10.2: SUMMATION OF DEVELOPMENT COSTS

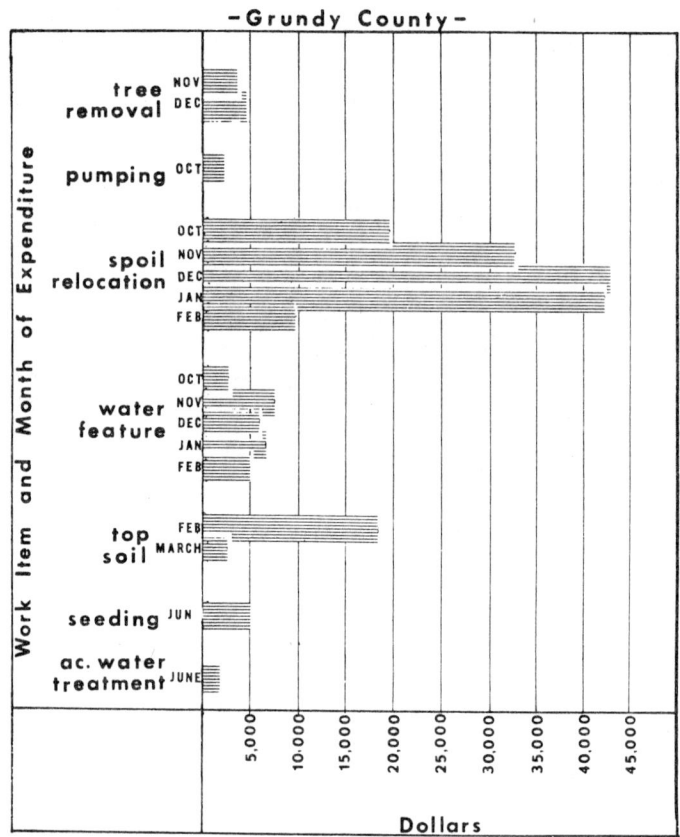

Source: PB 226 905

TABLE 10.3: SUMMATION OF UNIT COSTS BY WORK ITEM

Work Item	Total Units	Total Cost*	Cost per Unit
Tree removal, acres	15	$ 10,750.06	$716.00
Pumping, gal	885,600	2,223.56	0.0025
Spoil relocation, yd^3	605,230	146,733.50	0.2424
Water feature alteration, yd^3	41,400	27,672.64	0.6684
Topsoil, yd^3	75,689	20,687.45	0.2733
Seeding, acres	35	5,111.83	146.05
Acid water treatment, lb	79,000	1,869.70	0.0237

*Includes: equipment, materials, operating costs, and labor

Source: PB 226 905

As a general conclusion, it is clear that the project was successful from the time of its inception to its completion in the way that it conformed to the concept envisioned by the Department of Conservation; by its proof that environmentally significant problems can be solved; in the utilization and treatment of equipment; and in the ways in which equipment efficiency was affected by application in a surface mine area. In addition, work functions not normally performed have been identified, utilized, and proved to be both feasible and necessary for aesthetic reasons in future reclamation projects.

For example, the use of a small drag-line in the areas of shoreline alteration achieved an aesthetic effect (and at the same time provided a way of working adjacent to water shorelines that are not readily accessible by other types of heavy equipment). The biggest success of the project, however, is the fact that Illinois is now equipped to actually verify the cost of different procedures and operations in the effective reclamation of strip mine areas.

Since the prime purpose of the project was to verify actual cost of reclamation by outside or off-site interests, then the success is measured by the actual cost per dollar based on a low-risk or high-risk project. It should be understood that this particular project, due to the amount of grading changes and material moved for aesthetic reasons, is in effect a high-risk project. The cost for this particular project was something over $3,600.00/acre, excluding land cost. The ultimate use of the project as an active recreational area would indicate that the $3,600.00 per acre to transform an environmental disaster into a usable area was well worth the cost.

One other conclusion regarding reclamation costs can be made, which is also significant to a strip mine operation and having similar characteristics such as acid production of spoils, low pH ranges in existing lake bodies, and normal percentages of organic soils available on site; this conclusion is that acid can be controlled and acid water can be corrected economically by utilization of simple equipment that has versatility. A combination of vegetation establishment on spoil banks, reduced gradients of those spoil banks, and proper soil treatment prior to seeding begins to reduce the acid sources available on the spoil bank areas.

In addition, acid treatment or neutralization of the lake bodies themselves can be done efficiently with a minimum amount of materials if the primary source of acid production is reduced to a point of inactivity. This should not be construed as a panacea for all acid mine spoils. This particular site merely reflects the incorporation of organic soils into the consolidated material and some placement of topsoil above those consolidated materials in an effort to produce areas compatible for vegetative growth. The acid conditions that were prevalent on this site are also identified as a state that could ultimately correct itself if the length of time that it takes for the normal evolutionary processes is acceptable. The alteration of the pH ranges in the remainder of the lakes on site suggests that normal evolutionary processes can be speeded up by isolated neutralization treatments if required.

A primary consideration for any reclamation project is that the work be actually accomplished by contractors who are qualified and competent to perform the scope of work required. The success of this particular demonstration project was primarily based on the cooperation of the contractor and his ability to provide good organization in conducting many operations simultaneously and yet not have them interfere with one another.

SURFACE MINE RECLAMATION, MORAINE STATE PARK, PENNSYLVANIA

Purpose of Project

The Moraine State Park project was the first under the Appalachian Surface Mine Reclamation Program in which the Commonwealth of Pennsylvania and the Federal Government (represented by the Bureau of Mines) entered into a cooperative agreement to reclaim and restore surface-mined lands. The project involved approximately 177 acres of surface-mined lands adjacent to a 3,200-acre lake, Lake Arthur.

This project was designed to cope with conditions resulting from past surface and underground mining: elimination of acid mine water seeping or flowing from abandoned underground mine workings which would be detrimental to the ecological environment in and around Lake Arthur; backfilling and regrading of surface-mined areas to original contour to eliminate public hazards associated with open pits, highwalls, exposed deep mine openings, and water containments; reduction of erosion on barren outslopes and highwalls; and restoration of aesthetic values.

The report (1970) on the project was prepared to provide the general public, private industry, and Federal and State Governments with surface mine reclamation data on the methods and costs of a reclamation project where project specifications required that the disturbed lands be returned to their original contour.

Location

Moraine State Park, a 15,000-acre recreational facility, located in Brady, Franklin, Muddy Creek, and Worth Townships, Butler County, Pa., is east of the intersection of U.S. Route 422 and Interstate Route 79. This site was selected to serve the six-county Pittsburgh Region, and it is strategically situated within a 60-mile radius of approximately 4 million people. The park attendance is expected to exceed 1 million people annually.

The topography in Moraine State Park ranges from 1,170 feet along the major drainage channel in the park to 1,520 feet along the hill crests along the northern park boundary. The major drainage channel is Muddy Creek. This creek with its tributaries feeds Lake Arthur.

The restoration project was conducted in two main areas, the North Central Section and the Northwestern Section. These two sections are adjacent to the northern edge of the lake site.

Hydrology

Muddy Creek, the southernmost tributary of Slippery Rock Creek, is the main stream flowing through Moraine State Park. The Muddy Creek Watershed which drains 58.5 mi^2 will supply the water for the 3,200-acre Lake Arthur. The small tributaries of Muddy Creek are Bear Run, Big Run, Shannon Run, and Swamp Run.

The alkaline condition of the water flowing in Muddy Creek is normally favorable to the support of aquatic plant and animal life and to the development of

Area Reclamation Projects

the major recreational facility. However, during periods of unusually high rainfall and the winter and spring snow melts, the stream is subject to an increased inflow of acid mine drainage.

The acid water pollution in Muddy Creek is directly attributed to past surface and underground extraction of coal from approximately 66 mines which have been located on the five productive coal seams in the watershed. This pollution originates predominately from unsealed abandoned underground mine shafts, slopes, and adits. Surface mining is believed to contribute less than 20% of the total pollutant; however, surface mining is indirectly responsible for escape of acid water pollution because surface mining frequently intersects and exposes abandoned underground mine workings.

The water quality of Muddy Creek is inversely related to the amount of water flow in the watershed. The increase in acid content is attributed to greater inflow of acid mine water from underground mines predominately during periods of high surface water runoff. The acid load in Muddy Creek has been measured at 630 lb of acid per day.

Condition of Sites Prior to Reclamation

The two project sites, referred to as the North Central Section and the Northwestern Section, are located adjacent to the northern shore of Lake Arthur. These two areas which cover approximately 177 acres are but a small portion of the total 15,000-acre park; however, they are situated in areas designed to receive a high flow of park visitors.

Bituminous coal was extensively mined throughout the area using both underground and surface recovery techniques. Most of the mining occurred prior to World War II and in comparison with modern coal mining operations these mines were relatively small and of short duration. The last surface mining in and around the project sites was conducted about 1945. Surface mine regulations in the mid-1940s were minimal, and little effort was made by operators to reclaim or restore the disturbed acreages. Little or no backfilling of the surface excavations was required; the large percentage of rock and dirt existing in the pit bottoms was the result of sliding and spalling of rock from the highwall. The surface condition of the project areas was characterized by irregular mounds and pits, fill benches, vertical highwalls, and a variety of trees and plants.

The North Central Section, the smaller of the two project sites, covered about 57 acres. The area was comprised of two units with somewhat contrasting conditions. The Middle Kittanning coal was mined in the lower area to the southeast. Overburden from this operation has been cast over a wide area, including part of the adjacent valley bottom. In the larger area to the northwest contour strip mining took place in the Upper Freeport coal; the essentially flat-lying coal seam cropped out horizontally around the hill. As a result, a large isolated island was created within the stripped area. The highwall height averaged about 45' throughout this section.

The Northwestern Section consisted of approximately 120 disturbed acres. The surface configuration was very rough. The Middle Kittanning coal was mined in a lower cut and the Lower Freeport was extracted in a larger disturbed area. The highwall heights averaged 45' to 50'; however, a few heights in excess of

75' existed. Surface mining operations exposed several underground mine workings throughout the project areas. Many of these openings were not only a public safety hazard, but also a source of mine water drainage. The acidity of the mine water was measured to have a pH 3 to 4. Many of the stream beds were coated with "yellow boy," an iron precipitate from mine drainage which smothers bottom flora and fauna.

A substantial vegetative cover over the project areas had been established since mining terminated 25 years ago. Many of the trees had been planted under State programs, while many of the grasses and legumes were volunteer growth.

Project Specifications

The reclamation and restoration work in Moraine State Park involved the clearing, excavating, backfilling, and regrading of the disturbed lands in the project areas. This work was conducted in accordance with the rules and regulations of the Land Reclamation Board, Pennsylvania Department of Mines and Mineral Industries, and the Pennsylvania Bituminous Coal Open Pit Conservation Act, unless otherwise required. The following project specifications are excerpts from the project contract.

Preparation of Sites: All vegetation within and immediately adjacent to the project areas is to be removed or disposed of by the following procedure.

(a) All trees with a diameter of 1½" or greater, measured 12" above the ground, shall be uprooted or cut down and burned or removed from the site.

(b) All brush and trees with a diameter less than 1½" may be buried in the fill area. If inspector should determine that the accumulation of brush and/or trees to be excessive the contractor will be required to remove or spread the material over a greater area.

(c) In the fill area, all tree stumps with a diameter greater than 1½" which are not uprooted may be left in place.

(d) All brush and trees which are to be buried must be buried in the pit on the opposite side from the highwall and covered with at least 3' of fill.

Backfilling and Terracing: Because of the location of the project areas adjacent to Lake Arthur and other recreational and education facilities within Moraine State Park, the project specifications required that the disturbed lands be restored to approximately their original contour or to a contour compatible to the surrounding terrain. This involved moving a vast amount of spoil bank material back into the strip pits. Specific backfilling specifications had the following requirements.

(a) All rocks in excess of 1 yd^3 to be disposed of and other rough rock must be covered with at least 6" of the best available material.

(b) No depressions to retain water which may seep through the spoil and produce acid drainage will be permitted.

(c) All toxic material is to be buried in strip pits with at least 3' of cover.

(d) The best available fine spoil and topsoil is to be used for the surface cover in the reclaimed areas.

(e) A drainage channel will be constructed at the downstream end of the lower seam in the North Central Section according to the contract specifications.

Area Reclamation Projects

Contractor's Responsibility: Upon completion of the work the contractor was required to restore the land which he had affected outside the project area. He was to return this land to its original elevation, contour, and condition existing prior to commencing project work; restore or replace all improvements to the land (including structures, buildings, fences, landscaping, etc.) which may have been removed or damaged by or as a result of the project work.

Project Work

The actual project work consisted of site preparation, backfilling, and terracing the surface pits to original contour.

Preparation of Sites: The initial work effort was directed toward improving the project sites to provide access roads and clean the overburden spoil of the vegetative overgrowth for the larger earth-moving equipment. The work involved the clearing, grubbing, and disposal of brush and trees. This task was performed by a Series 3-T Caterpillar D-7 Bulldozer equipped with blade and rootrake. The rootrake was found to be ineffective because of the small size of trees and brush found on the project sites. The vegetation, however, was easily handled and removed by the D-7's blade.

The average time required to clear the sites ranged from 2.5 hours per acre in the North Central Section to 3.5 hours per acre in the Northwestern Section. This difference in operating time is attributed to the steeper terrain in the Northwestern Section and adverse winter weather conditions during work in the Northwestern Section. Table 10.4 presents data on site preparation time and costs.

TABLE 10.4: SUMMARY OF SITE PREPARATION COSTS

Section	Work, hours	Acres	Time/acre, hours	Average[*] D-7 cost per hour	Cost/acre
North Central......	148	57.3	2.58	$13	$33.54
Northwestern.......	424	120.2	3.52	13	45.76
Average..........	-	-	3.22	13	41.86

*1965 figures

Source: PB 225 165

Backfilling and Terracing: The two project sites were returned to their original contour by moving approximately 1.4 million cubic yards of overburden. The slope to which the disturbed area was returned averaged 10°. Locally, original slopes ranging up to 13° existed in the Northwestern Section. Figure 10.7 illustrates the profile of the two project areas prior to reclamation and the configuration of the areas after completion of the restoration work.

Project work in the North Central Section was conducted during the summer and fall months. The spoil material consisted predominately of weathered shale and siltstone with intermixed sandstone slabs. The only difficulty in the reclamation of this section was encountered on the lower seam in the southeast unit. Beneath the Middle Kittanning coal bed there is a clay bed which when water-soaked made it practically impossible for the equipment to maintain adequate traction while moving the overlying material in the outcrop area. The distance material had to be transported in the North Central Section averaged between 250' to 300' along the upper seam and greater than 600' along the lower seam (Figure 10.7).

FIGURE 10.7: BEFORE AND AFTER CONFIGURATION OF SURFACE IN PROJECT AREAS

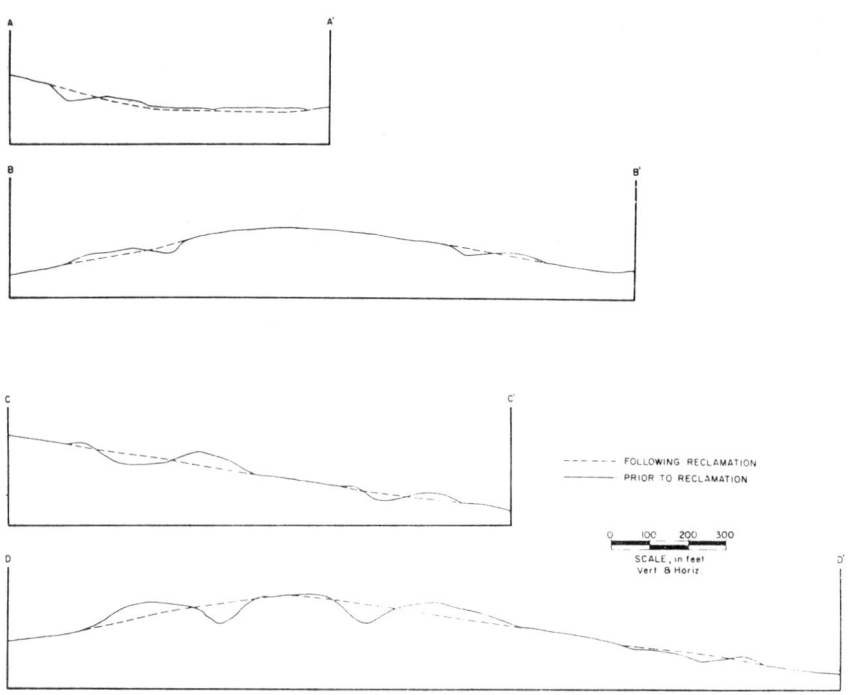

Source: PB 225 165

Reclamation in the Northwestern Section was done during the fall, winter, and early spring months. The spoil material was similar to that found in the North Central Section, except that there was a greater number of sandstone blocks distributed throughout the spoil. The work efficiency suffered somewhat in this phase of the project because the ground was frozen a substantial portion of time during the winter. This setback was offset to a large extent by using a bulldozer equipped with a ripper to break the frozen spoil. The average distance the spoil material had to be transported increased to about 400' while a few sections covered more than 600' (Figure 10.7).

In June 1967 the initial equipment which was deployed to the North Central Section included two Series 18-A Caterpillar D-9 bulldozers. A Series 66-A Caterpillar D-9 equipped with Boulderson blade and a Series 14-A Caterpillar D-8 were added to the work force late in July. A second Series 14 A Caterpillar D-8 arrived at the project site in August.

By mid-September all project work except the grading of a drainage ditch was completed in the North Central Section, and the equipment was transferred to the Northwestern Section. The final grading on the drainage ditch was delayed because the area was too wet to be effectively graded at that time. This work was completed in December.

In December, the contractor working under a codicil to the original agreement removed material from five coal refuse banks from areas to be inundated by the lake. This material was buried in the surface pits in the Northwestern Section. The volume of this acid-producing material was negligible when compared to the volume of strip spoil moved. Three Westinghouse scrapers and two Euclid trucks performed the work. This work did not interfere with the original reclamation work nor was the cost of this work borne by the original project contract.

Removal of the refuse banks was completed in March, at which time the Westinghouse scrapers were assigned to the original reclamation project. A Series 46-A Caterpillar D-8 was used to assist the scrapers. The reclamation work in the Northwestern Section was completed in May 1968. About 29 hours of total equipment time per acre were required to complete the reclamation objectives in the North Central Section as compared to the 51 hours of operating time per acre required in the Northwestern Section. Essentially two factors influenced this increase in working hours:

(1) To return the mined areas to their original contour the North Central Section required moving approximately 5,000 yd^3/acre while the Northwestern Section with its deeper excavations required about 9,000 yd^3/acre.

(2) The distance that the equipment had to travel was significantly greater in the Northwestern Section as indicated in previous paragraphs. Although there was a substantial increase in the quantity of material and the pushing distance, the operating time and costs were held to a minimum because of equipment modification and operating procedures. (See Discussion.)

A breakdown of total equipment time required to complete the backfilling and terracing phase of the reclamation project is presented in Table 10.5. The average cost of material moved was $0.16/$yd^3$ in the North Central Section and $0.15/$yd^3$ in the Northwestern Section (Table 10.6). These costs are based solely on operating equipment time and do not include site preparation costs, repair and maintenance costs, support equipment, and office facilities.

Adequate drainage provisions were made during the reclamation operation. A ditch was constructed in the North Central Section in the southern tip of the area mined for Middle Kittanning coal. A seal was placed (under a separate project contract) at the mine mouth. A second drainage was constructed in the Northwestern Section by laying a concrete pipe in an existing drainage channel from the Upper Freeport coal seam to prevent erosion of the reclaimed area by mine drainage seepage from the underground mine. The outlet was immediately above the reclaimed lower section, and consequently a channel was constructed to divert the water around the reclaimed area.

TABLE 10.5: SUMMARY OF EQUIPMENT TIME

Equipment model (series)	North Central Section		Northwestern Section	
	No. of pieces	Work, hours	No. of pieces	Work, hours
D-9 (66-A)	1	384	1	1,352
D-9 (18-A)	2	1,018	2	2,116
D-8 (46-A)	-	-	1	848
D-8 (14-A)	2	270	2	1,388
Scrapers	-	-	3	432
Total	-	1,672	-	6,136

TABLE 10.6: SUMMARY OF RECLAMATION COSTS

Section	Total work, hours	Acres	Time/acre, hours	Average* cost of equipment per hour	Cost/acre	Cost/cubic yard
North Central	1,672	57.3	29	$26.88	$780	$0.16
Northwestern	6,136	120.2	51	27.50	1,402	.15

*1965 figures

Source: PB 225 165

Vegetation of Project Sites

The establishment of a vegetative cover on the reclaimed areas was not included in the Federal-State cooperative contract for the restoration of the surface-mined areas. The revegetation operations were conducted by the Pennsylvania Department of Forests and Waters after the reclamation work had preliminary approval by the Federal and State project representatives. The reclamation work in the North Central Section was completed in the fall and the area was immediately prepared for planting. The North Central Section received chemical treatment (8¼ tons of 10-10-10 fertilizer and 110 tons of standard ground limestone) before grasses were sowed. The quantities of grasses applied to the area were as follows: 440 lb common rye grass, 220 lb timothy, 275 lb orchard grass, 110 lb common clover, and 110 lb bird's foot trefoil.

Grasses were not sown in the Northwestern Section; however, volunteer growth was observed at several locations during the May inspection.

In addition to the application of grasses, trees were planted on an 8' x 8' spacing in the North Central and Northwestern Sections. Evergreens were planted in May 1968; the type and number of trees included: 17,500 red pine, 16,800 white pine, 2,800 banks pine, 1,400 pilch pine, 2,100 Australian pine, 14,000 white spruce, and 9,800 Norway spruce.

A 70 to 75% survival was estimated in April 1969. At this time, deciduous trees, such as red oaks and red maple, were introduced to the Northwestern section.

Discussion

The fine texture of the spoil material facilitated its movement, and except for a few large sandstone slabs no major earthmoving obstacles were encountered. Although most of the project work was conducted during the winter months, only 5 work days were lost.

The reclamation work, which ran approximately $800/acre in the North Central Section and $1,400/acre in the Northwestern Section, is costly when compared to basic reclamation cost necessary to satisfy many of the State surface reclamation laws. This project, however, was to repair the affected acreage to a higher and more specific land use which is not the case in basic reclamation and was undertaken long after the original surface mining had ceased and the mining equipment had left the site. Consequently, these costs are applicable only when the reclaimed land is intended for multiple use or special objectives as outdoor recreation, residential, commercial, or industrial development, cropland or wildlife habitats, and/or when reclamation is accomplished at a late date, and/or when reclamation is performed by persons other than the mine operators.

Operating costs/yd^3 of material moved were less in the Northwestern Section than the North Central Section even though (a) the average distance the overburden had to be moved was greater; (b) the number of cubic yards of spoil necessary to reclaim an acre of disturbed land was almost doubled; (c) the spoil contained a greater number of sandstone slabs; and (d) the spoil was frozen during the winter months.

These lower operating costs in the Northwestern Section are attributable to:

(1) The project equipment arrived at the North Central Section at various intervals throughout the project work.

(2) The total work force had developed efficient working procedures by the time work began in the Northwestern Section. At the outset of project work the bulldozers operated individually; however, shortly after a month had expired a team approach was initiated which resulted in a more systematic reclamation effort. This proved advantageous in regard to support assistance and equipment maintenance; and the project inspectors could more easily evaluate the actual work accomplished.

(3) The Caterpillar D-9, equipped with a Boulderson blade, significantly increased the number of cubic yards moved per haul cycle as compared to the machines equipped with the normal or U-blades.

(4) The scrapers were instrumental in an early completion of reclamation work. These machines not only have a greater load capacity than the bulldozers but they also have the advantage of speed over the bulldozers.

It was found that the use of large equipment, such as the scrapers and bulldozers equipped with the Boulderson blade, expedited the project work. Similar results could be anticipated on any reclamation program which would be conducted under similar physiographic conditions and of comparable magnitude to the Moraine State Park project.

EVALUATION OF POLLUTION ABATEMENT PROCEDURES, MORAINE STATE PARK

Project Background

This project was an evaluation of the various mine drainage pollution abatement techniques completed during the construction phase of the Moraine State Park, Pennsylvania. The major objective of the pollution abatement project was to insure good water quality in Lake Arthur.

Information for the examination concerning the existing mining conditions in the park consisted of a reconnaissance of deep and strip mine areas, a search for old mine maps and contacts with persons who worked in or had knowledge of these mines. Water samples were collected weekly from May 23, until November 1, 1967 and monthly thereafter. The water samples were analyzed for pH, alkalinity, acidity, iron, manganese and aluminum. A total of 85 weirs were installed to measure flows of all known mine drainage discharges.

A diamond core drilling program was performed consisting of 23 holes and 1,858.6' of drilling. These holes were drilled at specific locations to determine the elevation and nature of the coal and associated strata, the extent of mined-out areas, and other geologic data. Additional field work consisted of surveys for mine drainage points and for the location, elevation and cross sections of mine refuse piles. Maps of the park area were obtained from the Department of Forests and Waters for use in the examination. Several mine drainage projects were being performed concurrently during the period of the mine drainage examination, one of these being the project described in the preceding section.

Pollution Abatement Measures

Pollution abatement projects performed in the watershed included strip mine reclamation, refuse pile removal, surface sealing, deep mine hydraulic sealing, well plugging, Jeep mine air-trap sealing and miscellaneous work incidental to restoration and abatement. Details of the strip mine reclamation work are discussed below.

Strip Mine Reclamation: Strip mine reclamation work included backfilling and regrading to terrace-type restoration, approximate original contour and special restoration as required. Most of the reclamation included installation of diversion ditches and slope drain flumes. Soil treatment and planting were also included. The strip mine reclamation consisted of work performed under seven separate contracts in nineteen strip pits located throughout the park in three general areas.

Approximate original contour restoration started at or beyond the top of the highwall and were regraded and sloped to the toe of the spoil at a maximum angle not exceeding the original contour of the land before mining, with no depressions to accumulate water. For the open pits backfilled by terracing, the steepest grade of the highwall and toe of spoil was limited to 45°. The terrace was regraded to a maximum descending gradient of 5° from the base of the highwall toward the top edge of the toe of spoil with no depressions to hold water. For backfilling in all but one project, the maximum limits of work were 100' above the highwall of the affected area and 50' below the toe of spoil.

Diversion ditches excavated above the highwalls in the strip mine areas have a maximum cross-sectional area of 10 ft^2 and have a uniform descending gradient. Slope drain flumes constructed across the backfill areas started at the point of intersection with the diversion ditch above the top of the highwall and extended down across the entire width of the graded backfill area to a point of discharge below the toe of spoil. The flumes were constructed with 36" bituminized fiber ½ section pipe.

All areas, unless designed otherwise, received soil treatment and seeding in accordance with the following:

(1) Apply ground limestone at the rate of 2.5 tons/acre.
(2) Apply fertilizer 10-10-10 at the rate of 300 lb/acre.
(3) Use a disc harrow and spring tooth harrow to thoroughly mix both applications with the soil.
(4) Apply seed using a tractor-mounted broadcast seeder as per the following formula. The seed shall be purchased already mixed. The seed shall be applied at the following rates: common rye grass, 8 lb/acre; common timothy, 4 lb/acre; orchard grass, 3 lb/acre; common clover 2 lb/acre; and bird's-foot trefoil, 2 lb/acre.
(5) After seeding, go over entire area with a disc harrow and spring tooth harrow.

Where specified, various species of trees were planted in the reclaimed areas. The deciduous and evergreen trees were planted on 8' x 8' centers which allows for approximately 700 trees/acre. The species included red oak, red maple, red pine, white spruce, hybrid poplar, white pine, hemlock, European black, alder and Norway spruce. The shrubs (arrowwood, viburnum and autumn olive) which are a source of wild game food, were not intermixed with the trees but were planted in rows along the edges of the reclaimed areas.

In addition to the 461.7 acres backfilled under the various contracts there were 265 acres of strip-mined lands backfilled by the previous mine operators. None of these areas were contributors of mine drainage pollution. In the North Corridor of the Moraine State Park there were 167.6 acres of strip mine reclamation. However, all of this area drains into Big Run and is not considered in this study.

Project ASMRP-1: Reclamation work for the strip mined lands in the project areas consisted of clearing, excavating, burying toxic materials, backfilling and regrading to the approximate original contour. Restoration work consisted of 57.3 acres in the North Central Section in Brady Township and 120.2 acres in Worth Township. Also included in this contract was the construction of a drainage channel in the lower seam of the North Central Section and the burial of 82,347 yd^3 of refuse pile materials in the strip mine pit during backfilling.

After completion of the restoration work, the reclaimed areas received soil treatment and planting with grasses and trees. This work was not included in the Federal-State contract and was conducted by the Pennsylvania Department of Forests and Waters. The project work was performed during 1967 and 1968 at a cost of $219,118. Details of this work are given above under "Surface Mine Reclamation, Moraine State Park, Pennsylvania."

Project MD-8C: The strip mine reclamation part of the project consisted of excavating, burying acid-producing materials including 134,721 yd^3 of refuse pile rock, backfilling and regrading to approximate original contour of 36.0 acres of affected area in a strip mine adjacent to the Lincoln deep mine and 11.0 acres in a strip mine adjacent to the Hilliard deep mine. Also included in this work was the installation of a diversion ditch above the highwall and a slope drain flume across the backfill. After restoration work was completed, the area received soil treatment and planting with grasses. This work was performed in 1967 and 1968 at a cost of $29,964.

Project SL-105-1: This project included three contracts for reclamation work in three separate sections of the park. This included work in five strip areas in the Northwest Section, four strip areas in the Southwest Section and four strip areas in the Eastern Section. Reclamation work in the Northwest Section included 23.9 acres backfilled to approximate original contour in the first area; 6.8 acres and 12.8 acres in the second and third areas both with terrace-type restoration; no backfill work was required in the fourth area, however, drainage ditches and slope drain flumes were required; and 2.3 acres of approximate original contour in the fifth area for a total of 45.8 acres of restoration.

Work in the Southwest Section included two areas of 6.3 acres and 25.2 acres with terrace-type backfill, one area of 17.0 acres backfilled to approximate original contour and an area of 11.6 acres with a combination of terrace-type and approximate original contour backfill for a total of 60.1 acres of restoration. The Eastern Section reclamation included one area of 4.8 acres with approximate original contour and three areas of 29.3 acres, 23.7 acres and 46.8 acres with terrace-type restoration for a total of 104.6 acres.

Diversion ditches were installed in four areas of the Northwest Section, two areas of the Southwest Section and all four areas of the Eastern Section. All areas received soil treatment, except for two areas in the Eastern Section. All areas were planted with grasses and/or trees except Area 4 of the Northwest Section which had adequate vegetative cover. The project work was performed in 1969 and 1970. The costs consisted of $109,628 for the 45.8 acres in the Northwest area, $120,200 for the 60.1 acres in the Southwest area and $156,900 for the 104.6 acres in the Eastern area.

Project SL-105-1A: The project consisted of reclamation of 8.7 acres of affected area in the Eastern Section of the park. Restoration included terrace-type backfilling and installation of 1,300' of diversion ditch above the highwall. No soil treatment was required and the area was planted in trees. Work in the project was performed in 1969 and 1970 at a cost of $8,775.

Project SL-105-1B: The project area is located on and adjacent to Interstate 79 in the Southwest Section of the park. Effluent from the drainage facilities installed as part of the highway construction discharged surface water and drainage from the highway into a strip mine area along the east side of the highway. This water, along with other drainage into the strip mine, produced acid discharges. Restoration work included 15.6 acres backfilled to approximate original contour and 2.4 acres, within the right-of-way of the highway, graded and backfilled to special restoration.

Drainage facilities installed as part of this contract included the installation of a storm sewer along the highway to carry the drainage formerly discharged into the strip mine to a natural drainage course below the strip mine area. After completion of the restoration work, the area received soil treatment and planting. The area within the highway right-of-way was planted with grasses and outside the right-of-way with grasses and trees. The project work was performed in 1970 at a cost of $27,623.

Method of Investigation

The periodic water sampling and flow measurements were started in May 1967 and continued through June 1971. Eighty-five weirs were installed in 1967 at all the known mine drainage discharge points. Since that time, many of the sampling points were either moved or eliminated due to the construction in the abatement projects and due to the inundation of the lake.

Water samples were collected at least once a month. These samples were analyzed and tabulated indicating the flow in gal/min, pH, and the alkalinity, acidity, iron and manganese in both mg/l and lb/day. The mine drainage data have been compiled into the various deep mine and strip mine areas, and this information divided into eight periods of six months' duration each, starting with July 1, 1967 and ending with June 30, 1971.

After examining the mine drainage data over a period of three years, acidity was considered to be the major component. The principal criteria for evaluating the effectiveness of the mine drainage abatement projects was the examination and comparison of the discharge flow rates, and average pounds per day of net acidity and iron before and after abatement. As supplemental information, the average pounds per day of alkalinity and acidity on a six months' basis for the various mining complexes in the watershed area were computed and tabulated.

The investigation and evaluation of the pollution abatement procedures were limited chiefly to the deep mine sealing and strip mine reclamation projects. Although it was not possible to compute the pounds of acid and iron that would have been generated by the refuse piles, the subsidence areas, the mining appurtenances and oil and gas well sites, it was estimated that serious and continuous pollution would have resulted if the refuse pile removal, surface sealing and well-plugging projects were not performed prior to the inundation of the lake. The data for Lake Arthur are an indirect indication of their effectiveness.

Strip Mining of Coal

The gates of the dam were closed on May 15, 1969. Starting in September 1969, water samples were taken from six locations in the lake and analyzed. This information was compiled and reported on a quarterly basis from the third quarter of 1969 through the second quarter of 1971. Visual observations and evaluations were made during the study relative to the erosion and turbidity, particularly in the areas of strip mine reclamation.

The flooded heights and water quality for the 24-mine observation holes were reported on a periodic basis starting in 1970 or after completion of the applicable mine seal. The last measurements and tests were performed during the fourth quarter of 1971.

Discussion of Results

General Considerations: Averages, analyses and flow volumes as associated with mine drainage conditions, are difficult to ascertain accurately and can be misleading due to variations in precipitation and infiltration from one period to another. Without continuous monitoring, the duration of both maximum or peak flows as well as minimum and no-flow periods were estimates and, as a result, have a certain margin or error. However, the periodic sampling on a monthly basis, supplemented with adjustments for abnormal variations or conditions have produced fairly reliable values which can be used for comparison in determining the effectiveness of an abatement program.

In the evaluation, acidity, alkalinity and iron in average pounds per day and mine discharge flow rates in gal/min are used as a measure for comparison. These averages, in many cases, are several times less than the maximum during high flows and do not indicate the effects of "slugging" which often accompany high flows. Seepage from most of the pile areas, some strip mine areas and the abandoned well sites were practically impossible to monitor. However, the pollution potential from these sites was recognized and abatement measures were performed to alleviate the possibilities of pollution, particularly after inundation of the lake. The average pounds per day of both alkalinity and acidity at six-month intervals have been computed for the various mine discharges. These values for the strip mine areas are shown in Table 10.7.

TABLE 10.7: AVERAGE POUNDS PER DAY ALKALINITY AND ACIDITY

	1967		1968			
	July – Dec.		Jan. – June		July – Dec.	
Strip Mine Areas	Alk.	Acd.	Alk.	Acd.	Alk.	Acd.
Northwest	<1	<1	38	<1	6	<1
Southwest	<1	<1	4	33	1	8
East	1	0	24	70	7	18
East (outside)	0	3	0	17	0	8

	1969				1970				1971	
	Jan. – June		July – Dec.		Jan. – June		July – Dec.		Jan. – June	
Strip Mine Areas	Alk.	Acd.	Alk.	Acd.	Alk.	Acd.	Alk.	Acd.	Alk.	Acd.
Northwest	10	<1	9	<1	6	1	2	2	2	1
Southwest	2	26	5	12	1	1	4	4	<1	<1
East	4	30	15	13	6	27	7	20	8	32
East (outside)	<1	22	0	6	<1	3	<1	11	1	16

Source: PB 221 337

Area Reclamation Projects

Rainfall Data: In order to correlate observed mine water pollutant loadings with atmospheric precipitation, official monthly rainfall data for the two nearest gauging stations (at Butler and Slippery Rock) were compiled for the years 1967 through 1971. The rainfall for the Moraine State Park area was estimated by averaging the monthly values from the two stations.

Strip Mine Reclamation: The strip mine discharges were far more erratic than the deep mine discharges; however, their contributions have always been less than 21% of the total pollution load. The production of both alkalinity and acidity in the strip mine discharges is directly affected by the climatic conditions, particularly precipitation. The production of acidity from strip mines has ranged from less than 1% of the total acid production during dry periods throughout the four-year study.

A comparison of average values before and after strip mine reclamation indicates a reduction in discharge flow rates from 142 to 136 gal/min (probably an insignificant difference), an overall reduction in net acidity from 50 to 22 lb/day and a minor increase in iron from 3 to 4 lb/day (probably not significant).

The Northwest Area, consisting of 166.0 acres of backfill reclamation, had an average reduction in discharge flow rates from 26 to 13 gal/min (50%). In this area, the test data indicate only a minor net acidity and iron production both before and after reclamation. Representative values for the "before" reclamation conditions in the area were not possible as the 120.2 acres in the ASMRP-1 project were in the process of being backfilled during the initial periods of the study.

A comparison of values in the Southwest Area (78.1 acres) indicates an overall reduction in discharge flow rates from 24 to 3 gal/min (88%), an overall reduction in acidity from 21 to 2 lb/day (90%) and a reduction in iron from 1 to <1 lb/day. The strip mine reclamation in the East Area consisted of 217.6 acres of backfill. A comparison of values in this area indicates an increase in discharge flow rates from 92 to 120 gal/min (30%), a reduction in acidity from 28 to 19 lb/day (32%) and a minor increase in iron from 2 to 4 lb/day.

Both terrace-type restoration and approximate contour backfilling methods were performed in the park area. Except for Project ASMRP-1, contour backfilling was restricted to relatively flat areas with limited surface drainage above the highwall. Strip mine reclamation performed under Projects MD-8C, SL 105-1, SL 105-1A, and SL 105-1B included the construction of diversion ditches above the highwall and slope drain flumes acorss the backfill at specific locations. This was the principal reason for the reduction in the slug discharge flow rates after reclamation in the applicable areas.

Terrace-type backfill with diversion ditches and slope drain flumes are the preferred methods for reclamation because they provided maximum control of erosion and turbidity. Erosion occurred at several locations in the backfilling, the most prominent being the Northwestern Section of ASMRP-1, an area of contour backfill without diversion ditches or slope drain flumes. Additional erosion occurred in several locations at the intersections of the diversion ditch with the slope drain flume. Installation of concrete or masonry construction at the intersections should correct these conditions.

Water Quality in the Lake Area: In examining the conditions prior to inundation, flows and water quality data were obtained on Muddy Creek at Weir #85 near Nealey from 1967 to July 1969. At that time, this sampling point was inundated due to the rising pool level. The water quality at this point ranged between pH 6 and pH 7 for the two-year period. During the years 1963 to 1966, Muddy Creek varied from a pH 5 to pH 7, as indicated by data from other sources.

Periodic water sampling was started in September 1969, at six locations in the pool area. From that time through June 1971, this information indicated the water quality in Lake Arthur remained alkaline, with little change from month to month. The range in analysis, as compiled from this data, indicated the following: pH 6.0 to pH 7.6, alkalinity 10 to 86 ppm, acidity 0 to 6 ppm, iron 0.2 to 3.0 ppm and manganese 0 to 3.5 ppm.

Water quality in the lake has remained good since the initial inundation in 1969 and aquatic life is flourishing. The Pennsylvania Fish Commission has stocked the lake with largemouth bass, catfish, musky, black crappies and alewives. The lake has been opened for boating, fishing and swimming.

Comparison of Effectiveness and Costs: The average discharge rates for both deep and strip mines before abatement were nearly equal: 146 gal/min (deep mine sealing) to 142 gal/min (strip mine reclamation). The deep mine sealing indicated a reduction in the discharge rates from 146 to 57 gal/min (60%) while strip mine reclamation had a slight reduction of 142 to 136 gal/min (4%). The overall reduction for total abatement was 288 to 193 gal/min (33%).

Values for net acidity indicated that the deep mines were producing an average of ten times more acid than the strip mines before abatement (501 lb/day from deep mines to 50 lb/day from strip mines). After abatement, the average net acidity from the deep mines was 160 lb/day indicating a reduction of 68%, and the strip mine reclamation reduced the average acidity to 22 lb/day indicating a reduction of 56%. Total overall reduction in net acidity varied from an average of 551 lb/day (before) to 182 lb/day (after), indicating a reduction of 67% for the combined abatement.

Average total values for both the deep mine sealing and strip mine reclamation indicated increases in production of iron after abatement. Deep mine sealing increased iron from 34 to 42 lb/day and the strip mine reclamation from 3 to 4 lb/day. As in the case of acidity, the deep mines were producing about 10 times more iron than the strip mines. However, the total increase from 37 to 42 lb/day of iron did not cause any serious problems in the watershed area.

Iron Chemistry in Abated Groundwater: Analyses of many observation hole samples taken after abatement construction has been completed indicate an increase in the total iron content of the abated waters involved. These increases could be due to three different mechanisms:

(1) The total iron reported includes suspended insoluble iron compounds, such as hydrous iron oxides, ferric hydroxide sols, iron oxide scale inadvertently knocked off the sides of the 6" steel casings which line the observation boreholes, iron corrosion products which accumulate at the air-water interface in the borehole, etc. Unfortunately, any or all of these compounds are determined in

the total iron analysis, which is the standard iron method for reporting mine water quality. Obviously, this method does not permit one to distinguish between the iron being produced in the mine waters in situ (from the decomposition of pyrite) and various other forms of iron introduced locally at the sample point by mechanical means (agitation, scuffing, air-water interface accumulation, etc.).

(2) Once an effective abatement structure is functional, the waters controlled or influenced by it normally undergo an increase in pH, and bicarbonate content. They are also excluded from contact with air, which helps to retain all soluble iron in the ferrous state. Under these conditions, the situation is ideal for the formation of ferrous bicarbonate, which has a solubility ranging from 25 to 710 ppm depending upon the bicarbonate iron concentration and the partial pressure of carbon dioxide in the mine atmosphere. (Solubility product of ferrous carbonate is 4×10^{-3}).

Where iron concentrations were very low before abatement was instituted, it would be logical to expect that ferrous ion concentrations would increase. The chemical mechanisms responsible for the presence of iron, however, are entirely different from those involved in the decomposition of pyrite to produce both acid and iron sulfate.

(3) With waters upgraded to the pH range of 6 to 8 and excluded from contact with air, normal ferrous ion can attain very high solubility levels, 200 to 300 ppm being very common values experienced in the neutralization of ferrous sulfate solutions at pH 6.

In summary, it is perfectly logical to expect that the iron content of "abated" waters will increase slightly, but this does not indicate that acid mine water generating processes are still at work. In short, the foregoing discussion points up the fact that iron analyses per se as determined by the standard total iron method are quite meaningless for evaluating the effectiveness of an abatement method especially where the iron contents are less than 10 ppm. A more significant index would be to make ferrous iron determinations on filtered samples, which method would eliminate the interference of mechanically entrained iron compounds and give more positive indications of the iron-producing mechanism involved.

Effect of Land Inundation (Lake Filling) on Pollutant Generation: There is a potential for an increase in pollutant generation as the result of ultimate inundation. The inundation was that of the land being covered by the impounded waters to form Lake Arthur with an average depth of 10' to 15'. When this hydraulic head is finally developed, the water table surrounding the lake rises an equivalent amount, inducing a higher groundwater flow rate in the abandoned mines. In the case of the sealed mines, the open sections in front of the barriers are still producing an amount of mine "make" water proportional to the mined area remaining unsealed.

This is related to the apparent trend of an increase in acid production of the abated mine water sources during 1970 and 1971. The increase in groundwater flow (due to water table rise) plus an increase in rainfall during 1970 and 1971 would both contribute to increased flow in the unsealed portions of the mined areas, thus increasing the total acid loading. These two meterological effects should also increase the flow and alkalinity loading of the alkaline streams in the area, so that the two effects should easily counteract each other, leaving the lake unimpaired.

Treatment Trends and Stability: Sometimes effective responses to abatement treatment can be observed from a study of the data obtained during the relatively short period of operation. These can be subdivided according to attendant causative factors.

Rainfall Effect — There is an interrelationship of monthly rainfall and average total mine discharges of the two major classes. Greatly diminished discharge acidities for the deep mines follow the rainfall intensity pattern very closely as would normally be expected, since the mine "make" water in the unsealed sections should be proportional to total rainfall. However, the total amounts of acid being generated remain at a low level (less than 200 ppd) even during periods of excessive rainfall (July to December 1970).

The reclaimed strip mine areas do not show any direct correlation to rainfall intensity, primarily because the sampling of these areas was not timed to correspond to periods of rainfall. The obvious (but unquantified) effect of abatement treatment was the diversion of large volumes of surface run-off from these sites, thus greatly reducing the magnitude of acid slugs normally produced. After abatement treatment, acid generation in these areas remains very small and undergoes very little change under varying conditions of atmospheric precipitation. This demonstrates the establishment of a significant degree of pollutant generation control.

Rising Water Table Effect — As mentioned previously, the filling of Lake Arthur to a maximum depth of 15' would be accompanied by a corresponding rise in the groundwater table level. This would be sufficient in several places to increase the average groundwater flow through untreated portions of unsealed mines and through the lower levels of restored strip mines. The end result would be a slight increase in acid generation rate from both of these sources. The annual increases are not very great.

Natural Alkalinity Generation and Storage — Of paramount importance to the overall abatement concept for the area is the ability of the regional streams to generate compensating alkalinity.

Calculations indicate that there is a tremendous alkalinity reserve in Lake Arthur (about 2,200 tons) and that the net average alkalinity generation rate of the Lake Arthur stream system is about 5 tons per day. This large compensating alkalinity supply is in great excess of any known future acid source and provides a high degree of stability insofar as water quality in the lake is concerned.

In summary, there are indications that a slight increase in acid generation (above the maximum abatement level attained in July to December 1969) has occurred. This, however, is to be expected for the reasons cited above and should peak out at a level of about 200 ppd. The increase is very small and is essentially insignificant in view of the large alkalinity reserves available in the lake and its alkaline feed streams.

Cost Effectiveness of Abatement Methods: Comparison in terms of acid reduction versus costs indicates the deep mine hydraulic sealing work to be 6.5 times more effective than strip mine reclamation in the watershed area.

In making a comparison of effectiveness of the abatement projects, several factors have been considered. The benefits of the deep mine hydraulic sealing is limited primarily to acid reduction. In strip mine reclamation, other benefits such as land use and esthetics plus erosion and turbidity control have been regarded as important for the recreational usage of the park. The covering, recontouring and draining of the strip mine areas has also eliminated their greatest environmental impact effect, the generation of strong acid slugs during periods of heavy rainfall.

SURFACE MINED LAND RECLAMATION IN GERMANY

The material in this chapter is excerpted from:

ORNL–NSFEP–16

To help design effective mining and land conservation measures, it is useful to examine programs which have been adopted by other industrial nations faced with similar problems. ORNL–NSFEP–16 describes the planning, technological, and regulatory procedures which are used in the Federal Republic of Germany (West Germany) to ameliorate the harmful environmental consequences of large-scale surface mining of brown coal.

In general, the land restoration program has been largely successful and strip-mining is no longer a controversial public issue in Germany. Of particular interest are the institutional arrangements that have been worked out to assure comprehensive planning for land restoration before the start of mining. Many features of the German approach could be applied to strip-mining problems in the United States.

THE IMPORTANCE OF LIGNITE IN WEST GERMANY

The Federal Republic of Germany is fortunate to possess large reserves of brown-coal (lignite) to serve in electric power production and possibly as a raw material for producing synthetic fuels. The total brown coal reserves of West Germany are estimated to be about 60 billion tons. Of these, some 55 billion tons are located in the Rhineland coal fields alone, making them the largest continuous deposit of lignite in Europe. (The lignite reserves of the world are estimated to amount to some 2,100 billion tons, as compared to 155 billion tons for all of Europe.) Using present-day mining techniques, approximately eight to ten billion tons of Rhineland lignite lie close enough to the surface for economical recovery.

As improved mining methods will almost certainly be developed in the future, it is clear that brown coal will continue to play a vital role in the West German economy far into the next century. The total West German production of brown coal in 1970 amounted to some 108 million tons, with 92.6 million tons being mined in the Rhineland alone. About 81 million tons of lignite were burned in thermal power stations to produce 60 billion kilowatt-hours of electricity; 24.7 million tons were used for briquette manufacture; and the remainder was used for miscellaneous purposes.

In view of the declining market for briquettes, new applications for brown coal are under study. It is believed that favorable physical properties of brown coal, such as high chemical reactivity, porosity, and low sulfur content, may make it suitable for a number of different production processes. Some of these potential processes are the reduction of iron ore, the gasification of lignite using nuclear process heat, and the production of metallurgical coke.

About one-third of all power generated in German thermal electric power stations comes from lignite-fueled plants. The present and potential uses of brown coal, both as an energy source and as a raw material, assure that it will eventually be mined to depletion. In Germany, the need to treat effectively the environmental brown coal mining has been fully recognized. Furthermore, this necessity will remain even if brown coal is eventually replaced by an alternative fuel for electric power production.

The vital role of lignite in the German economy can be seen from the intensive efforts which have been made to modernize the mines. This modernization process has required an enormous financial investment, more than one billion dollars since the end of World War II, and has led to a consolidation of the brown coal mining industry. The many shallow surface mines which were formerly common in the Rhineland have been gradually replaced by larger, more efficient open-pit mines. (While the number of active mines in the Rhineland decreased from 23 in 1950 to 6 in 1970, the annual production of lignite increased from 64 to 93 million tons.)

The post-war lignite industry in the Rhineland was shared by four major companies: Rheinische Braunkohle AG, Roddergrube AG, Braunkohle-industrie AG (BIAG), and Neurath AG. In 1959, these companies merged to form the present-day Rheinische Braunkohlenwerke AG, which today dominates the brown coal mining industry. This single company, a subsidiary of Germany's largest electric power utility—the Rheinisch Westfälische Elektrizitätswerke AG (RWE)—employs nearly 16,000 workers and produces 85% of the Geman lignite. Its large size and many resources have helped the Rheinische Braunkohlenwerke AG to institute enlightened land reclamation practices which have received worldwide recognition.

BROWN COAL MINING TECHNOLOGY

Open-Pit Mines

The Rhineland brown coal field lies in flat, plains country in the 1,000 square mile triangular area formed by the cities of Aachen, Cologne, and Mönchen-Gladbach. The lignite is deposited in highly faulted seams that are from 65 to 350 feet thick, with varying overburden up to 650 feet. The coal bed lies on a slightly-folded, inclined plane, with the shallower seams located at the base of the triangle, in the vicinity of Cologne and Aachen. Mining began in shallow surface mines near Cologne and has moved steadily northward, becoming progressively more complicated as the deeper coal deposits were reached. Prior to World War II, the mining of lignite in the Rhineland area was confined to sites where the brown coal lay sufficiently close to the surface (<100 ft) that conventional surface mining equipment could be used. Since the number of such suitable locations was limited and would be exhausted in the foreseeable future, it was clear that new mining methods needed to be developed to exploit the deeper lignite deposits.

Attempts to develop suitable new mining techniques began in 1938 when tests were made to explore the possibility of using deep mines to extract the lignite. These experiments were delayed by the war and were finally abandoned in 1953 when it became apparent that the high groundwater level of the region and the unconsolidated nature of the overburden precluded the large-scale introduction of deep mines on an economical basis.

At about this time, however, modern, massive excavating machines were developed that made the deeper brown coal seams accessible for the first time to surface mining techniques. Because the capacity and efficiency of these bucket-wheel type excavators were significantly greater than that of the older equipment, it became possible to design new open-pit lignite mines on a scale not possible heretofore.

Thus, in the years 1953 to 1955, a new epoch in the history of the Rhineland brown coal industry began. The result is such mines as the Fortuna-Garsdorf open-pit mine located near Bergheim. This mine is the largest material handling operation on earth, nearly twice as large as its closest competitor—the Kennecott Copper Bingham pit in Utah. In 1970, a total of 86.8 million cubic yards of spoil material, together with 36.2 million metric tons of lignite, was taken from the Fortuna-Garsdorf open-pit mine alone.

Mechanization and Automation of the Lignite Mines

The modernization of the Rhineland brown coal mines, beginning with the introduction of giant wheel-excavators in 1955, has greatly increased productivity and helped to make brown coal the cheapest source (next to hydroelectric power) of energy in Europe.

The process of mechanization did not end with the massive, new digging machines. A transportation system, capable of matching the prodigious capacity of the wheel excavators, had to be developed to haul away the spoil material and lignite from the mines. Since full restoration of the land disturbed by the mining operations was planned, giant spreader machines—similar to the wheel excavators—were designed and built to spread the overburden from active mines back into mined-out pits.

In addition, because the depth of the new surface mines extends well below the groundwater level, methods had to be developed to prevent flooding of the pits. That these problems were solved successfully is shown by the fact that lignite has not only held its position with respect to competitive fossil fuels, but has actually expanded its market.

Bucket Wheel Excavators: Each of the large, open-pit, brown coal mines is equipped with several wheel excavators for stripping off the overburden and extracting the lignite. A single machine costs up to ten million dollars and consumes as much as 10.4 Mw of electrical power in operation. Excluding maintenance personnel, only two operators are needed to operate the huge excavator. The lignite or loose overburden is carried by conveyor belt from the excavator wheel to the discharge boom, where it falls into waiting railroad cars. Because of their ability to excavate selectively and deliver the loose overburden, lignite, or topsoil to a separate, interfacing transportation system, the wheel excavator is especially suitable for use in areas where land reclamation is planned.

Transportation of Bulk Materials: Transporting the massive amounts of spoil material and lignite from the mine is accomplished with a specially designed system consisting of conveyor belts, heavy-duty trains, and slurry pipelines. Much of this equipment is automated or remotely controlled, thereby contributing to the high productivity of the overall mining operations. Seven-foot-wide conveyor belts moving on steel rollers at speeds up to 15 miles per hour are used to transport the lignite out of the mine pit. These conveyor belts can be installed in a straight line and operate satisfactorily on relatively steep inclines, thus eliminating ramps which would be required if trains were employed in the pits.

The total installed length of the conveyor belt network comes to about 70 miles. The belts are used to haul both lignite and spoil material. A crawler, equipped with special handling devices, is used to move the skid-mounted conveyor belt sideways as mining progresses. This can be done very rapidly, with little interruption of the mining operations.

Bulk material from the mines which must be transported over long distances is hauled on a company-owned railway system. The rail network consists of 310 miles of tracks connecting the active brown coal mines, the mined-out areas undergoing restoration, the briquette factories, and the power generating stations. The spoil material is hauled in eight-axle-gondolas, each with a capacity of 125 cubic yards and a gross weight of 240 metric tons. The gondolas can be emptied in a matter of seconds by a hydraulic system which tips the cars sideways.

Raw lignite is transported in four-axle, ninety-ton rail cars. Specially profiled, heavy-duty rails, with a linear density of 43 pounds per foot, have been designed to accommodate the enormous, 30-ton-per-axle loads which are encountered. Electric locomotives weighing up to 139 tons are used to pull loads of as much as 2,000 tons. This private railroad network carries a larger annual tonnage (not ton-miles) than the whole German Federal Railway system.

Ground and Surface Water Control

To extract the lignite using surface mining techniques, it is necessary to lower the groundwater level to prevent flooding of the mine pits. This is accomplished by pumping water from some 1,850 deep wells which have been drilled in the Erft river basin to an average depth of nearly 600 feet. Submersible motor pumps, some nearly 33 feet long and weighing more than 12 tons, pump water at rates as high as 33 tons per minute. At the present time, 1,100 wells in continuous operation provide sufficient water removal capacity for the six surface mines in the Erft region. On the average, 14 tons of water must be pumped out of the ground for each ton of lignite mined.

The water from the deep wells is discharged into the Erft, Inde, and Merzbach waterways and into a special drainage canal which connects the coal fields to the Rhine River. The canal can also be used to provide supplementary flood control benefits to the region north of Cologne. During periods of high water, a pumping station can divert 10 cubic meters per second from the Erft into the Rhine River. The Rheinische Braunkohlenwerke AG mining company is also studying the possibility of converting some of the huge mine pits into a freshwater reservoir to improve the supply and quality of water for industry in the region. Thus, the brown coal mining activities provide unforeseen community benefits.

LAND REHABILITATION

Extent and Costs of Land Reclamation

All lignite mining operations to date have affected less than one-tenth of the 620,000 acre Rhineland brown coal area. The total land area disturbed by brown coal surface mining from the turn of the century to January 1, 1969, amounted to 36,750 acres. Of this, 16,480 acres were not yet restored, representing either active mine sites or depleted pits being reclaimed. The remainder, 20,270 acres, has been restored for forestry (10,290 acres), agriculture (7,390 acres), and artificial lakes (2,590 acres). Restoring land to full agricultural productivity is the most expensive type of reclamation, costing from $3,000 to $4,500 per acre.

In recent times, rising land prices and lower reclamation costs, due to improved, more efficient methods, have resulted in arriving at an economic break-even point. Today, the market value of the restored farmland compares favorably with the expenditures for reclamation. In the United Stages, the cost of full land restoration would, in most cases, greatly exceed the value of the land. However, it is interesting to note that land reclamation was required in Germany long before it became marginally profitable.

A Panoramic View of Land Reclamation: The steady, northward progression of mining operations during the past fifty years occurred as the shallow, southern lignite deposits were gradually exhausted. Because of this, today the various stages of the mining and land restoration cycle are open to view, spread out in sequential order. At the active mines in the northern and central regions of the lignite field, the giant bucket-wheel excavators selectively strip off and save the top layer of loess. (Loess is an extremely fertile type of loam which covers most of the Rhineland region and has gradually come to be regarded as an important mineral in its own right.)

The excavators next peel off thick layers of sand, gravel, and clay overburden before extracting the loose, black layers of exposed lignite. Some commercial exploitation of the sand and gravel has begun, thus turning the extraction of brown coal into a total mining operation.

Further southward, near Quadrath-Ichendorf and Berrenrath, the huge surface mines have been exhausted of lignite, and restoration of the land is underway. Brought in by trains from the north, the discarded spoil material is filled back into the mined-out pits by mammoth spreader machines and leveled off by bulldozers. The leveled areas are subdivided into 5- to 10-acre tracts, or polders, by six-foot-high dikes of loam. These polders will eventually be filled with a loess slurry, which leaves behind a one- to two-meter-thick top layer of loess when it dries. Still further to the south, near Berrenrath, fields of grain and hay can be seen thriving on restored land which is less than 5 years old. The sequence in the forested areas is similar: To the north are newly planted stands of young trees, and in the south are forested areas reclaimed fifty years ago. The latter are nearly indistinguishable from natural forests and are superior to the stands of scrub timber which originally grew there.

Meticulous Advance Planning: The German land restoration program actually begins, with the preparation of detailed plans for the relocation of populated settlements and for the restoration of the land, years before the first shovel of brown coal is mined. Land-use patterns are proposed in advance, and the new landscape is designed accordingly—the topography, the drainage system, lakes, and the designation of areas to be restored for forestry and for agriculture. This comprehensive early planning enables the mining operations to be coordinated with concurrent land restoration work. New towns for the displaced people are designed according to modern urban requirements and are more compact than the former, unplanned settlements.

The basic resettlement costs are borne by the mining company with local and state governments providing supplementary funds to pay for the incremental costs of better schools, sewer systems, and other community services than those which existed at the former town site.

This comprehensive approach reflects an acceptance of the fact that surface mining affects not only coal, but also trees, buildings, people, and the land itself. In Germany, the State of North Rhine Westphalia and the lignite mining industry have accepted the challenge of finding acceptable solutions to the entire set of social and environmental problems created by brown coal surface mining. This approach makes it possible to treat the overall problem as a whole rather than dealing with separate aspects of the problem on a piecemeal basis.

Forests and Lakes

The restoration of brown coal mining lands began in the Rhineland some fifty years ago. Today, extensive tracts of both first and second generation forests (and some thirty-nine lakes of varying size) can be seen in the area. Historically, the forestland reclamation program divides naturally into three main periods. These are the "greening" action of the 1920s, the extensive planting of poplars and alders after the second World War, and the current phase of reforestation which began about 1958. The current program is distinguished by the planting of commercially valuable trees, with some poplars planted merely to provide a measure of protection against the weather. The planting program is also concerned with replanting areas which did not take well in earlier actions and the planting of trees and shrubs on slopes to reduce erosion.

The planting of commercial trees directly—without first having to prepare the way by establishing hardy but worthless quick growth tree types—has been made possible by applying a loess-improved topsoil to the areas which are to be reforested. Instead of simply grading the spoil banks and planting them to pioneer trees, a special layer of loess and overburden mixture is applied in depths of 3 to 5 meters. This mixture forms a loose, porous soil with good physical properties for tree growth. Because of the presence of limestone in the loess, the pH value of the soil ranges from 6.8 to 7.4. At first, the humus and nitrogen are at rather low levels, but this condition is improved by sowing lupine at the time of, or prior to, tree planting. Later, fallen leaves and organic debris resulting from forest thinning activities provide a rapid buildup of the humus content.

The Rheinische Braunkohlenwerke AG Mining Company maintains some 16,000 acres of forests, extending from Bonn in the south to Grevenbroich in the north. During 1970, some 670 acres of land were reforested—three million trees planted—and 9,500 cubic meters of wood harvested. The mining company has established a Forestry Division, with a staff of about 40 foresters and technicians, to plan and supervise these sizeable operations. During peak periods in the planting season, the mining company supplements its forestry division staff by contracting with local firms.

Much of the effort of the Forestry Division has been devoted to establishing the large forests in the southern part of the coal field. The focus is gradually shifting as mining operations continue to move northward—into areas primarily used for agricultural purposes. In the future, activities of the Forestry Division will center increasingly on forest maintenance, thinning, and harvesting.

A special problem to be dealt with is the upgrading of nearly 5,000 acres of poplars and alders planted during the second reforestation period. Extensive thinning and planting of more valuable trees are needed. More recently, there is a notable diversity in the kinds of trees planted. The tree types selected for a particular location are chosen in accordance with soil conditions and the expected exposure to sun and wind. To promote efficiency and economy in harvesting, the tree mixture in a given section is limited to one or two types, but the mixture varies strongly from section to section.

In all, the reforestation program employs twenty-two different types of deciduous trees (including basswood, oak, ash, beech, willow, alder, locust, poplar, elm, and maple), eleven types of conifers (including pine, fir, and hemlock), and eighteen different varieties of shrubs, such as hazelnut, dogwood, mountain ash, and wild roses. This mixture of trees and shrubs serves to provide not only an ecologically sound forest, but also the diversity appealing in a recreational forest. Forests in Germany have traditionally been used for recreation by the general public as well as for timber production.

In the past, most public and many private forests were open to all for biking, hiking, picnicking, and other recreational uses. In recent times, the accessibility of forests for relaxation and enjoyment has been extended by the 1969 Federal Forestry Law which requires that all private forests be open to the public. Commonly, all agricultural lands as well are open to people wishing to take walks along field paths and farm roads.

Winning New Farmland

In 1970, the Rheinische Braunkohlenwerke AG Mining Company restored some 470 acres of mining wasteland to agricultural productivity and an additional 670 acres as forest land. In view of the high costs of farmland reclamation, it is interesting to examine the considerations which have led to the adoption of this policy. Before improved reclamation techniques were introduced in 1960, the land restoration costs were far higher than the worth of the reclaimed land. Even today there are few economic incentives for rehabilitating the land. It is difficult for German agriculture to compete with neighboring European Common Market countries, and in many instances direct government subsidy is required to keep it viable.

Furthermore, the surface mining industry is not even the principal destroyer of farmland in North Rhine Westphalia. While the brown coal mining industry consumes about 600 acres of farmland annually, some 300 acres of farmland are lost each week to city growth, highway construction, and industrial expansion. These considerations clearly demonstrate that the decision to reclaim mined-out areas as farmland was not at all a "foregone conclusion."

The incentives for restoring mining areas to agricultural productivity in the Rhineland are diverse. First, the brown coal deposits are located in the richest, most fertile farm country of Germany which is favorably situated near the large population centers of the Ruhr valley. Thus, a ready, nearby market is on hand for the agricultural produce, affording a significant savings in transportation costs. Second, the problem of dispossessing and resettling farmers who are in the path of the mining operations is greatly eased by having restored land available as an acceptable substitute. Many of these farmers would be loathe to part with long established family farms were not satisfactory restored areas, fully commensurable in fertility and productivity with their former holdings, available as compensation.

Thus, the farmland restoration policy aids greatly in reducing social and political tensions and contributes to public acceptance of the temporary disruptions caused by the mining industry. Probably the most compelling reason for farmland restoration, however, is the prevailing conviction that to allow valuable soil to be irrevocably destroyed by a strictly temporary land use—mining—would represent extreme folly. Saving the loess has become one of the highest priority items of land planning in the Rhineland, reflecting a basic land ethic which cannot be evaluated or explained in purely economic terms.

Structuring the New Landscape: The spoil material and soil from active mines is transported to the site selected for restoration and is used to fill in the deep pits left by earlier mining operations. There, a completely new landscape, with a topography specified by prior design, gradually takes shape. Mammoth spreader machines, quite similar to the wheel excavators used in extracting the brown coal, are used to distribute the spoil material evenly over the area being reclaimed. Each spreader weighs 2,300 tons and requires only two operators.

A single machine is capable of handling up to 200,000 cubic yards of material daily. The transportation network of conveyor belts and trains on heavy duty tracks described earlier brings a steady stream of spoil material to the spreaders, enabling continuous operation. Ordinary bulldozers are used to level off and compact the overburden in preparation for applying the top layer of loess.

Before the loess is applied, the surface of the prepared area is deliberately furrowed, or deeply roughened, to prevent the formation of a clear interface between the top layer of loess and the substrata. In the past, the same spreader machines used to distribute the spoil material were also used to apply the loess. Again, final leveling of the surface is performed with bulldozers, which produce some undesirable compacting of the top layer to depths of about one foot.

Harrow disks are employed to break up the compacted loess before planting. In recent times, an alternative method of applying the top layer of loess has been developed. After the spoil material has been distributed, the surface is divided into small diked areas, or polders. The dikes consist simply of loosely piled loess and are about two yards in height. Loess and water are mixed in a one-to-one ratio and pumped through pipes into the polders, which are from 5 to 10 acres in size. The application of slurry is carried out in successive steps; the polder is flooded with slurry to a depth of about 2 feet and allowed to dry before more slurry is applied. This process is repeated until the final six-foot-thick top layer of loess has been obtained. Normally, a month or two is required for the drying process, but in unfavorable weather up to 8 months may be required.

The slurry technique of applying loess, adapted from methods developed in Holland to reclaim land from the sea, is faster and more economical than mechanical methods. Furthermore, it has been found to possess other superior attributes as well. For example, the pore volume of land prepared by the slurry method is higher than that of either virgin soil or mechanically prepared land.

The method has found increasing application in Germany, and much practical experience has been gained. To a large extent, the ease with which the slurry is prepared depends upon the condition of the loess. If the loess is strongly wetted, it does not readily form a suspension and loess-to-water ratios of 1-to-3 may be required.

On the other hand, if the loess was mined and stored during relatively dry weather, ratios as low as 1-to-0.7 suffice. The slurry discharge pipe must be relocated several times during the filling of a polder to prevent the formation of zones with different loess particle sizes. During the first slurry application the loess settles out and partly seals the bottom. In subsequent applications, less water penetrates into the subsoil and consequently the drying process can be greatly speeded by simply draining off the water after the loess has settled.

Once the land has received the top layer of loess, it is important to establish a vegetation cover as quickly as possible. Such a cover prevents hardening of the ground from heavy rains and siltation and makes it difficult for undesirable weeds and other plant growth to gain a foothold. Working land which has been formed by the slurry method is initially difficult because of its softness.

At first, the land cannot sustain the heavy loads of normal farm machinery, and special lightweight equipment must be used to work the ground and sow the first crop of alfalfa. Once established, the vegetation assists in drying out the polder, and the root system penetrates deeply into the fresh land. If reclamation occurs in late fall or early winter so that crops cannot be planted at once, special measures must be taken to prevent surface hardening due to winter precipitation. Experience has shown that the land should be deeply furrowed after the loess is applied. The ground then tends to freeze in rough clumps during the winter, promoting the drying process and allowing early spring planting.

Initial Cultivation and Interim Management: The newly reclaimed farmland is retained by the company for an interim five-year period. During this time it is subjected to intense management by agricultural experts of the brown coal mining company. The land improvement methods are based on experience and the results of scientific research which has in part been carried out in cooperation with the Agronomy Institute of the University of Bonn. The field work is conducted by trained personnel, working from centrally located company farms to assure that uniform methods of soil preparation and fertilization are used. The company farms also serve for conducting diverse experiments in soil physics and agricultural chemistry.

A primary objective of the interim management program is to build up the humus and nutrients in the soil. The nutrient level can be regained rather quickly by the use of fertilizer, or by planting leguminous crops such as alfalfa or lupine. Restoring the humus level which prevailed before mining, however, is a much slower process.

The humus content of the newly restored land averages about 0.5% while that of undisturbed areas in the same vicinity ranges from 1.3 to 1.8%. For this reason, alfalfa is prized not only for its ability to fix nitrogen, but also because of its deep root system which contributes to the buildup of humus when the plant is harvested. Research has shown that normal crop rotation increases the humus level by only 0.04 to 0.05% annually.

Therefore, other methods are employed to speed the process of soil conditioning. The stubble from winter rye, which follows the alfalfa crop, is disked and plowed under. Sewer sludge, composted garbage, and other organic wastes have been used experimentally to increase the humus level. Rape, a plant of the mustard family used for fodder, is sometimes sown after the grains have been harvested and the straw and stubble disked into the soil. The plant is fertilized to achieve rapid growth and the plant foliage is then plowed under as a form of green manure. Since rape possesses a flexible planting date and is relatively inexpensive, the method has proved attractive.

The reclaimed land requires higher than usual applications of fertilizer for at least the first 10 years of cultivation. Experiments have shown that the optimal amounts on the newly restored land are: 135 to 180 pounds of P_2O_5 and K_2O per acre, and 180 pounds of nitrogen per acre. Controlled experiments were carried out by the Agronomy Institute of the University of Bonn to determine the crop yields attainable on restored mining lands during the 5 year period of interim management.

Experimental plots of land which had been restored by the slurry technique were selected near Inden, Germany and control experiments were conducted on nearby, similar land which had not been disturbed by surface mining. The restored land was first conditioned by planting a crop of alfalfa. In succeeding years, identical crops were planted on the restored and the undisturbed land. The results are shown in the following table.

TABLE 11.1: CROP YIELDS ON RESTORED AND UNDISTURBED LAND*

		Yield (pounds per acre)	
Year	Crop	Old Land	New Land
1962	Winter Rye	3,822	410
1963	Rape (dry mass)	3,349	3,456
	Oats	3,411	3,367

(continued)

TABLE 11.1: (continued)

Year	Crop	Yield (pounds per acre)	
		Old Land	New Land
1964	Sugar Beets		
	Roots	44,025	54,652
	Sugar, %	16.7	16.7
	Foliage	34,291	43,489
1965	Sugar Beets		
	Roots	46,168	46,168
	Sugar, %	15.4	16.5
	Foliage	65,010	60,278
1965	Winter Wheat	5,537	5,572
1965	Winter Wheat	4,688	5,019

*Number of pounds to the bushel: rye (56), wheat (60), and oats (32).

SOCIAL AND ECONOMIC IMPROVEMENTS

Agriculture

The principle of primogeniture—or the exclusive right of the eldest son to inherit his father's land—is not embodied in the inheritance laws of North Rhine Westphalia. As a result, once-large farms have gradually become splintered and subdivided over the centuries, as the property passed from one generation to the next. Today, it is not uncommon for a farmer to own, or to have to lease, numerous small parcels of land which may be widely separated from one another. Such a land-holdings pattern is highly inefficient because it precludes the application of modern, mechanized methods of farming.

Consequently, a consolidation of the many small holdings to form larger, economically viable units has long been a prime objective of governmental planning. Obviously, the handling of such a sensitive issue requires much care, if the reform is to be accomplished in an equitable manner. Hence, in settled areas, the land consolidation program must proceed slowly and cautiously to avoid arousing dissatisfaction. On the other hand, in the resettlement of reclaimed brown coal lands, it was recognized at the onset that a unique opportunity existed for quickly accomplishing the desired land consolidation.

Resettlement consists of two distinctly separate transactions: the indemnification of the farmer for property confiscated and the purchase of new, reclaimed land from the mining company. In the sale of restored land, the mining company favors buyers who were dislocated by the mining operations. The purchaser is granted a $190 per acre rebate by the mining company to compensate for the extra fertilizer and seeding costs initially required in cultivating reclaimed land.

An intricate method has been worked out to estimate fairly the worth of farms confiscated by the mining company. State assessors appraise the property and judge the value of the land not only on the basis of its area, but also according to its fertility as established by past records of crop productivity. The farm buildings and such external improvements as woods, orchards, and wells are appraised by taking into account their current replacement cost and their present depreciation based on age. The value of each item is added to obtain the final settlement sum. If the farmer is dissatisfied with the negotiations, he may appeal through the courts. However, this happens in less than one out of six cases. The settlement payment, supplemented by savings and loans, enables the farmer to purchase new land and buildings. How high the replacement costs may be is revealed by a recent survey, which found that the total investment in buildings on an average farm in North Rhine Westphalia amounts to about $70,000.

The nearly 3,000 acre agricultural community developed on reclaimed land near Berrenrath, in the southern sector of the brown coal field, illustrates the new socioeconomic structure which is being attained. The new community will comprise some 27 separate farms, each with 40 to 80 acres in a single tract of land. The community is being built according to plans specified by the winning entry to a landscape design contest sponsored by the state of North Rhine Westphalia. The plan envisages 70% of the land being used for cultivation, 20% for forested tracts, and the rest for village growth and industry.

The farmers are clustered together in small hamlets of about six to eight families each. Contrary to usual German practice, the farm dwellings are located amidst the cultivated fields, reducing unnecessary traveling time. The hamlet pattern reduces the isolation of an individual farm family and makes it possible for several farmers to pool resources in purchasing expensive farm equipment since the machinery can then be used communally. The size of the farms, the spacious modern buildings, and the economies afforded by cooperative endeavor make it likely that the Berrenrath agricultural community will remain economically resilient for many decades to come.

Villages

Whole villages, caught up in the path of the brown coal mining operations, must be torn down and relocated. Thus, the mining juggernaut uproots people and institutions as well as the landscape itself. The number of people evacuated from their homes and resettled in fully new locations reached a total of 19,552 at the beginning of 1971 and is expected to grow to 30,000 by the end of this century. In all, nearly 5,000 homes, farms, and places of business have been forced to yield to the brown coal industry and move elsewhere. Moving an entire village requires extensive preparations and takes a relatively long time.

For this reason, the decision must be reached and made known well in advance of the planned start of mining. Most of the villages which have been relocated are fairly small in size, with a population of about 350 to 2,000 inhabitants. In some few cases, particular villages have been spared because of their historic interest or because they are located close to the edge of the brown coal field.

The villagers participate directly in selecting a new town site, choosing one from a dozen or so possibilities which have been presented to them. All of the sites being considered have previously been approved by the regional planning commission. Past experience has shown that the villagers prefer to move en masse, retaining some portion of the community identity. This practice tends to avoid excessive social upheaval and the creation of a condition of rootlessness. The village may simply be reconstructed at another site, or the community may decide to join another already existing town. The latter is desirable because the connecting arteries of transportation already exist, and the consolidated population pattern which results makes it possible for the new, merged community to afford improved public services. The new community invariably provides better schools, parks, sports facilities, and playgrounds than the old one. Because of unsatisfactory conditions in the old village, the resettlement in many cases amounts to nearly a complete urban renewal action.

Mining company representatives negotiate directly with the individual villagers to determine a fair settlement payment for the confiscated property. The money paid by the mining company enables the villager to replace his home at the new town site. Generally the new homes are substantially improved—larger than before, and with central heating and indoor plumbing. The added improvements are paid for privately, not by the mining company. The new village is designed according to modern principles by professional architects and planners commissioned by the town council. By-pass roads separate the residential section from areas of heavy traffic, and more space is allotted for playgrounds, ball parks and kindergartens.

The businesses are centrally located, the village is provided with a sewer system, more public green areas, and is generally more compact and efficient than the former one. Given this type of choice, the villagers apparently prefer to trade the picturesqueness of their old towns for a new environment with more modern conveniences.

A sociological survey of some resettled communities reveals that the end result has been generally satisfactory. According to the survey 85% of the resettled people are happier in their new homes, and 57% likewise prefer the new location of the town. Some 83% of those interviewed believe that the advantages outweigh the disadvantages in resettlement. This does not imply, however, that the resettlement proceeds without a certain quota of worry and concern. The typical evacuee experiences feelings of anxiety about his future job, income, the new school and community accommodations, his social position in the new community, and the unavoidable problems involved in constructing a new house. On the plus side, however, he gains the following:

> (1) He receives a desirable building lot located on an improved residential street away from areas of heavy traffic.
> (2) His modern, new home is equipped with central heating and indoor plumbing and is considerably more spacious, nearly one-third larger than his former dwelling.
> (3) His new house is worth more, will last longer, and has lower maintenance costs than his old one.
> (4) In building a new home, he receives substantial federal tax advantages which are provided in Germany to encourage the construction of residential housing.

Recreation

The recultivation of mined lands furnishes a unique opportunity to sculpture the new landscape to meet the recreational needs of the people as well as to serve the interests of commerce, industry, and agriculture. The growing population and changing life styles in Germany have combined to create an unprecedented demand for outdoor recreational facilities at the same time that ever greater amounts of land are being consumed for highways, cities, industry, and military uses. For this reason, land reclamation in the Rhineland brown coal fields is planned in such a way as to provide for multiple land usage. The forests are planted not only to meet the commercial need for lumber production, but also to provide restful settings for weekend relaxation.

The agricultural landscape as well is prized for its potential recreational value. Trees and hedges are planted by the mining company on reclaimed farmland to divide the fields and protect the land against moisture loss from action of the sun and wind. These windbreaks and groves of trees amidst the cultivated fields provide a scenic variety which attracts countless city dwellers on Sunday afternoons. The well maintained farm roads, with side paths and benches, enable the elderly as well as the young to undertake leisurely strolls and enjoy a closer contact with nature than would otherwise be possible in their urban life.

The oldest and most impressive restored region lies in the southern section of the coalfield near Brühl, where a forest-and-lake landscape of distinctive charm has been created. The numerous ponds and lakes, formed from deep pits left over from the lignite mining, have been artfully fitted into the landscape and are scarcely distinguishable from natural bodies of water. The lakes have sloping shores and shallow edges which are planted to reeds and grasses to provide wildlife habitats. The restored region comprises some 3,000 acres and serves both as a recreation park and a wildlife preserve.

Because of the many opportunities for water sports and hiking, the park attracts some 20,000 visitors each weekend from the greater Cologne area. It is maintained jointly by the Forestry Division of the mining company and the State Forestry Service in cooperation with civic organizations of neighboring communities. The achievements here, and elsewhere in the Rhineland brown coal area, demonstrate that, with proper planning and effort, the needs of industry for raw materials and power can be met without producing excessive environmental damage.

GOVERNMENT REGULATION AND SUPERVISION

Historical Development

The present form of the brown coal industry and the adoption of enlightened land restoration practices in Germany evolved gradually over the past several decades. Although reforestation of land areas disturbed by lignite surface-mining has been carried out in Germany since the early 1920s, the practice of restoring land for agricultural use was not instituted until much later.

The location of the Rhineland lignite deposits, in the midst of rich, fertile farmland, provided a strong incentive for this type of land reclamation. Public concern over the large tracts of unreclaimed land left over from World War II began to appear after the war, particularly in the vicinity of Cologne, where the proximity of the mining operations made the disturbed lands highly visible to great numbers of people. As a result, new surface-mining control legislation was enacted in the year 1950 to assure orderly, well-planned mining practices.

On March 11, 1950, the state legislature of North Rhine Westphalia passed

Germany's first Regional Planning Law. This law, modified in May 1962, established a Land Planning Commission charged with the responsibility of developing overall guidelines for land use within the state. The main purpose of the commission is to help coordinate the diverse social, economic, and industrial activities of the region. The commission designates specific land areas for use by agriculture, forestry, and industry and sets the boundaries of populated settlements. It develops long range plans for transportation arteries and networks, the preservation of historic sites, and the construction of recreational facilities to serve the entire region.

On April 4, 1950, the state legislature enacted two additional laws applying specifically to the brown coal producing areas of the region. These are The Law for Overall Planning in the Rhineland Brown Coal Area (Gesetz Uber die Gesamtplanung im Rheinischen Braunkohlengebiet - GS NW. S. 450) and another law establishing a Community Fund to finance land restoration. The first of these laws created the Brown Coal Committee which develops detailed plans for exploiting the lignite resources of the state within the framework and spirit of the overall regional planning law.

The Brown Coal Committee

The basic responsibility of the Brown Coal Committee is to safeguard land areas temporarily used for brown coal mining from long-term damage and from being rendered unsuitable for more lasting uses. This responsibility encompasses more than merely preventing the creation of desolate areas by requiring that the land be restored for forestry or agriculture. Rather, in light of the general objectives of the overall regional planning, the Brown Coal Committee assures that the land is restored in such a way as to harmonize with the social, cultural, and industrial interests of the rest of the region.

The Brown Coal Committee is composed of 27 members especially selected to represent the societal interests most affected by the impact of the mining operations. Members of the committee are the district governors in Cologne, Aachen, and Düsseldorf; the head of the state mining agency; the Rhineland land planning commissioner; the minister of agriculture; a representative of the Rhineland Agriculture Association; the director of the State Land Settlement Office; three representatives from the brown coal mining industry; three representatives of mining unions; five representatives from county governments, three representatives of the farmers; one representative of Crafts and Trades; one representative from the power industry; one representative of the stoneware industry; one representative from the industrial union for chemistry-paper-ceramics; and one representative from the Erft Basin Conservation Club.

This broad base of representation on the committee affords an opportunity to resolve conflicts long before actual mining activities begin. The committee formulates land restoration requirements based on the future use of the land. These requirements are determined as early as possible to enable the mining company to design its mining operations accordingly.

The primary function of the committee is to act as a review body to consider proposals for extending mining operations to new land areas and to make appropriate recommendations to the minister-president of North Rhine Westphalia. As can be expected in view of the composition of the Brown Coal Committee, the final recommendation to the state government is based on considerations of overall land use, conflicting local issues, and national coal requirements. The Brown Coal Committee has gradually emerged as a powerful force, defining the conditions under which the brown coal industry must operate. Its existence subjects the brown coal industry to public scrutiny and has been instrumental in bringing about the conservation practices of the industry. The Brown Coal Committee serves as a quasi-public forum where the divergent interests of society can be considered before mining commences. Public hearings and the signature of the state chief executive are required before the recommendations of the committee become legally binding.

The adoption of requirements that a certain portion of the land disturbed by the surface mining of brown coal be restored to agricultural productivity illustrates the importance of such a planning and review body. Shortly after World War II, a coalition of agricultural groups within the Brown Coal Committee became concerned over the destruction of fertile farmland by the mining operations.

In the late 1950s, this coalition of agricultural interests, known as the "green front," successfully campaigned within the committee to require that the valuable top layer of loess, often 15 to 20 feet thick, be saved, and that a portion of the land disturbed by surface mining be restored to agricultural productivity.

Implementing the Mining and Reclamation Plan

After the mining and land rehabilitation plan is adopted, the State Mining Office is responsible for supervising its implementation and assuring that the mining and land restoration activities are carried out in accordance with its stipulated provisions. The mining company is required by law to cooperate by providing all information which the state enforcement agency needs to carry out its regulatory function.

For example, the brown coal mining company routinely submits aerial survey photographs every six months to document its mining and land restoration progress. The planning and enforcement process, with participation of nonmining interests, affords flexibility in resolving the social and environmental problems posed by surface mining. The recommendations of the Brown Coal Committee serve as a living law which changes in accordance with the requirements of specific situations. Since the deliberations take place well in advance of actual mining, amply sufficient lead time is available for full consideration of all of the issues and problems.

By virtue of its representation on the Brown Coal Committee, the state enforcement agency is fully cognizant of the spirit and intention behind provisions of the operations plan and is able to draft supplementary regulations accordingly.

State Mining Office

The State Mining Office (Bergamt) of North Rhine Westphalia is the agency which oversees mining operations and enforces the provisions of the land restoration plan. Since most of the German lignite deposits are located within this single state, nearly all government control of brown coal surface mining is on the state, rather than the federal, level. The task of setting adequate reclamation standards is facilitated by the lack of significant economic competition from neighboring states. The legal authority to regulate the extraction of minerals in West Germany derives from a general mining law based on an older Prussian model written June 24, 1865. The law reserves nearly all mineral rights to the state which may grant mining concessions to private companies. The concession confers on these companies the right to commercially exploit the mineral resources by state-approved methods. Traditionally, the concession has generally been granted to the discoverer of new mineral deposits in order to encourage prospecting.

The State Mining Office of North Rhine Westphalia has thirteen regional offices to oversee the mining activities in the state, including the mining and reclamation operations of the brown coal industry, with its annual production of nearly 100 million tons. Each regional office is staffed with about ten mine inspectors, usually trained mining engineers, who spend most of their time in the field observing the progress of operations and checking for compliance with regulations.

The cost of administering and enforcing a strip-mine reclamation bill can constitute a substantial part of the overall restoration costs. In spite of the expense, however, a well-supported enforcement agency is vital to the success of any reclamation program. The agency not only enforces the explicit provisions of the law, but also, by writing supplementary regulations as part of its interpretive role, largely sets the tone of the program.

In Germany, the brown coal mining industry was at first reluctant to undertake the highly expensive reclamation of farmland. The regulatory agency, backed up by a strong law, played a highly important role in bringing about an acceptance of the practice, today an even slightly profitable venture.

RELEVANCE TO U.S. SURFACE MINING

Some elements of the German surface mining and land reclamation techniques are applicable to U.S. strip-mining in spite of important differences in the climate, terrain, and geological features of the coal-bearing regions of the two countries. Bucket wheel excavators of the type used in the Rhineland have already been used in North Dakota and Illinois to remove soft, unconsolidated overburden. These machines can operate continuously and deliver the loose overburden by conveyor belt to a separate, interfacing transportation system. Because of its digging selectivity, the bucket wheel excavator is especially suitable for separating the fertile topsoil from the remaining overburden material and saving it for later use in land reclamation.

It is not well-suited for hard rock digging or for the handling of drilled or blasted materials. Similarly, the extremely heavy weight and limited mobility of the bucket wheel excavator make it wholly unsuitable for Appalachian contour mining. Nevertheless, the coalfields of the United States, especially those in the interior, northern great plains, and western provinces, contain vast expanses of gently rolling or flat land where the German technology could be applied if there were sufficient incentives for full land restoration.

Whether or not legislation exists, requiring that quality land restoration be integrated into the mining cycle can change the economics of mining and thereby influence the selection of a specific excavation technology. If the digging machines do not have to be coupled to a separate transportation system, to haul away and save the topsoil and to transport massive amounts of spoil material to refill areas, the operational costs of excavating would tend to dictate choosing the giant shovels and draglines used in southern Ohio and Illinois. If full restoration of the land after mining is planned, it may well be that some adaptation of the excavation and transportation system used in the Rhineland is more economical and efficient. In other words, different conclusions are reached depending upon what portion of the overall mining cycle is included in the optimization process.

In the main, the methods of land reclamation which are adaptable from those employed in the Rhineland lignite fields apply to the rolling plateau country and the flat lands described earlier. In such topography, the slurry method of spreading loam on graded, filled-in areas is technically feasible.

Of course, as in the case of choosing an excavation technology, economic considerations may favor some alternative method of accomplishing the same purpose. The large amount of directed research carried out at the University of Bonn to determine suitable plants and trees for revegetation of the mined lands, and the factors affecting their growth rates, should be valuable and useful in the United States.

Much of the environmental degradation from surface mining can be prevented by making a conscientious effort in land reclamation. Nevertheless, it is still too early to exclude the possibility of long-range, adverse effects from surface mining. For example, the altered ground strata and mineral content of mined lands could unfavorably affect groundwater movements or percolation characteristics. These could conceivably lead to undesirable long-range results such as increased soil salinity in the mined land areas or elsewhere.

Although there is no reason to suspect that such events are actually occurring in the Rhineland, the possibility of subtle, but ultimately harmful changes cannot be dismissed. For this reason, it is important to gain a better understanding of possible geological effects and to continue to develop improved mining techniques. The restoration achievements in the Rhineland clearly demonstrate that a meticulously planned, well-funded program can produce impressive results in land reclamation.

In summary, there is much that can be learned from the German experience in restoring surface-mined lands. Their program has been in effect for some twenty years and has helped to minimize social dislocations and environmental damage from brown coal surface mining.

The land restoration program in North Rhine Westphalia embodies four main principles which have made it viable and effective. First, the regulation of surface mining is incorporated within an overall regional development plan. This makes it possible to protect the larger interests of the whole region. Second, a planning body composed of representatives from diverse interest groups participates in formulating detailed requirements for mining and land restoration long before the actual mining begins.

Thus, a broad spectrum of society is consulted and untimely haste is avoided. Third, the recommendations of the planning body are submitted for public review before being adopted and implemented. This provides a political pressure relief valve as well as a mechanism for detecting possible adverse side-effects which had escaped consideration. Fourth, an enforcement agency is empowered to enforce the plan which is finally approved and adopted. The German program offers visible evidence that, with detailed advance planning, striking achievements in reducing environmental damage from surface mining are possible at a price that can be borne by the consumer.

RECLAMATION COSTS

The material in this chapter is excerpted from

PB 203 189
PB 226 905
PB 238 538

RECLAMATION COSTS SUMMARY

The cost of reclamation can vary widely, depending on the primary objectives of the restoration activities. Reclamation of orphan lands is generally considered a public burden and constitutes an economic problem. All work reported in this section was completed under government contract, by the lowest bidder, with money coming from State and/or Federal funds. Cost would have been substantially lower had the reclamation been concurrent with mining.

It has been estimated that costs could probably be reduced by at least one half, if the reclamation had been conducted along with the mining. The expense of clearing and grubbing of volunteer vegetation, disposal of buried trees and brush, loosing of compacted spoil, and reestablishing access to areas could be saved. In addition, a contracting firm doing the work for a government agency would have mobilization costs and receive a profit. Mining companies also are profit conscious and would consider these costs in anticipating their profits (e.g., overhead would be less if reclamation is integrated with mining).

The data presented in this section can serve as a guide for estimating and determining cost ranges, however, it should be recognized that variations exist. Adjustments may be necessary from the standpoint of physical conditions, economic conditions, price changes, and more restrictive requirements of recent surface mining laws. The cost figures are taken from PB 238 538, published in Oct. 1974.

Reclamation conditions, procedures and successes in the eastern United States, particularly in Appalachia, have no bearing on conditions to be expected in the western United States. The situations are wholly different. Reclamation in the West differs from that in the East, primarily because of aridity especially during the summer months. Because of the many variables and differences between reclamation in eastern and western areas, they will be discussed separately.

Eastern Surface Mining Reclamation

Since cost considerations are different between orphan lands and active mines, each type of reclamation is discussed separately.

Orphan Lands: The cost analysis for Pennsylvania (Table 12.1) was prepared from information in the State files at Harrisburg. Selected projects were evaluated for the various mine drainage pollution control techniques completed during the construction phase of the Moraine State Park.

Ohio information (Table 12.2) was obtained from State files in Columbus. Twenty projects were evaluated dating back to 1965. All projects were reclaimed with money collected as a result of bond forfeitures. Expenditures on a given tract of land are limited to the amount of bond forfeited on that land. Where the bond forfeited on any given area of land is insufficient to pay the cost of doing all the reclamation work, the State was required to pursue reclamation only to the extent that such money permitted. Tree planting had top priority, and if any money remained, other pollution control measures were included. In most cases, only sufficient funds for tree planting were available.

Kentucky furnished the information (Table 12.2) for 5 projects that were reclaimed with bond forfeitures and State money from a special reclamation fund. The Kentucky law requires that bond forfeitures be spent on the land for which the bonds were forfeited. However, they can also spend additional money from a special reclamation fund to do total reclamation for pollution control.

The information for West Virginia (Table 12.2) was furnished by the State for eight projects. These projects were reclaimed with money from a special reclamation fund that could only be spent on surface mine problems. Tables 12.3 and 12.4 are based on data collected from several Federal Government publications, an environmental impact statement (Palzo Project), Myles Job (Breck & Brooks), and the TVA.

Active Mines: Available information for active operations is sketchy and probably not very accurate. For example, a survey by the U.S. Bureau of Mines of reclamation work conducted in 1964 by the major surface mining industries showed that in the principal coal-producing areas, average costs of completely reclaiming coal lands ranged from $169 per acre in the South Atlantic States to $362 in the Mid-Atlantic area. Partial reclamation costs ranged from $74 per acre in the East South Central region to $261 in the Mid-Atlantic. Details are lacking as to the exact type or degree of reclamation represented by the costs reported but the level was probably influenced by legal requirements of the seven States that had surface mining laws. The cost also might have been influenced by the fact that reclamation work was conducted with the mining operation, and the extra expense of repairing access roads to move heavy equipment back to the site was avoided.

TABLE 12.1: PENNSYLVANIA RECLAMATION COSTS

Pollution Control Measures	Unit[1]	Cost Maximum	Cost Minimum
Backfilling and grading:			
1. Approximate original contour:	Acre	$1,522.00	$1,000.00
2. Terracing	Acre	1,500.00	700.00
Revegetation:			
1. Trees only --700/acre	Acre	500.00	90.00
2. Grasses and legumes - 19 lb/acre	Acre	220.00	180.00
3. Grasses, legumes, and trees	Acre	500.00	386.22
Diversion ditch:			
1. Cross section, 10 sq ft	LF[2]	1.00	---
2. 6 ft. Bottom, side slopes 1½ to 1	LF	14.90	7.93
3. Rock protection for ditch	Sq Yd	12.00	---
Reconditioning stream bed	LF	1.50	1.00
Curtain grouting of outcrop	LF	11.87	5.80
Mine seal, bulkhead type	Each	7,000.00	6,000.00
Coal refuse pile (gob):			
1. Removal and grading	Cu Yd	1.06	1.00
2. Contouring and grading pile	Cu Yd	1.00	0.33
3. Top soil:			
a. Clearing and grubbing borrow area	Acre	700.00	187.50
b. Excavation and covering refuse	Cu Yd	2.00	0.26
4. Drainage:			
a. Ditch	LF	2.60	1.67
b. Pipe and laying	LF	20.00	10.00
5. Revegetation	Acre	900.00	100.00

[1] Acre = 0.40 hectares
Foot = 30.48 centimeters
Square yard = 0.84 square meters
Cubic yard = 0.76 cubic meters

[2] Linear foot

Source: PB 238 538

TABLE 12.2: OHIO, KENTUCKY, AND WEST VIRGINIA RECLAMATION COSTS

		Cost	
Pollution Control Measures	Unit[1]	Maximum	Minimum
Ohio:			
Backfilling and Grading			
1. Strike-off	Acre	$181.23	$169.86
2. Terracing	Acre	214.09	47.34
Revegetation			
1. Trees only	Acre	50.00	22.07
2. Grasses and legumes	Acre	50.14	38.09
Kentucky:			
Backfilling and Grading			
1. Approximate original contour	Acre	1,200.00	171.00
2. Terracing	Acre	185.00	167.00
Revegetation			
1. Grasses, legumes and trees	Acre	150.00	40.00
West Virginia:			
Backfilling and Grading			
1. Approximate original contour	Acre	641.23	
2. Georgia V ditch	Acre	600.00	211.57
Revegetation:			
1. Grasses, legumes, and trees	Acre	287.69	90.00

[1] Acre = 0.40 hectares

Source: PB 238 538

TABLE 12.3: COSTS OF MINE RECLAMATION CONTROL MEASURES[1] (DOLLARS)

Pollution Control Measures	U.S. Bureau of Mines #6772	U.S. Bureau of Mines #8456	MYLES JOB	U.S. EPA ELKINS	T V A	PALZO PROJECT	U.S. EPA TRUAX-TRAER
Surface Backfilling by grading:							
1. App. Orig. contour	11.70 to 15.73/LF	780 to 1,402/acre	---	---	472/acre	---	---
2. Terracing	5/18/LF	---	250 to 400/acre	---	650/acre	---	---
3. Swallow tail	---	---	---	582/acre	400/acre	---	---
4. Pasture	---	---	---	568/acre	---	---	---
5. Final pit only	---	---	---	---	7,300/acre	600/acre	---
Surface backfilling by using explosives:							
1. Terracing	8.84 to 14.08/LF	---	460/acre	---	---	---	---
Scalping	---	---	---	---	75/acre	---	---
Clearing and Grubbing	---	33.54 to 45.76/acre	---	25 to 164/acre	---	100/acre	---
Revegetation	---	---	---	114 to 282/acre	---	---	---
Municipal waste sludge, liquid:							
1. Irrigating	---	---	---	---	---	500/acre	---
2. Incorporating -12"	---	---	---	---	---	100/acre	---

(continued)

TABLE 12.3: (continued)

Pollution Control Measures	U.S. Bureau of Mines #6772	U.S. Bureau of Mines #8456	MYLES JOB	U.S. EPA ELKINS	T V A	PALZO PROJECT	U.S. EPA TRUAX-TRAER
Dry							
1. Hauling	---	---	---	---	---	---	12/hour
2. Application	---	---	---	---	---	---	.12/ton
Masonry seals:							
1. Dry	---	---	---	2,212 each	---	---	---
2. Wet	---	---	---	4,076 each	---	---	---
Clay Seals	---	---	---	950 each	---	---	---
Treatment for refuse piles and slurry ponds:							
Soil Cover:							
1. 4" cover	---	---	---	---	---	---	1.00 cu yd
2. 12" cover	---	---	---	---	---	---	1.00 cu yd
3. 24" cover	---	---	---	---	---	---	1.00 cu yd
Straw mulch application	---	---	---	---	---	---	30/ton 27/acre
Limestone	---	---	---	---	---	---	5.50/ton
Fertilizer, 6-24-24	---	---	---	---	---	---	55.30/ton
Rototilling 8"	---	---	---	---	---	---	6/acre
Discing 8"	---	---	---	---	---	---	3/acre
Handraking	---	---	---	---	---	---	3/acre

[1] Acre = 0.40 hectares; Foot = 30.48 cm; Sq yd = 0.54 sq mi; Cu yd = 0.76 cu m; Short tons = 0.907 metric tons.

Source: PB 238 538

TABLE 12.4: DENTS RUN PROJECT, WEST VIRGINIA, RECLAMATION COSTS

	Cost per acre		
Item	Job 1,[a] section G, strip R (16 acres)[c]	Job 2,[b] section G, strip A (10 acres)	Job 3,[b] section C, strip B and C (22.8 acres)
Description of work:			
1. Grading	$3300	$2820	$3825
2. Lime	25[d]	85	92[d]
3. Fertilizer	48	51	49
4. Seeding and planting	241	219	216
5. Mulch	173	192	192
Total/acre	$3787	$3367	$4374
Total Cost[e]	$60,592	$33,670	$99,727

[a] Job 1 = Construction consisted of: Modified contour backfill, diversion ditches, rip rap outslope, compacted backfill (auger holes), 1973.

[b] Job 2, 3 = Construction consisted of: Same as above except for grading which was pasture backfill, 1973.

[c] Acre = 0.40 hectares

[d] Cost includes water treatment of impounded mine water.

[e] Grant total (3 jobs) = $193,989

Source: PB 238 538

In a 1971 study of surface coal mining in West Virginia, Stanford Research Institute concluded that the total reclamation costs of complying with existing surface mining laws and regulations range from about $500 to $1,000 per acre, not including sedimentation costs. When these are added, the total would be raised to about $650 to $1,200 per acre for a nominal range of conditions. More difficult reclamation terrain would require additional costs over and above these, which could raise the total to about $2,500 per acre. The variations in costs are a result of the displaced overburden being rehandled in northern West Virginia while spoil cast downslope in the south was not graded.

Mathematica Inc. found that reclamation costs of active mines are virtually unknown, even to the mine operators themselves, and results of past studies have varied widely in many cases. Reclamation requirements in the 1971 West Virginia law are quite similar to those in force in Kentucky. However, West Virginia does require that highwalls be reduced to 30 feet and that topsoiling be provided where acid-producing materials are present. Thus backfilling costs will be higher than those in eastern Kentucky, where high-wall reduction and topsoiling are not required. Major variables that have a decided effect on reclamation cost according to Mathematica Inc. are listed in Table 12.5.

TABLE 12.5: VARIABLES AFFECTING BACKFILLING AND GRADING COSTS

(1) Geographic location.
(2) Topographic setting (original, prereclamation and final ground slopes).
(3) Type of strip mine:
(a) area, (b) contour, (c) area-contour, (d) other.
(4) Coal seams mined and thickness.
(5) Inclination of coal seams in back of highwall:
(a) dip, (b) rise, (c) horizontal.
(6) Condition of coal seams in back of highwall:
(a) not mined, (b) auger mined, (c) drift mined (entries opened or caved), (d) mine workings exposed by stripping operation.
(7) The probable hydraulic head that could develop if coal in back of highwall was mined.
(8) Strip mine area information:
   (A) Length, width, and area (acres) covered by spoil before reclamation
   (B) Highwall height (maximum and average height)
   (C) Highwall length
   (D) Number of cuts
   (E) Total area affected during reclamation in acres (including area above highwall and outside of slopes)
   (F) Volume of spoil to be moved (cubic yards)
   (G) Average haul distance for backfilling and grading
   (H) Texture of spoil
   (I) Amount of large rock and material requiring special handling (mining timbers, machinery, and debris, junked cars, and other solid waste)
   (J) Amount and reactivity of pyritic material (minerology and mode of occurrence, for example, finely dispersed; single crystals or crystal aggregates; coatings on joint surfaces; in form of lenses, layers or modules; "sulfur balls"; pyritic shales, etc.)
   (K) Clearing and grubbing requirements.
(9) Type of backfill:
(a) contour, (b) pasture-reverse slope, (c) swallowtail, (d) head of hollow, (e) submergence, (f) other.
(10) Physical sealants for covering toxic material:
(a) none, (b) clay, (c) bituminous material, (d) plastic material, (e) other.
(11) Compaction desired:
(a) none, (b) only toxic materials, (c) all spoil material with exception of upper layer (1 to 3 feet).
(12) Accessibility factors:
   (A) Right-of-way problems
   (B) Ingress and egress construction (include clearing and grubbing for access and postconstruction revegetation)
   (C) Other factors affecting access.
(13) Surface and subsurface ownership of strip-mined area. Also, ownership of properties for ingress and egress:
(a) public, (b) private, (c) in process of being acquired on line placed on property, (d) abandoned, (e) temporary easement, (f) other
(14) Time of year reclamation performed.
(15) Weather conditions during reclamation period(s).

Source: PB 238 538

Table 12.6 shows estimated production costs based on an average stripping ratio of 8:1 and 0.5 acres disturbed per thousand tons of coal produced. The 8:1 ratio is representative of surface mines in eastern Kentucky at today's coal prices. Total production costs, under the stated assumptions, are $4.17 per ton. It is interesting to note that the stripping costs account for 58% of the total per ton production costs; and reclamation costs, when totaled, account for about 8%.

TABLE 12.6: ESTIMATED AVERAGE PRODUCTION COSTS*

Cost Element	Cost ($/ton)	Cost (% of total)
Sediment structure	0.12(1)	2.9
Acreage fees	0.01(1)	0.2
Bonding	0.00	0.0
Scalping	0.08(1)	1.9
Stripping	2.40	57.5
Overburden haulage	0.05	1.2
Coal loading	0.10	2.4
Coal haulage	0.50	12.0
Backfilling and grading	0.08(1)	1.9
Revegetation	0.03(1)	0.7
Royalties	0.50	12.0
Severence tax	0.30	7.3
Total	$4.17	100.00

*Average stripping ratio = 8:1 and 0.5 acres disturbed per thousand tons of coal produced

(1) Reclamation costs. These costs are equivalent to $0.32 per ton or 7.6% of the total costs.

Source: PB 238 538

Summary: The cost figures presented are indicative and show the importance of preplanning reclamation and incorporating it with the mining cycle. Reclamation costs can be reduced significantly if restoration is concurrent with the mining. It has been estimated that the cost of contour backfilling could be reduced by ⅔ if done immediately following mining.

Early reclamation avoids the cost of removing vegetation, burying toxic materials, providing access and moving heavy equipment back into the area. If mined land is allowed to remain bare for any length of time, landslides can develop on steep slopes; erosion and sedimentation can become excessive. Thus, prompt reclamation is essential to reduce not only reclamation costs but more importantly, environmental degradation.

Costs for reclamation of orphan land varies considerably, depending primarily on the condition of the land, and the desired result. To obtain averages, mediums, etc. from the data presented in this section would be misleading. The following ranges of cost are presented for reclamation in 1974 where reclamation is performed by contractors under bid cost. (See Table 12.7 on the following page.)

TABLE 12.7: RECLAMATION COST RANGES

Desired results and condition of land:	Range/acre*
1) Trees only -- land does not require grading or soil amendments and is not toxic............	$ 50 - $ 150
2) Grasses and legumes-- land does not require grading, but does require liming, fertilizer, seedbed preparation, and seeding...............	$ 100 - $ 400
3) Complete reclamation-- land requires grading, water control, soil amendments, mulching, seedbed preparation, seeding, etc.............	$1,800 - $4,000

*Acre = 0.40 hectares.

Source: PB 238 538

Variables affecting cost of reclamation for active mines have been mentioned. Table 12.8 has been prepared to place these costs in perspective to the coal being strip mined on a tonnage basis.

Western Surface Mining Reclamation

Strippable coal reserves of the West are becoming increasingly important because of their magnitude and low sulfur content. Stringent air pollution regulations are causing coal-using industries to seek the western low-sulfur coal. Western coal lies in seams up to 100 feet with overburden depths up to 200 feet.

Considering these facts alone, it is safe to assume that in the immediate future there will be a tremendous expansion of the surface mining industry in the West. Obviously reclamation costs on a per-ton-of-coal-mined basis in the West will be much lower than the East (Table 12.8). However, the reclamation costs per acre could reasonably be higher because of the semiarid to arid conditions that require more sophisticated restoration techniques than those practiced in the East.

Costs estimated for reclamation are scarce, mainly because of the small scale of coal strip mining in the past and the lack of State requirements. The Burlington Northern Railroad is reclaiming approximately 1,000 acres of orphan land in eastern Montana. These acres were surface mined between 1923 and 1958 and reclamation work began on September 13, 1974. Currently more than 580 acres have been contoured, seeded, or prepared for seeding at a cost of $600 per acre, for a total of $390,000 to date. Remaining contouring and seeding is estimated to cost another $400,000.

The Ozarks Regional Commission is sponsoring a regional project to demonstrate that mined land can be restored to productive use. This project is known as "Mined-Land Redevelopment", and in 1973 it included Kansas, Missouri and Oklahoma. Demonstration sites vary in size from 20 to 150 acres and are orphan areas. Although the majority of acreage reclaimed was for grassland, other uses such as catfish farming, recreation, housing, industrial parks, and solid waste disposal were also demonstrated.

TABLE 12.8: APPROXIMATE RECLAMATION COSTS PER TON OF COAL MINED

State	Average Thickness (feet)	Calculated Production Per acre 80% recovery (tons)	Cost per ton at reclamation costs per acre* of									
			$200	$300	$400	$500	$600	$800	$1,000	$1,200	$1,500	$2,000
Illinois	5.0	7,200(a)	.028	.042	.056	.069	.083	.111	.139	.167	.208	.278
Indiana	4.6	6,624(a)	.030	.045	.060	.075	.091	.121	.151	.181	.226	.302
Kentucky												
Eastern	3.1	4,464(a)	.044	.067	.090	.112	.134	.179	.224	.269	.336	.448
Western	5.1	7,344(a)	.027	.041	.054	.068	.082	.095	.136	.163	.204	.272
Ohio	3.4	5,328(a)	.038	.056	.075	.094	.113	.150	.188	.225	.282	.375
Pennsylvania	3.2	4,603(a)	.043	.065	.087	.019	.130	.174	.217	.260	.326	.434
Tennessee	3.2	4,176(a)	.048	.072	.096	.120	.144	.192	.239	.287	.359	.479
West Virginia	4.9	7,056(a)	.028	.043	.057	.071	.085	.113	.142	.170	.213	.283
Virginia	4.1	5,904(a)	.034	.051	.068	.085	.102	.136	.169	.203	.254	.339
Montana												
Subbituminous	30.0	42,240(b)	.005	.007	.009	.012	.014	.019	.024	.028	.036	.047
Subbituminous	50.0	70,400(b)	.003	.004	.006	.007	.009	.011	.014	.017	.021	.028
Lignite	16.0	22,448(c)	.009	.013	.018	.022	.027	.036	.045	.053	.067	.090

*Acre - 0.40 hectares; short tons - 0.907 metric tons; cu ft = 28.82 liters; pound = 0.453 kilograms

(a) Based on specific gravity of 1.32 = 82.64 pounds per cubic foot or 1,440 tons per acre-foot at assumed 80% recovery, bituminous.

(b) Based on: 1,409 tons per acre-foot at assumed 80% recovery, sub-bituminous:

(c) Based on: 1,403 tons per acre-foot at assumed 80% recovery, lignite.

Source: PB 238 538

All the grassland sites have been reclaimed to the following specifications:

(1) Spoil banks are graded until slopes on 90% of land area are 10% or less. Remaining land can have slopes up to 15%.
(2) The entire area adjacent to the water pits slopes to within 4 ft of the water, except for the highwall side. Slope specifications are the same as (1).
(3) Soil testing is done on the grid pattern, with four samples per acre taken and composited for pH testing.
(4) Soil treatment (lime and fertilizer) is furnished as recommended by the agricultural extension agent.
(5) Wood and brush control management is employed.
(6) Annual applications of fertilizer are made if needed.

One-tenth of the orphan land in Kansas has been reclaimed to productive use by this project. Costs are shown in Table 12.9.

TABLE 12.9: RECLAMATION COSTS PER ACRE FOR KANSAS MINED LAND DEMONSTRATION SITES, MAY 1973

Item	Number of Sites	Acres*	Range ($/A)	Weighted Average
Grading	68	1,307	$120 - $508	$158
Lime (all sites)	61	1,188	$ 0 - $ 42	$ 9
(lime users)	38	676	$ 1 - $ 42	$ 17
Fertilizer	61	1,188	$ 0 - $ 27	$ 11
Seedbed preparation	61	1,188	$ 1 - $ 62	$ 15
Seeding	61	1,188	$ 4 - $258	$ 13
Total	6		$136 - $551	$208

*Acre = 0.40 hectares

Source: PB 238 538

The Kansas figures would have been higher if the reclamation had been performed by contractors under bid costs instead of by local persons with a vested interest. The leveling of spoil piles is the major cost factor in reclamation. Costs vary greatly for backfilling and grading of overburden because of the various degrees of leveling that are performed.

For example, grading to a rolling topography does not stipulate the minimum grade that is to be attained. Thus, there is no standard to follow in determining the quantities and distance that the overburden must be moved. To be meaningful, any cost data must state the type of backfilling and leveling to predetermined grade. In summary, there is insufficient information from orphan and active western mines to provide data for analysis. Rough estimates can be obtained by using eastern mining data, but even these data are questionable.

Strip Mining Economics

For any organization, including the nonprofit ones, the objective of economic performance is supreme. The economic decision-making process involves many factors some within and many outside the mining company's control. The physical and chemical attitudes of the coal seams and their overburdens are more easily obtained. Selection of the right equipment, and method for shipping and coal recovery, though difficult, can be achieved. The legal and social outlooks, on the other hand are more unpredictable. Their effects on costs are more critical, and therefore, more important to evaluate. It has been contended that reclamation requirements not only add directly to the mining cost but indirectly escalate the cost by decreased productivity.

Capital cost considerations can hardly be overemphasized. However, equipment costs vary widely, and are a function of the amount of steel and the fabrication in the design and construction of the equipment. Cost involved for other surface facilities (e.g., storage, office space, buildings, etc.) are also subject to great regional variances. Also the financing and accounting procedures of companies differ, thereby making it difficult to arrive at meaningful comparisons.

Estimation of the mining costs must be by necessity, based on the company's experience. The labor and material cost must be estimated for drilling, explosives, overburden removal, reclamation, pit cleaning, coal loading, haulage, road building, fuel, oil, grease, maintenance, supervision, depreciation, etc. Additionally, costs for transporting, erecting, dismantling, and moving the primary stripping and other equipment must be considered. Since the viability of a project must be determined over the mine life, these have to be projected into the future taking into account the inflationary and productivity trends.

A factor clouded with more uncertainties is the selling price of coal. It is a complex function of the demand and the availability of other energy resources and their prices. The correlation between the selling price and the mining and preparation cost, on one hand, and the attractiveness of investment in stripping on the other, is strong. The most important decision-criteria in strip mining is the stripping ratio, defined as the amount of cubic yards of overburden to be removed to recover a ton of coal. It relates the selling price of coal with the cost of mining the coal and stripping the overburden.

In literature, sometimes the calculations are based on average overburden depths, though in reality, the break-even stripping ratio is a point-value, beyond which the coal seam cannot be economically stripped; i.e., as the overburden depth increases, more money is spent on exposing the coal seam till a limit is reached when the value of the recovered material (clean coal) is just enough to pay for all the cost involved in mining, preparation and selling the material.

It is the improvement in technology more than any other single factor that has not only sustained the coal mining industry but extended the technique to deeper coal seams. The importance of technological evolution in the ability to strip coal seams not heretofore possible must be borne in mind. Strip ratios of 25:1 and greater in 3 and 4 ft thick coal seams have been achieved in recent years. To give some idea on the capital investment in modern day strip mines, reference is made to Table 12.10, which shows significant increases in capital investment, mining costs and interest rates for opening a mine, 2 million tpy (short tons) capacity, under identical conditions in 1973, as compared to opening one in 1958. Thus, for the same return-on-investment before tax, a ton of coal must realize $7.56 in the market in 1973.

TABLE 12.10: COMPARISON OF TWO STRIP MINES 1958 vs 1973

	Year	1958	1973
1.	Capital Investment	$9,700,000.00	$17,500,000.00
2.	Capital Investiment/ton	4.85	8.75
3.	Mining Cost	2.05	5.18
4.	Interest*	0.15	0.54
5.	Total Cost (3 + 4)	2.20	5.72
6.	Realization	3.22	7.56**

*40% equity in both cases, 1958 interest 5%, 1973 interest 10%.
**Required realization for maintaining the 1958 ROI (Return on Investment).

Source: PB 238 538

The U.S. Bureau of Mines provides cost estimates for twelve hypothetical mines with a 20-year life. Coal seam and overburden data are considered typical for the hypothetical mine area. The analyses are based on the use of new equipment, the prevailing wage scale, and the payment of all costs including UMWA welfare fund, royalties, license and permit fee. Tables 12.11 and 12.12 provide a summary of the study.

TABLE 12.11: PHYSICAL DATA USED IN COST ANALYSES

Production, million tons per year	Mine location	Average coal-seam thickness, inches	Average overburden thickness, feet	Stripping ratio (feet to feet)	Estimated Average Btu per pound[1]
BITUMINOUS COAL: EASTERN PROVINCE--APPALACHIAN REGION					
1	Northern West Virginia..	72	60	18:1	13,200
3	........do.............	72	60	18:1	13,200
BITUMINOUS COAL: INTERIOR PROVINCE					
1	Western Kentucky........	66	100	18.2:1	12,000
1(2 seams)	........do.............	120	100	10:1	12,000
3	........do.............	66	100	18.2:1	12,000
1	Oklahoma................	16	32	24:1	12,500
SUBBITUMINOUS COAL: ROCKY MOUNTAIN AND NORTHERN GREAT PLAINS PROVINCES					
1	Southwestern United States	96	60	7.5:1	10,600
5	........do.............	96	70	8.8:1	10,600
5	Montana.................	300	75	3:1	8,500
5	Wyoming.................	300	75	3:1	8,500
LIGNITE: NORTHERN GREAT PLAINS PROVINCE					
1	North Dakota............	120	40	4:1	7,200
5	........do.............	120	50	5:1	7,200

[1]As-received basis for raw (unwashed) coal. Average calorific values for bituminous coal are taken from analyses of face, tipple, and delivered samples of mostly underground-mined coal and applied to strip coal.

Source: PB 238 538

TABLE 12.12: SUMMARY OF COST ANALYSES

Production, million tons per year	Estimated capital investment	Operating cost Dollars per year	Operating cost Dollars per ton	Operating cost Cents per million Btu	Selling price, 12-percent DCF Dollars per ton	Selling price, 12-percent DCF Cents per million Btu
BITUMINOUS COAL: EASTERN PROVINCE--APPALACHIAN REGION						
1	$12,727,500	4,146,400	4.15	15.7	5.40	20.5
3	28,005,000	9,167,100	3.06	11.6	4.01	15.2
BITUMINOUS COAL: INTERIOR PROVINCE						
1	$13,709,800	3,900,100	3.90	16.3	5.35	22.3
1(2 seams)	8,280,100	2,984,300	2.98	12.4	3.81	15.9
3	24,870.100	7,748,400	2.58	10.8	3.46	14.4
1	15,998,000	5,267,000	5.27	21.1	6.95	27.8
SUBBITUMINOUS COAL: ROCKY MOUNTAIN AND NORTHERN GREAT PLAINS PROVINCES						
1	$7,898,100	3,025,900	3.03	14.3	3.83	18.1
5	28,656,700	12,030,800	2.40	11.4	3.03	14.3
5	13,879,100	6,943,400	1.39	8.2	1.64	9.6
5	13,921,100	7,892,500	1.58	9.3	1.83	10.8
LIGNITE: NORTHERN GREAT PLAINS PROVINCE						
1	$6,381,800	2,373,200	2.37	16.5	3.01	20.9
5	20,749,700	8,384,600	1.68	11.7	2.12	14.7

Source: PB 238 538

Table 12.13 gives average percent breakdown of cost for seven strip coal mines as were experienced in 1969. Labor and supplies each accounted for nearly ⅓ of total cost.

TABLE 12.13: PERCENTAGE BREAKDOWN OF COSTS

	% Total Cost
Labor	32.0
Supplies	32.0
Power	3.0
Payroll taxes	1.2
Compensation insurance	1.7
Welfare fund	12.9
Other employee benefits	0.4
Property & other taxes	1.8
Insurance	0.3
Direct administrative	2.8
Total operating	88.1
Selling	1.6
General adminstration	2.6
Royalties	0.8
Total other cash costs	5.0
Total cash cost	93.1
Depreciation	2.1
Depletion	0.4
Amort., development	0.2
Amort., capital	4.2
Total noncash charges	6.9
Total Cost	100.0

Source: PB 238 538

Table 12.14 presents estimated per ton production cost for 5,000,000 tpy (short ton) mines. The figures in these tables are to be taken as indicative rather than conclusive and require adjustment with cost increase indices for use today.

TABLE 12.14: ESTIMATED PER-TON PRODUCTION COST

Direct Cost		Indirect Cost	
Production:		15% labor, maintenace,	
Labor	$0.150	supplies	$0.135
Supervision	0.037	Fixed Cost	
	0.187	Taxes and insurance (2% of	
Maintenance:		plant cost)	0.107
Labor	0.047	Depreciation	0.242
Supervision	0.005	Deferred expenses	0.133
	0.052		0.482
Total Labor and supervision	0.239	Annual production cost,	
Operating supplies:		$11,673,306	$2.33
Spare parts	0.400		
Explosives	0.136		
Lubrication	0.014		
Diesel fuel	0.025		
Tires	0.035		
Miscellaneous	0.050		
	0.660		
Power	0.160		
Union welfare	0.400		
Royalty	0.175		
Payroll overhead	0.084		

Source: PB 238 538

ILLINOIS RECLAMATION COST-EFFECTIVENESS ANALYSIS

This section is included as an example of one state's approach to the problem of reclamation of "pre-law" lands. In 1972, the Center of Environmental Studies at Argonne National Laboratory entered into a cooperative project with the Illinois Institute for Environmental Quality to develop plans for the reclamation of 109,000 acres of land that was strip mined in Illinois prior to 1962, the year the first state reclamation law was passed.

The objectives of the project were to develop cost-effective alternative plans for the reclamation of these "pre-law" lands; to design and develop demonstration projects that would establish and verify such important inputs as reclamation costs and technological feasibility; to establish and develop a base of practical experience with alternative methods of reclaiming the land; and to develop and apply data-collection and management systems that would allow the large quantities of information required to be assembled and analyzed efficiently.

Several critical questions regarding pre-law disturbed lands, include: (1) what can be done with the land? (2) how much will it cost? (3) who would conduct reclamation projects? and (4) how should reclamation activity be financed? The approach used to analyze these, and related questions, involved the steps shown on the following page.

Reclamation Costs

(1) Obtain a data base in a computer-readable format, describing current condition and utilization (if any) of pre-Law lands. (A data base prepared by Southern Illinois University's Cooperative Wildlife Research Laboratory was used for this purpose.)

(2) Develop a general data management system to retrieve specified information from the land condition data base as required.

(3) Categorize the land in terms of current condition as either environmentally degraded or nonutilized.

(4) Prepare reclamation cost estimates as functions of land condition descriptors and desired end-use categories.

(5) Compute and display for analysis purposes cost functions for specific current-condition/end-use combinations.

An existing commercial data management system (MARK IV), which was already operational at Argonne, was tested and deemed satisfactory for use as the retrieval program. The data base was coded in MARK IV format and stored in the system. Criteria for environmentally affected lands were established and a report of such lands was prepared using the MARK IV. Data retrieved from the land condition file was also fed to computational cost routines for the purpose of calculating reclamation cost functions. The operation of the system is diagrammed schematically in Figure 12.1. The results of this analysis are described in the remainder of this section.

FIGURE 12.1: USE OF ENVIRONMENTAL DATA SYSTEM IN STRIP MINE RECLAMATION PLAN EVALUATION

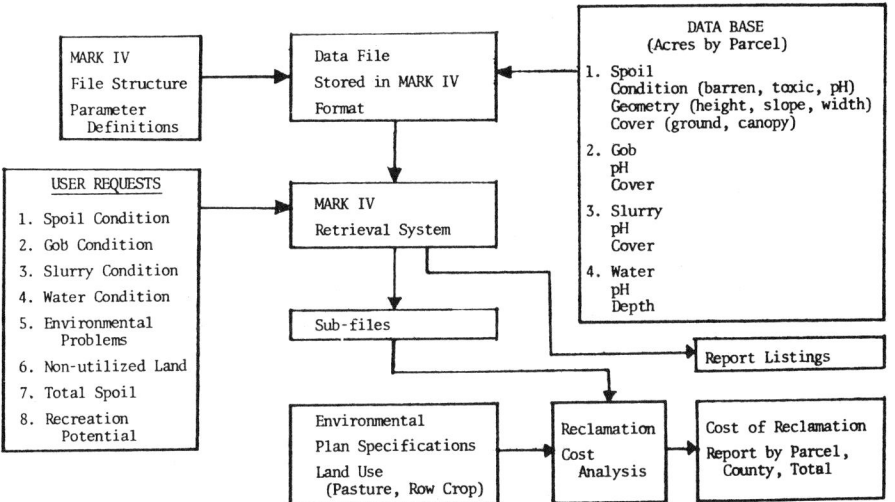

Source: PB 226 905

Land Condition

Because Illinois strip-mined areas differ from place to place due to their age, the equipment and procedures used in the mining process, soil chemistry and topographic factors, the reclamation costs vary as well. Some areas have undergone a degree of natural reclamation, thanks to natural seeding and watering, while others are so contaminated that they are completely barren. Reclamation costs also vary depending on the type of land being reclaimed, spoil, water, gob (refuse), or slurry. Therefore, for the purpose of conducting a cost analysis, affected lands have been categorized into environmentally affected and nonenvironmentally affected lands. The breakdown of environmentally affected lands and the criteria for selecting sites with environmental problems are shown in Table 12.15.

TABLE 12.15: AFFECTED LANDS SUMMARY

CATEGORY	ACRES ENVIRONMENTAL PROBLEMS	ACRES NON-ENVIRONMENTAL PROBLEMS	TOTAL
• SPOIL Utilized	2317	47584	49901
Non-utilized	8707	29077	37784
TOTAL	11,024	76,661	87,685
Environmental Criteria: Toxic (pH < 5.0) or Ground Cover < 25% and Canopy Cover < 10%			
• WATER Environmental Criteria: Acid (pH < 5.0)	767	11072	11839
• GOB & SLURRY Environmental Criteria: Toxic (pH < 5.0) Cover < 75%	5077	205	5282
• TOTAL	16,868	87,938	104,806

Source: PB 226 905

A large fraction of the affected lands has already been put back to some productive use. If these lands are not environmentally affected, they are omitted from further consideration in this analysis. Nonenvironmentally affected lands that are still idle are retained as part of the existing affected-lands "problem." The breakdown of these lands is also shown in Table 12.15.

Land-Use Alternatives

Reclamation costs also vary as a function of desired end use. Since a majority of surface mining in Illinois has been done in rural areas, it is reasonable to expect that most reclaimed lands will be returned to agricultural production, forest, pasture, or cropland. It is assumed that isolated instances where affected lands may be particularly suited for recreation or residential development will not significantly alter the results of the analysis. For purposes of this study, two types of pastureland and two types of cropland are designated as possibilities for agricultural end use.

> Spoil land where the tops of the spoil piles have been graded, where acid conditions (if any) have been neutralized, and to which fertilizer, mulch, and seed have been applied are designated strike-off pasture.
>
> Spoil land that has been graded to rolling contours (slope $\leqslant 10°$), where acid conditions (if any) have been neutralized, and to which fertilizer, mulch, and seed have been applied are designated general grade pasture.
>
> Spoils which have been graded to fine tolerance (slope $\leqslant 5°$), where acid conditions (if any) have been neutralized, and where fertilizer, mulch, and seed have been applied are designated neutralized row crop.
>
> Spoils which have been graded to fine tolerance, with 2 ft of topsoil put in place, and fertilizer, are designated topsoil row crop.

Reclamation Alternatives—Cost Analysis

Each of the existing land condition categories described above has been evaluated in terms of the total expenditures required to reclaim pre-Law lands under each of the desired land-use categories (less acquisition costs). On the basis of data items from the master file that describe the current condition of each parcel, a set of computer programs has been developed that calculates reclamation costs as a function of these conditions. Costs include tree removal, recontouring, soil neutralization, covering, revegetation, and water treatment.

The results of these cost calculations are displayed in Table 12.16, which shows both a total cost (1973 dollars) and a required annual rate of expenditure over the time periods indicated (assuming a cost escalation of 5% per year). Several conclusions are immediately evident from this table:

> Gob and slurry areas are a significant part of the environmentally affected lands problem and a significant part of the cost.
>
> Virtually none of the disturbed lands can be economically returned to row crop land.
>
> Some small percentage of nonutilized lands may be economically returned to good pastureland, but the more likely end use is strike-off pasture.
>
> It may be possible to economically return some environmentally affected spoil lands to good pasture.

Cost-Function Analysis

Environmentally Affected Lands: Before acid spoils can be returned to pastureland, they must be graded to rolling contour and neutralized. Thus, the only feasible and potentially economical agricultural use for these is $300/acre pastureland. Some environmentally affected sites also contain acid water, which must be neutralized and gob and slurry areas, which must be covered and vegetated.

TABLE 12.16: RECLAMATION ALTERNATIVES

Alternative Set	Type of Affected Land and Condition	Reclamation and End Use	Total Cost Data			Approximate Cost per Year at 5%, M Dollars			
			Current Total Cost (M$)	Total Acres	Average Cost ($/A)	5 yrs	10 yrs	20 yrs	30 yrs
Environmental	a. Gob and Slurry	a. Vegetation	13.6	5,077	2,674	3.13	1.76	1.09	.88
	b1. Spoil b2. Gob, Slurry and Water	b1. Pasture b2. Vegetation	29.7	16,868	1,759	6.85	3.84	2.38	1.93
	c1. Spoil c2. Gob, Slurry and Water	c1. Row Crop (neutralization only) c2. Vegetation	49.2	16,868	2,916	11.36	6.37	3.95	3.20
	d1. Spoil d2. Gob, Slurry and Water	d1. Row Crop (with 2 ft topsoil) d2. Vegetation	101.1	16,868	5,993	23.35	13.09	8.11	6.58
Non-Environmental Non-Utilized	a. Spoil	a. Strike Off Pasture	13.0	29,077	446	2.99	1.68	1.04	.84
	b. Spoil	b. Good Pasture	21.2	29,077	731	4.9	2.75	1.70	1.38
	c. Spoil	c. Row Crop (neutralization only)	66.5	29,077	2,456	15.36	8.61	5.34	4.33
	d. Spoil	d. Row Crop (with 2 ft topsoil)	221.3	29,077	7,611	51.12	28.66	17.76	14.40
Non-Environmental Utilized	a. Spoil	a. Strike Off Pasture	30.7	76,661	401	7.1	3.98	2.46	2.00
	b. Spoil	b. Good Pasture	50.5	76,661	658	11.66	6.54	4.05	3.28
	c. Spoil	c. Row Crop (neutralization only)	164.6	76,661	2,147	38.02	21.31	13.21	10.71
	d. Spoil	d. Row Crop (with 2 ft topsoil)	572.7	76,661	7,471	132.29	74.17	45.96	37.26

Source: PB 226 905

Reclamation Costs

Because it was originally felt that areas with potential surface water problems could be given a high priority in the reclamation program, cost functions for environmentally affected lands were calculated separately for sites with acid surface water and those without acid surface water. The unit-cost functions for each of these site categories are shown in Figures 12.2 and 12.3, respectively.

FIGURE 12.2: ENVIRONMENTAL LANDS WITH AFFECTED SURFACE WATER

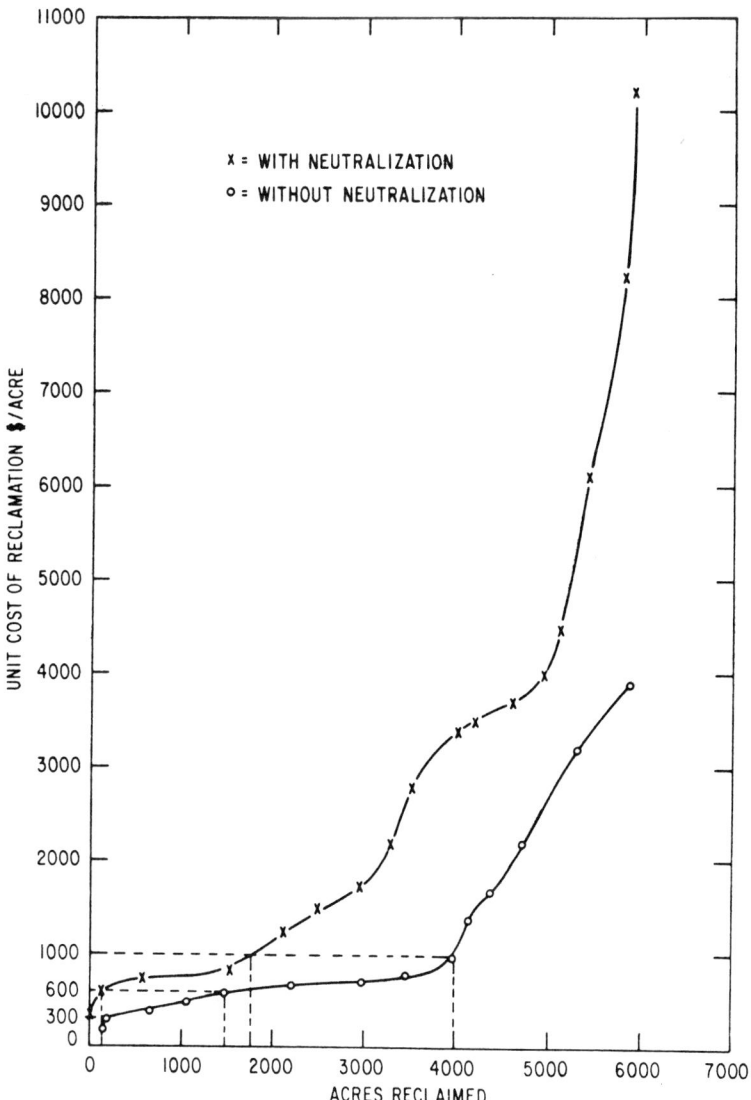

Source: PB 226 905

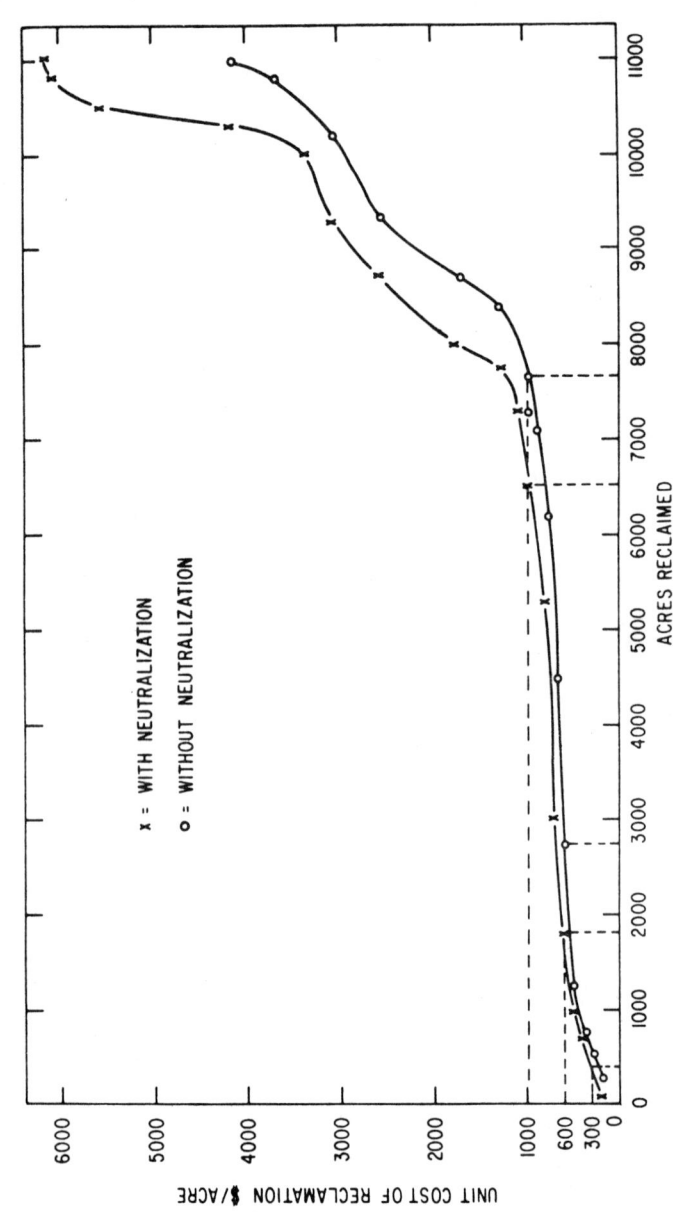

FIGURE 12.3: ENVIRONMENTALLY AFFECTED LANDS WITHOUT AFFECTED SURFACE WATER

Source: PB 226 905

It can be concluded from Figure 12.2 that virtually no sites with acid water will be reclaimed by the private sector, assuming a current market value for pasture of $300/acre. As illustrated in Figure 12.3, a few sites without acid water have the potential for being reclaimed. However, for this latter land category, it is also evident that the unit cost curve is extremely elastic in the area from $400 per acre to $900 per acre, indicating a strong latent market potential for private reclamation efforts with a minimal incentive program.

This is particularly true if the price of pastureland can be expected to rise in the next few years. Thus, if the neutralization costs borne by the private owner could be subsidized, a substantial incentive toward reclamation may result in the private sector. This form of subsidy is particularly efficient, since there would be no incentive to encourage the reclaimer to overneutralize.

Unit cost curves were generated, assuming that such a subsidy would be in effect, and the results are displayed in Figures 12.2 and 12.3. Some benefit is derived from this subsidy for lands without acid water, but a dramatic improvement can be noted for lands with acid waters. It can be concluded that if neutralization costs are subsidized, at least at the lower ends of each cost curve, a strong market potential for private reclamation exists for both lands with and without acid water. Therefore, as long as water runoff is contained on site, both types of land can be treated equally and an effective rate of reclamation attained with a combination of a neutralization subsidy and increased land prices relative to reclamation costs.

It remains to consider the high ends of each cost function. It will be noted that each cost curve displays a kink at approximately $1,000 per acre, at which the slope of the cost curve increases dramatically. In each case, these tails are due to sites containing gob and slurry areas that are extremely costly to reclaim and which dramatically increase the average cost per acre for the entire site. If reclamation of these areas were also subsidized or they were reclaimed by the State, in addition to the neutralization subsidy, a strong market potential would exist to reclaim most of the environmentally affected lands, provided that land prices double in the next 5 to 10 years relative to reclamation costs, as shown in Figure 12.4.

This analysis leads to the conclusion that an effective and efficient reclamation program for environmentally affected lands need not include the purchase or condemnation of these private properties, except possibly for those sites that cause off-site damages and for which the owner refuses to take voluntary corrective action. Rather, direct subsidies, in the form of neutralization-cost reimbursement and gob and slurry covering and vegetation, would, in a reasonable period of time (10 to 12 years), be sufficient to reclaim most of these lands, with the private sector bearing the costs of tree removal, earthmoving, fertilizing, seeding, and project management.

Nonutilized Lands: Except for specific instances, the two most economically attractive end uses for nonutilized lands are low grade (strike-off; $200 per acre) and medium grade (rolling contour; $300 per acre) pasturelands.

Unit-cost functions for each of these end uses is shown in Figure 12.5.

FIGURE 12.4: LANDS WITHOUT GOB AND SLURRY, NEUTRALIZATION NOT INCLUDED

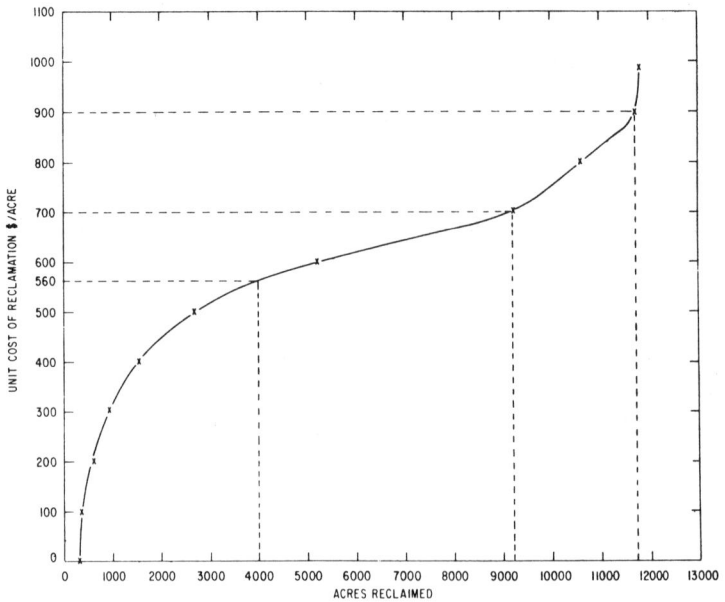

Source: PB 226 905

FIGURE 12.5: NONUTILIZED LAND

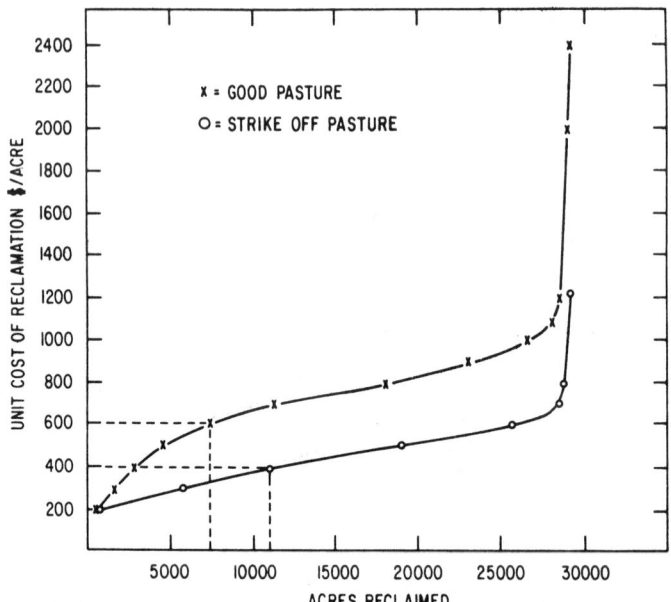

Source: PB 226 905

Reclamation Costs

As in the case of environmentally affected lands, little reclamation activity in the private sector will take place at current land prices and reclamation costs, unless incentives are provided. However, as before, the nonutilized land cost function is extremely elastic in the price range up to $1,000 per acre for good pasture and up to $600 per acre for strike-off pasture.

Thus, for this land category, incentive programs should be a very effective means of encouraging private sector reclamation. This conclusion indicates the potential payoff of demonstration projects that would lead to additional cost-saving techniques and that would significantly reduce reclamation costs to the State.

Gob and Slurry Areas: From the previous discussion, it was concluded that existing gob and slurry areas will not be reclaimed by private interests, nor are they of signficant productive value after reclamation has taken place. Therefore, if they are to be reclaimed, government will have to complete the entire task. For these reasons, the total cost function is of interest; that is, the total cost of reclaiming a given number of acres. This cost function (in 1973 dollars) is shown in Figure 12.6.

FIGURE 12.6: RECLAMATION COST FUNCTION—GOB AND SLURRY

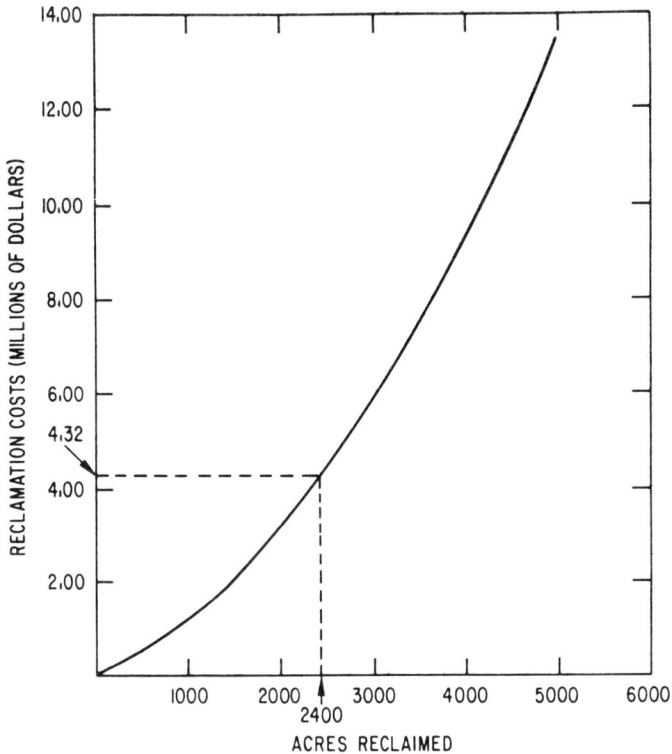

Source: PB 226 905

Sites have been ranked on a least-cost-per-acre-first basis in generating the curve. The cost function can be used to estimate the number of acres that can be reclaimed for a given annual expenditure over a specified planning horizon. For example, a $1M per year expenditure for five years has a value of $4.329M. This expenditure would allow the reclamation of 2,000 acres, or 47%, of the total gob and slurry areas in Illinois. This cost function has been used to develop cost-effectiveness estimates for the recommended program.

Ownership of Disturbed Lands

An indication of the potential success of private sector incentive programs can be gained from ownership figures and rates of change of ownership in the past ten years. Ownership of disturbed lands is summarized in Table 12.17.

TABLE 12.17: OWNERSHIP OF DISTURBED PRE-LAW SURFACE-MINED LANDS

	Coal Companies	Private	Corporations	Total Owned	Total Affected	% Owned of Total Affected
Non-utilized	16,283	8,142	3,344	27,769	29,077	96%
Environmentally with affected water	1,747	2,736	932	5,415	5,872	93%
Environmentally without affected water	4,639	3,535	2,099	10,273	11,046	93%
Totals	22,669	14,413	6,375	43,457	45,945	95%
Total Owned	47,000	34,000	10,000	81,000		
% Unreclaimed of Total Owned	48%	42%	64%	54%		

Source: PB 226 905

Of the 47,000 acres still retained by coal companies, 22,669 (48%) are considered to remain in an unreclaimed state, while only 42% (14,413 acres) in private ownership is unreclaimed; but 64% (6,375 acres) of affected lands mined by corporations are still unreclaimed or idle. 34% (2,186 acres) of these lands are held by farm corporations. The remainder are owned by banks and trust companies, or miscellaneous corporations (mostly materials manufacturers).

While coal companies have apparently restored a significant portion of their mined lands to use, they have also divested themselves of considerable holdings. Corporate holdings have taken up a large amount of these divested lands, particularly in the northern portion of the State. Private, recreational interests have acquired a considerable amount of land in the areas near Chicago, while the Metropolitan Sanitary District has purchased large areas of Fulton County. Pri-

Reclamation Costs

vate holdings have increased somewhat, particularly in the southern portion of the State, but the percentage increase has been slight. From these data, it can be concluded that, although the coal companies have divested themselves of a considerable amount (38%) of previously mined pre-Law land during the past ten years, they still retain a surprisingly large amount of land (22,669 acres) in an unreclaimed or idle state.

Market analyses indicate that there is a potential profit available to anyone who is willing to take some portion of this land as a gift, unless no market exists at the prices assumed. If the latter is not the case, then some barrier must exist to transfer of this land back to productive use.

There are indications that at least a partial barrier does exist. This barrier takes the form of the real estate tax assessment practices used by the counties. Coal company versus private tax assessments were sampled in selected areas of the State, and it was found that, in general, large coal companies are paying artificially high tax rates on mined land. Moreover, coal producers are reluctant to sell lands to private interests because they are, in turn, indirectly blamed when the new owner applies for low tax rates on his mined-out land.

The net effect of this barrier would appear to be a withholding of lands from the private agricultural land market by coal companies, who, in turn, themselves return the least costly land to agricultural use. Thus, coal companies, instead of selling the land to individual farmers, are turned into secondary farm corporations.

ELKINS DEMONSTRATION PROJECT

An acid mine drainage reclamation project near Elkins, West Virginia was funded by Congress as a demonstration project in 1964. A summary of the work performed is presented in Table 12.18.

TABLE 12.18: RECLAMATION WORK PERFORMED

Reclamation	
Surface mines reclaimed	12.5 miles
Backfill, total	3.6 million yd^3
Subsidence holes filled	450
Mine seals	101
Grass planted only	322 acres
Grass hydroseeded only	16 acres
Trees planted only	57 acres
Hydroseeded grass and trees planted	195 acres
Grass and trees planted	120 acres

Source: PB 207 189

Cost Analysis Procedures

The reclamation contract was entered into on June 30, 1966, at an estimated cost of $1,640,382. This contract did not include revegetation nor the filling of subsidence holes beyond 100 feet from the highwall. Due to the many unknown conditions existing in the heavily mined-out areas, the contract was a cost plus fixed fee type. Daily records of labor and equipment were kept by the contractor for work performed on each work area in the project. These data were later transferred onto computer cards and a computer program developed to obtain the desired cost breakdown.

Indirect Costs: Indirect costs included everything not directly applied to the work areas, such as office work, supplies, etc., and were distributed to the various work areas on a cost basis. For example, if 10% of the direct costs were charged to Area 2, then 10% of the indirect costs would also be charged to that area.

Cubic Yards and Acres: Approximately 650 acres of surface mine were reclaimed during the reclamation contract. Aerial photographs were taken of the project area during the planning stage and were used to develop contour maps showing the finished grade, acreage, and cubic yards of earthen material to be moved for specified types of backfill on the work areas. Upon completion of the contract, a land survey was made of each work area to determine the total acreage and cubic yards of material moved.

Accuracy of the backfilling quantities is somewhat limited because of the necessity of moving backfill material two or three times in an attempt to separate the toxic spoil from the nontoxic fill material and burying it in the strip pit.

Revegetation: A contract in the amount of $205,911 was awarded to the Tygarts Valley Soil Conservation District on a cost reimbursable basis in September 1967 to revegetate the reclaimed work areas on the project. In the spring of 1968 approximately 710 acres of land disturbed during reclamation were revegetated. Soil samples were taken and analyzed as a guide to the fertilizer and lime requirements and for choosing the best type of vegetation. The District completed the revegetation of the project in one growing season instead of two as originally planned, reducing the contract cost for revegetation to $177,727.

The contractor was required to make a cost analysis at the completion of the contract, therefore accurate and complete records were kept on all phases of work as it progressed. Actual labor and equipment hours expended each day were recorded by work areas. In addition, a daily record was kept of all lime and fertilizer applied and all grass seed and tree seedlings planted in each work area. This was further broken down as to method of application, for example, truck spreading or box spreading of fertilizer and conventional method or hydroseeding of grass seed. In addition, a record was kept of the species of grass seed and tree seedlings planted in each work area.

Each month a summary was made of all data compiled during that month and a cumulative total made of labor and equipment hours and material applied to each work area. Foremen's time and overhead costs were distributed to the different work areas on a basis of direct labor hours worked in each area during the month. Vehicle rental distribution was based on actual hours equipment was used on each work area during the month.

Discussion and Results

Cost of surface mine reclamation, mine sealing, and revegetation is presented for purposes of estimating cost of future reclamation work. An average overall cost, including both direct and indirect charges, is calculated for surface reclamation and mine sealing. Since equipment rental was a main item of expense (40% of the total cost) on the reclamation work, equipment costs were analyzed to determine the best and most economical equipment utilization for each type of work.

Equipment Summary: During the period of the reclamation contract, twenty-six pieces of equipment were leased by the contractor on a monthly basis to perform the reclamation work on the project. The lessor was to be notified by letter thirty days prior to terminating the lease on any of the equipment. Table 12.19 lists the equipment that was utilized during reclamation and shows the work hours, cost per hour, and range of cost per hour for each particular type of unit used.

TABLE 12.19: SUMMARY OF EQUIPMENT TIME AND COSTS

Type Equipment	No. of Pcs.	Work Hours	Total Cost	Avg. Cost Per Hour	Range in Cost Per Hour
600 Motor Grader	1	481	$ 10,385	$21.59	0
TD-25 Dozer	2	2,678	29,636	11.06	$ 7.47 - $17.14
D-7 Dozer	1	2,492	28,259	11.34	0
D-8 Dozer	2	2,851	28,358	9.94	8.50 - 10.31
D-9 Dozer	6	10,859	237,360	21.86	12.63 - 79.86
Koehring Shovel	1	951	23,024	24.21	0
Compactor	1	560	16,100	28.75	0
977 Traxcavator	3	5,615	63,315	11.27	7.71 - 16.97
DW-21 Pan	2	3,818	55,883	14.64	14.30 - 14.99
Scraper Pan	2	1,048	25,195	24.04	11.81 - 30.55
John Deere Crawler	1	1,892	4,162	2.20	0
Air Tract Carrier & Attachments	1	396	10,363	26.17	0
Compressor	2	2,294	17,478	7.61	6.27 - 8.94
105 LeRoi Air Compressor	1	16	1,600	100.00	0
Totals	26	35,951	$551,118		

Source: PB 207 189

Range of cost varied considerably for the D-9 dozers due to the necessity for keeping certain dozers on rental during periods of adverse weather. For example, the dozer which showed the highest cost per hour ($79.86) was on rental during four winter months and, because of bad weather, was utilized only 144 hours during the rental period. If this equipment had not been kept on rental, the lessor would have moved it from the project making it unavailable for spring operation. The LeRoi air compressor was rented for one month but after only 16 hours of use, it was found to be insufficient for the job; therefore the average cost was extremely high at $100.00 per hour. The 977 traxcavators were used as a combination hi-lift to explore the strip pits for buried deep mine openings and as a root rake to clear areas prior to backfilling. Utilizing this equipment during the winter months was difficult which resulted in considerable variation in the cost per hour as shown on Table 12.19.

The Koehring shovel was operated at an average cost per hour of $24.21 and was used for stream channeling and estalishing drainage from work areas. In February 1967 the shovel was damaged by a highwall fall and was down for repairs for the remainder of the project. The scraper pans were used mostly in work areas requiring compacted backfill and thus had limited use. The grader was used exclusively to maintain haulage roads to and from the work areas at an average cost of $21.59 per hour. The compressor, air tract, and crawler were used mostly for underground work pertaining to masonry seals.

Clearing and Grubbing: The first work actually performed on the project was the clearing of certain areas which were covered with volunteer trees and other vegetation established over the 25 years since stripping. This was done to prepare the land for the backfilling and sealing operations and was designated as Clearing and Grubbing. The following work was performed during this operation:

(1) All trees with a diameter less than four inches, measuring 12 inches from the ground, were uprooted, cut and burned.
(2) All trees with a diameter greater than four inches were cut, trimmed to saw log lengths, and stockpiled at a convenient location for the property owner.
(3) All stumps and brush were uprooted and burned.
(4) Boulders and rocks large enough to impede revegetation were buried in the spoil near the outer slope.

Average overall cost for clearing and grubbing was $330/acre or 16.6% of the total cost for surface mine reclamation (excluding revegetation). An average of 32 labor hours/acre was required to clear and grub (Table 12.20).

TABLE 12.20: CLEARING AND GRUBBING COSTS FOR 651 ACRES

	Cost in Dollars	Hours
Direct labor, total	72,662	21,468
Direct labor, average/acre	112	32
Equipment, total	38,329	3,461
Equipment, average/acre	59	53
Direct cost, total	110,991	–
Indirect cost, total	103,518	–
Total cost	214,509	–
Average cost per acre	330	–

Source: PB 207 189

These costs were higher than originally estimated, partially due to the dense forest in some areas and the extra handling to cut pulpwood for the landowners. Average direct cost varied considerably with respect to type of backfill performed on the work areas. For example, the average cost/acre for clearing and grubbing prior to contour backfilling on three selected areas was quite high and ranged from $127/acre to $367/acre. High costs were incurred in areas containing a fractured highwall. A portion of the highwall was unsafe and had to be cleared so it could be pulled down. Also the material was needed for fill. Generally, low costs were noticed in pasture and swallowtail backfill operations and in stripped areas where toxic spoil had prevented dense foilage and where it was not necessary to disturb vegetation on the highwall.

Surface Mine Reclamation: The average cost for surface mine reclamation was $1,658/acre. Cost of moving earth (3,060,000 yd^3) was $0.35/yd^3 (Table 12.21). These costs are higher than those reported by the U.S. Bureau of Mines for surface mine reclamation at Moraine State Park in Pennsylvania. In their report, the cost/acre for two areas was $780 and $1,402. The average earth moving cost was $0.16/cubic yard. Labor hours (39/acre) were the same for both projects.

TABLE 12.21: SURFACE MINE RECLAMATION COSTS FOR 651 ACRES

	Cost in Dollars	Hours
Direct labor, total	96,884	25,558
Direct labor, average/acre	149	39
Equipment, total	457,706	26,028
Equipment, average/acre	703	40
Direct cost, total	554,590	–
Direct cost, average/acre	852	–
Direct cost, average/cubic yard	0.18	–
Indirect cost, total	524,984	–
Total cost	1,079,574	–
Average cost/acre	1,658	–
Average cost/yd^3	0.35	–

Source: PB 207 189

The average direct cost for surface mine reclamation varied from a low of $472/acre on contour backfill to a high of $1,130/acre for a combination of pasture-contour backfill. Average direct cost per acre for pasture backfill reclamation was higher than contour backfill costs, an unexpected result. Further studies showed that, in general, the spoil was more highly toxic in the pasture backfill areas than in the contour areas. Because of its toxic nature, the spoil had to be moved several times, thus increasing the cost. Swallowtail backfill, because of additional earth work, was slightly more costly than pasture backfill. High costs for all phases of reclamation for a combination of pasture and contour are due to complex problems that existed in the work areas including the six conditions given below.

(1) Unknown interrelated conditions between the strip and underground mines which made it necessary to spend considerable time opening up the pit to locate fractures and openings into the underground mine.

(2) The contractor was required to separate the toxic spoil from the nontoxic backfill material where feasible and bury the toxic material in the strip pit. This required moving the material two or three times in some areas. As a result, the amount of earthen material actually moved greatly exceeded the 3,060,000 yd^3 determined from before and after cross sections.

(3) Approximately 17% of the total backfill material moved was used for excavation material to fill subsidence holes on top of the highwall and as clay compacted material for seals.

(4) It was necessary in many work areas to establish drainage by re-channeling streams from strip mines prior to reclamation.

(5) Adverse weather conditions during the winter months hampered the reclamation work on the project and necessitated payment of rent on equipment which could not be utilized.

(6) The highwall, in many instances, was fractured to the extent that it could not be left standing. In such cases, the wall was pulled down and the material used to complete the backfill.

Revegetation Costs: The overall cost for revegetating the reclaimed work areas is summarized in Table 12.22. Average direct cost was $200/acre and total cost $248/acre. Cost varied considerably depending on the type revegetation work performed. Higher revegetation costs were incurred in steep areas where it was necessary to use a hydroseeder. This also increased cost in contour backfill areas. The more level areas on which conventional equipment could be used were revegetated at a much lower cost.

TABLE 12.22: REVEGETATION COST FOR 709 ACRES

	Cost in Dollars	Hours
Direct labor, total	31,860	9,539
Direct labor, average/acre	45	14
Equipment, total	17,493	4,365
Equipment, average/acre	25	6
Material cost	45,190	–
Hydroseeding contract cost	47,475	–
Direct cost, total	142,018	–
Direct cost, average/acre	200	–
Indirect cost, total	33,709	–
Total cost	175,727	–
Total average cost/acre	248	–

Source: PB 207 189

Masonry Seals: Forty-three dry masonry seals and 12 wet seals were constructed in the entries to abandoned drift mines at an average cost of $4,138/seal. High equipment cost was attributed to the exploration of the strip pit to locate mine openings and to clearing debris from openings at the face of the highwall. Preparation of seal sites, such as timbering and clearing debris from the seal sites in the mine, was performed manually.

The average direct cost (SWA) for dry seals and wet seals shows that wet seals cost about twice as much as dry seals. Cost of dry seal on Work Area 8 was considerably higher than cost of other seals due to high labor cost involved in opening and timbering the portal prior to constructing seal.

Clay Seals: In areas where the highwall was badly fractured and the stripping operation had intercepted the deep mine workings, openings were sealed by compacting clay against the openings and the highwall with a vibrating sheeps foot compactor. Although 41 openings were sealed this way, data were recorded only for Work Areas 1 to 9 and 10. These data are summarized in Table 12.23. The cost/seal in Work Area 10 was higher than in Areas 1 to 9 due to haulage distance from the borrow pit to the seal site.

TABLE 12.23: CLAY COMPACTED SEALS

Work Area	No. of Seals	Compacted Backfill, yd^3	Total Cost	Cost per Seal	Average yd^3/Seal	Cost/yd^3
1-9	10	10,490	$ 9,500	$ 950	1,049	$0.91
10	6	11,670	14,160	2,360	1,945	1.21

Source: PB 207 189

AMD CONTROL FOR A SMALL COMPANY COSTS AND EFFECTS

The material in this chapter is excerpted from:

AD 740 157

The purpose of the study reported in AD 740 157 was to investigate the possibility of using economic analysis in the study of water pollution control in general, and specifically to apply economic analysis to the water quality program of a small, privately owned strip mining company located in southwestern Ohio.

The objectives were to:

(1) Measure the quantity and quality of acid mine drainage attributable to the stripping operations of the company. Data were acquired directly from information compiled by the coal company over time and from state and federal agencies concerned with rainfall, runoff, and water quality.

(2) Measure the costs of controlling acid mine drainage from stripping sites and to separate these costs from other reclamation costs of the firm. The data were developed from the coal company's records as well as from information compiled in other studies concerning the costs of controlling mine drainage pollution.

(3) Estimate the effect of these costs on the total costs of stripping coal for the firm and the ability of the firm to pass these costs on to customers or to landowners in the form of reduced mineral rights fees, or to absorb them and thus measure the effect on its competitive position of absorbing these costs. To do this, an attempt was made to estimate the demand curve and price elasticity to demand for the industry and firm.

(4) Estimate the costs and the effect of the costs associated with proposed alternative drainage control programs on the economic position of the firm in the market area where it operates.

AMD Control for a Small Company—Costs and Effects   321

The Target Mine: The data for this study were generously provided by a coal stripping firm located in southeastern Ohio. Because of a promise not to identify the firm in any way, it was possible to examine its books as well as other aspects of its operations. Because of the promise of confidentiality, figures, locations and other information were not reported in a specific manner.

Consequently, some of the information was disguised and proxy variables used where identification would be possible. These changes do not affect the results of the calculations nor do they change the conclusions of the study. Changes in cost since the publication date (1971) should be taken into account.

LAND AND WATER IN TARGET MINE AREA

Geography and Geology

The target mine is a small stripping company operating in a limited geographic region within Hocking and Perry Counties. Perry and Hocking counties are located in the hilly country of southeastern Ohio on the western edge of the Appalachian Coal Basin.

In the vicinity of the target mine the topography is sharply rugged with steep-walled valleys and narrow ridges. The slopes average 20 to 30% with upper limits of 50%. The mine operates in the Hocking River watershed along minor tributaries of Monday Creek. Some of the streams in the immediate vicinity of the stripping are intermittent.

The rock strata are of the Allegheny Series of the Pennsylvania System. Four different coal beds have been deep or strip mined in the area. They are the Lower Kittanning (No. 5), Middle Kittanning (No. 6), Lower Freeport (No. 6a), and the Upper Freeport (No. 7). The coal beds are often characterized by shale and flinty partings. Pyritic material is present in the coal and adjacent shale in nodules and dispersed particles.

The No. 6a coal, stripped by the target mine, at the time of the study, occurs high on the narrow ridges, so the overburden is completely removed and all of the coal is taken. The coal is 24 to 30 inches thick with a shale parting, so the coal must be mined in two stages. The overburden and the upper layer of coal are removed. Then the parting is separated for future burial and the lower layer of coal is removed. The No. 7 coal is also high on the ridges, where present, and is removed in the same manner as the No. 6a. The seam's thickness varies from a few inches to 48 inches.

As a broad generalization, the toxic materials necessary for forming acid mine drainage are intimately associated with the coal. The black shales immediately above and below the coal, especially the underclay, are pyritic and reactive.

The pH analysis of three spoil samples indicated rapid improvement of the material as the in-place distance from the coal increased. A sample of coal and immediately adjacent clay had a pH of 2.06. Recently exposed soil, from which most of the black material had been separated, had a pH of 3.77. Spoil that had weathered for about 3 months had a pH of 4.12. A variety of grasses had already begun to volunteer on the latter spoil without any attempt at reclamation by the mining company.

Since there is little limestone in the area, the natural buffering action common to much of the eastern Ohio coal fields is largely lacking. Consequently, the water and spoil are locally quite toxic. However, the Forest Service has found that reclaimed spoil (which essentially means separation and burial of black materials and some grading) consistently has a pH of 4.5 or slightly above and it will support growth of acid loving conifers.

Water Quality

The Division of Engineering of the Ohio Department of Health found the following conditions at the mouth of Monday Creek in 1967. During the low flow period, the dissolved solids concentration was 1,560 mg/l. The pH index was 2.9, sulfate concentration was 1,070 mg/l, chlorides 45 mg/l and zero bicarbonates. Field surveys by the U.S. Forest Rangers in Wayne National Forest confirm the aforementioned findings and that the conditions prevail continuously and are of long duration.

Lost Run Creek, a tributary of Monday Creek, runs through the property on which the target mine is presently operating. The stream's source is a mine tunnel leading into the No. 6 coal. Additionally, water seeps or flows from hundreds of mine openings, pits, and spoil banks to feed Lost Run upstream from the target mine.

The stripping is pre-law, so no reclamation has occurred. Vegetation has volunteered on portions of the disturbed areas, but in many areas the spoil banks are bare and continuously eroding. Without an extensive and expensive remedial program, there is no hope that the heavy mineral and sediment load entering the area's streams will be reduced in the foreseeable future.

Water quality data indicates that the quality of the water coming from the mine is poor, but is not as bad as the quality of water in Lost Run. Field readings by U.S. Forest Service personnel and State officials have consistently recorded pH levels of about 3.0 for all instream waters in the vicinity of the target mine.

Target Mine's Water Quality

The pH is generally below 3.0. The acidity ranges from 170 to approximately 1,000 mg/l, total iron ranges from 2.0 to 180 mg/l, aluminum ranges from 30.0 to 145 mg/l, manganese ranges from 20.0 to 145 mg/l, and sulfate content is 500 to 2,000 mg/l.

Water from the mining activities can be separated into three different categories because most water goes through three different stages before entering the streams. Water trapped in the pits has a pH of 3.0 or below, acidity of about 900 mg/l, and relatively high levels of sulfate, iron, aluminum, and manganese.

Water from the pits then seeps, flows or is pumped over the hill to a series of ponds about 50 feet below the stripping. The ponds were created by spoil from previous stripping operations, so the high level impoundments do not reflect deliberate planning. However, they act as silt basins and the pH, acidity, and mineralization contents improve somewhat relative to water in the pits. This is probably due to the natural buffering action of minerals in the spoil and dilution by nonacid water.

From the high level ponds water seeps and flows to a lower level pond formed by slippage of a spoil bank which blocks a narrow valley. The pond is elongated and retains run-off from disturbed and undisturbed areas. The low level pond is an effective silt retention structure and the acidity of the retained water is considerably improved. The pH is around 4.0 and the acidity (less than 200 mg/l) and iron (less than 10 mg/l) content are also greatly reduced.

Overflow from the low level pond forms the northerly flowing stream which enters Lost Run Creek upstream of the target mine's tipple. The water is free of sediment and the water quality measurements are better than those of the larger streams.

Ohio Water Quality Standards

Ohio water quality standards at the time of the study, did not deal directly with quality and quantity of coal mine effluent. However, the Ohio Department of Health's Water Pollution Control Board under certain sections of the Ohio Revised Code did have criteria for determining the quality of stream water that must be maintained for specific uses, such as municipal, industrial, recreational, aquatic life, and agricultural purposes.

The Revised Code further stated that no water which exceeds minimum standards established for specific uses shall have its quality reduced. Of the specific minimum standards established for various water uses, those provided for aquatic life are most appropriate for the study area because it is primarily national forest and sparsely populated. The dissolved oxygen must average 5.0 mg/l per calendar day and not less than 4.0 mg/l at any time. The pH must range between 6.0 and 8.5, except for fluctuations with photosynthetic activity.

However, within the Hocking River Basin, Rush Creek, Monday Creek, Sunday Creek, Federal Creek, and those minor tributaries polluted by acid mine drainage from underground mines and pre-reclamation law strip mines were exempt from Ohio water quality legislation, at the time of the study, except that no further degradation of stream quality from mining was permitted.

Water Quantity

Monday Creek is the largest stream in the area in which the target mine operates. In 1967 the Federal Water Quality Administration measured the flow in Monday Creek downstream from the mouth of a tributary that runs through the target mine's property. The average flow was 7.6 cfs with a maximum of 15.0 cfs.

Lost Run, the tributary flowing through the target mine's property, originates in the tunnels of abandoned deep mines of the No. 6 coal seam about 2 miles upstream of the target area. Two other small streams, one with a northerly flow and the other flowing in a southerly direction empty into Lost Run on the mine property. Each stream drains areas previously or presently being mined.

Both streams are said to be intermittent. However, during the months of August to December, 1970, there was some flow in each stream.

It had been anticipated that the quantity of water would be available from indirect sources. The limited duration of the study precluded any meaningful analysis through direct measurements. No evidence was found of any previous records being made of stream flow on Lost Run or its tributaries.

On October 8, 1970, Federal Water Quality personnel attempted two measurements in Lost Run, but the low flow makes the results somewhat suspect. One measurement was taken in Lost Run immediately downstream from the confluence of the northerly flowing tributary. The reading was 0.16 cfs, or 72 gpm. The flow in the tributary was estimated to be approximately 15 to 20 gpm.

A second reading was taken on Lost Run downstream of both tributaries. The flow was 0.17 cfs, or 76.5 gpm. The flow in the downstream tributary was estimated to be 20 to 25 gpm.

The rainfall in the area averages about 40 inches per year. The mining of the target company is near the top of the ridges, so there is relatively little water directly associated with the stripping. During observation of two heavy rains, sheet wash was not extensive. Apparently the spoil banks and ponds absorbed much of the water and retarded run-off.

No hydrographic studies were attempted, but the volume of water being carried by Lost Run was not drastically increased after the storms. Where the grade of the stream is low, the channel is ill defined because of the sediment load, so over-bank flows at those points occur readily. Sediment sources are numerous and plentiful in quantity because of previous mining activities. However, the target mine does not appear to be a significant contributor because the ponds on the site act as effective silt basins.

According to the Hydrology Division of the Corps of Engineers the conditions present in the Monday Creek portion of the Hocking River Basin leads to average annual run-off of 0.00134 cfs/acre. The run-off is based upon 40 inches average annual rainfall and given soil and vegetative conditions. The 0.00134 cfs per acre is equivalent to 868.32 gallons per day per acre or 316,936.8 gallons per year per acre of run-off.

Since the total run-off from an average acre of land is only 29.2% of the annual rainfall, these figures are not representative of the run-off conditions from recently stripped acreage. Studies by the U.S. Forest Service of three disturbed watersheds in eastern Kentucky showed the run-off to be about 50% for the 1968 water year. Assuming the higher run-off rate, an average yield per acre annum would be 543,124 gallons.

To get a better measure of short term quantities of water the company might be forced to handle, it is assumed that a 24 hour storm with a 10 year frequency results in 3.8 inches of rainfall on land where the antecedent conditions are normal. The vegetative cover will have been completely removed, so it is further assumed that the run-off rate would be slightly greater than would occur with row crops of similar soil characteristics but less than the run-off from a paved surface. The run-off from an acre of land meeting the above criteria would be 81,463 gallons.

In the vicinity of the target mine, the water problems are greatly complicated if coal is stripped from any seams lower than the No. 7 or 6a. The No. 6 seam has been extensively deep mined and the maze of tunnels have become channels for good-sized streams surfacing at points where tunnels were cut into the No. 6 coal outcrop along the hillsides. Stripping the outcrop may result in additional breakthroughs into the rooms and tunnels of the abandoned deep mines. Acid water then flows or seeps from the deep mines into the strip pits, creating expensive handling problems.

An even more complicated problem for the target mine existed in 1968 and 1969 when it stripped 12.5 tons of coal from the No. 5 seam. The No. 5 seam is about 20 feet below the No. 6 coal which had previously been deep mined and then the outcrop was stripped until numerous breaks into the deep mine had occurred.

The target mine stripped the overburden from the No. 5 coal until only a narrow ledge was left from the earlier No. 6 stripping. The water coming into the No. 5 strip pits was from rainfall as well as flows or seeps from the bench of the No. 6 stripping. Much of the water coming off the bench appeared to flow from the breaks in the highwall leading into the deep mine.

During periods of little rainfall the water entering the No. 5 strip pit was limited to seeps whose contribution to any water already in the pit was minimal. However, during and after heavy rains the volume of water coming down the highwall was plentiful and sustained. No quantitative measures of the marginal contribution of water from the previously mined area to the total volume of water per unit area of No. 5 stripping were attempted, but the extra water was significant, possibly doubling the water that would have to be treated.

Since the area disturbed was less than an acre, the total volume of water was limited. Mining under such conditions should not be attempted unless the per acre coal yield is profitable enough to offset greatly increased treatment costs.

AMD PREVENTION AND ABATEMENT PROGRAMS USABLE BY TARGET MINE

Small mining companies are faced with technical and economic considerations that are not present for large companies, at least not in the same relative magnitudes. Probably one of the most important constraints placed on small companies is the lack of managerial skill and technical knowledge needed to preplan and manage a program that gives low cost water handling and effective reclamation.

Large firms can hire specialists on a full and part time basis to analyze their problems and supervise effective programs to alleviate the situation. Large companies also appear to make better use of information and services made available by public agencies.

In this section, four methods which the target mine might use in treating water before allowing it to enter the surrounding streams are discussed. Since all of these methods are hypothetical at this time, no attempt will be made to analyze their relative effectiveness. It is assumed that each method will achieve the goal of improving the pH and reducing mineralization to meet Ohio water quality standards. Where appropriate, obvious problems of a given method are discussed.

Water Treatment Facilities

Water treatment facilities, as a general rule, require large capital outlays as well as skilled operators and involve economies of scale. Neutralization plants are particularly effective for large deep mines that have relatively controlled water sources and work the same site for many years. Acidity of the mine drainage is not generally as much a problem for strip mines but the difficulties of trapping and treating the water are greater because of the magnitude of the surface area involved.

The target mine disturbs only 10 to 15 acres of land per year and it is continuously moving, so a neutralization facility would have to be small and mobile. Assuming that all run-off water from an acre of land could be trapped for treatment, an upper limit of approximately 543,000 gallons per annum could be expected, or 1,488 gallons per day. If 10 acres of land are involved, the annual volume of water would be 5,430,000 gallons, or 14,880 gallons per day.

Since a treatment facility would be forced to have a capacity greater than the average daily run-off, assume a capacity large enough to handle the run-off from a 24 hour storm with a 10 year frequency. Ten acres of land could be expected to yield 814,628 gallons of water from such a storm. The capital cost of such a structure, considering the water quality at the target mine, would have been approximately $340,000 in 1965. A capital outlay of this magnitude by such a small firm is unrealistic.

Had the company acquired a sophisticated water treatment plant utilizing hydrated lime and including sludge disposal capable of treating 200,000 gallons per day of water containing 1,000 ppm acidity, its capital cost in the mid-sixties would have been $84,000. Using only a 5% per year rate of price increase would have resulted in a 1970 capital cost of approximately $107,217.

The total operating costs (including capital costs) were $0.48 per thousand gallons of water treated in 1968. Inflating $0.48 by 5%, the 1970 operating cost per thousand gallons would have been $0.53.

Such a treatment facility, hereafter would have cost the company $0.09 per ton of coal in 1970. The cost per ton was derived by assuming that 5,430,000 gallons of water would have been treated for 10 acres of disturbed land, yielding 32,000 tons of coal.

This treatment method is somewhat inappropriate for small mines for several important reasons. The initial capital outlay would discourage capital-short small firms. The facility would probably prove inflexible for contour stripping. The very sophistication of the plant would require considerable attention and training, unnecessarily complicating the operation of small firms.

To attempt to take a simpler approach such as retaining water temporarily within the pits in relatively small ponds and using a small daily capacity treating facility that is highly mobile would probably be fairly effective. A device such as this might involve a pump bringing water to a flume over which a hopper containing a screw feeder would feed a limestone slurry into the polluted water at a rate necessary to achieve neutralization. Because the neutralization efficiency of limestone is low without aeration, it is assumed that two tons of limestone would be required to neutralize one ton of acid. The acidity of the water being treated is assumed to average 900 mg/l. The water could then be gravity fed into a settling basin to remove the sludge.

For this simple treating facility, hereafter referred to as Method 1, capital costs involved would be pumps and hoses, flume, pipe and a hopper and mechanical screw feeder, plus construction cost of a settling basin or basins. Total capital costs would vary with the distances involved, but a reasonable assumption might be $2,000.00. Of this, $1,000.00 would be for motors, hose, pipe, screw machine, flume materials and assembly and an additional $1,000.00 for a settling basin of two or three acre feet capacity retained behind a semi-compacted earth dam with an overflow channel that would prevent erosion of the dam. Steeply sloping terrain would increase construction cost of the dam. Because of the movement of contour strippers along the hillside, it is assumed that at least three basins would be required over a ten year period.

Operating costs would involve $30.00 per 8 hour shift in labor, $5.00 per ton of pulverized limestone, plus gasoline and oil and maintenance of the facility. The sludge would be left in the pond which would also serve to catch sediment during its useful life.

The total cost of treating water during an average year would be $2,333.00. The cost assumes that the company would treat water 50 times each year. The equipment would have a useful life of 10 years and three different sediment basins would have to be constructed during that period. Assuming that the company recovers 32,000 tons of coal on 10 acres of land that annually yield 5,430,000 gallons of run-off, the cost per ton for acid neutralization would be $0.073. The total annual costs are shown in Table 13.1.

TABLE 13.1: NEUTRALIZING FACILITY COSTS (METHOD 1)

Item	Annual Cost
Capital, plus 3 basins	$ 533.00
Labor, 50 days at $30.00 per day	1,500.00
Limestone, 40 tons	200.00
Fuel and Maintenance	100.00
Total Annual Cost	$2,333.00

Source: AD 740 157

The primary drawback to this approach is that it probably would fail to handle portions of the water because of sheet wash that would go off immediately and the ponded water would percolate through the spoil and into the streams, or overflow during heavy rains. Standing water would have an opportunity to continuously react with the highly toxic material previously separated for burial.

An alternative technique available for preventing stream pollution from mine drainage is to utilize gravity flow and diversion around the pit. Hereafter this technique is referred to as Method 2.

The topography above the highwall is steeply sloping, yet somewhat undulating with natural drainage ditches that, before mining, carried run-off down the hill to the streams. By cutting diversion ditches along the contour above the highwall until the natural channel is intercepted water could be diverted around the pit, or across the pit at control points.

When the water has to be directed from above the highwall through the strip pit, a ditch should be cut in the pit to quickly and directly convey the water to a discharge point on the hillside below the spoil material from the first cut. Since spoil from successive cuts and the need to maintain haul roads would fill in the ditch, it would be necessary to use plastic or tile pipe to conduct the water from the open pit area to the discharge area. Extensions would be placed on the pipe with each new cut and the pipe beneath the spoil would be permanently buried.

In order to prevent accumulation of water in the pit, the underclay should be graded so that a slight slope in the direction of the ditch is created. In this way, the natural flow would carry all rainwater immediately into a ditch and, through the pipe, over the hill.

Since the underclay and toxic material which had already been separated during the stripping process would contain pyritic matter which would form acid, the pit floor and toxic piles should be covered with pulverized limestone after each rainfall. The neutralizing agent could be applied with a regular farm lime spreader.

A highwall cut one mile long usually disturbs approximately 30 acres of land. That linear distance could reasonably be expected to have several points where natural drainage patterns would necessitate conducting water through the open pit. Therefore, it is doubtful that the discharge from any single point would be very large or sludge-laden. It appears unlikely that an operation the size of the target mine would generate enough acidic water under the above described conditions to necessitate settling basins for sludge removal at each discharge point.

The target mine estimates its cost of installing a diversion ditch to be $1.00 per linear foot, plus an upper limit of $100.00 per acre for scraping the underclay and installing the ditch across the pit to utilize gravity discharge of rain water.

This includes cleaning an area 25 feet wide above the highwall. The ditch would be 3 feet deep with one to one grade on downside and two to one grade on up side. Hard rock would be hit at one foot depth.

Plastic pipe cost would vary with the diameter and length. Assuming three inch diameter, the cost would be $0.60 per linear foot. Spreading of limestone in the pit would be between $10.00 and $15.00 per acre per application. U.S. Weather Bureau information indicates that it rains an average of 40 days per year in which measurable run-off would occur in the study area.

Assuming that one mile of highwall is created for every 30 acres of disturbed land in contour stripping, 10 acres of disturbed land would result in 1,760 linear feet of highwall. The total annual cost would be $7,060.00, and the cost per ton of coal for handling water using Method 2 would be $0.22. The costs are listed in Table 13.2.

TABLE 13.2: WATER DIVERSION (METHOD 2)

Item	Cost
Highwall, 1,760 linear feet at $1.00/ft	$1,760.00
Scraping/acre at $100.00	1,000.00
Liming at $10.00/acre 40 times each year	4,000.00
Pipe 3 inch PVC at $0.60/ft and 50 ft/acre	300.00
Total Annual Cost	$7,060.00

Source: AD 740 157

If a ponding arrangement were necessary, it should be one so constructed that it would handle sediment from erosion as well as sludge from the treated water. The cost of constructing a debris basin meeting the U.S. Soil Conservation Survey requirement and capable of handling 30 acre feet of water is estimated to be $4,000.00. This cost assumes sloping terrain, but with a natural hollow, and some hard rock movement. Addition of the 30 acre foot pond would add $0.031 per ton of coal. Combined with Method 2 to form Method 3, the cost per ton of coal would be $0.25.

Reclamation Costs

Reclamation costs include separating the toxic matter when removing the coal, grading the spoil to permit reforestation and reduce erosion, backfilling the final cut to a level of 6 feet above the coal while burying the toxic material and planting seedlings in accordance with Ohio law. Reclamation definitions are variable and most do not include separation of toxic material while removing coal. It is an extra cost incurred by miners since the passing of reclamation legislation. The target mine has been effectively complying and has integrated it into its costs. The target mine estimates its per month cost of separating toxic material from coal to be $400. The company is mining an acre per month. The additional reclamation of grading, backfilling, burial of toxic materials, and reforestation is estimated to be $150 to $200 per acre. The reclamation cost of $150 would be $0.047 per ton if 3,200 tons per acre were mined. To blast the highwall to a 45° angle and grade to the original contour and seed with minimum erosion would cost $500 per acre, the per ton cost would be $0.156.

INDUSTRY AND TARGET MINE SUPPLY AND DEMAND RELATIONSHIPS

From recent experience it would seem that the demand for coal is expanding. Coal prices appear to be rising, suggesting that even with expanded production of coal, the demand is rising faster than production. This situation will continue until energy production facilities are expanded or the demand for energy subsides.

Coal production figures show that the Ohio proportion of total U.S. output remained relatively constant over the decade 1960-1969. This was also true for the southeast Ohio proportion of Ohio's output except for several years.

Market Area of Target Mine

One of the advantages of studying an individual mine is the fact that a relatively identifiable and isolated market area can be defined. The target mine operates in several counties. This is true of several producers in Ohio, particularly the larger mines in northeastern Ohio. The market area of the firm encompasses an area of a circle of approximately a 50 mile radius from the mine.

There are overlapping markets involved here. For instance, the target firm has shipped coal as far away as Dayton, Ohio. Moreover, the dominant firm in this market recently began trucking coal to Cincinnati, a distance of approximately 140 miles. In a period of rising prices, longer hauls will occur, although during the study period long trucked shipments were rare. This determination of the market was based upon several factors.

(1) The target firm indicated a lack of capability of making a profit in hauling farther than 50 miles, which would be beyond Columbus, Ohio.

(2) The firm does not ship by rail so trucking is its means of delivery on every contract. All of the producers in the Hocking-Perry county area ship almost exclusively by truck and are assumed to be subject to the same limitation as the target mine. The large firms in the area ship largely by rail and thus a market exists for truck sales and for rail sales and this study was restricted to truck sales and to rail deliveries within the 50 mile radius. Most producers in the area ship exclusively by truck. It also shows the target mine's relative market share during the study period.

(3) The 50 mile radius is applicable to the northwest and southwest from the mine since the market to the northeast and southeast is dominated by larger firms in Ohio, West Virginia and Kentucky, all of whom ship by rail, water, pipeline and truck and therefore do not truly represent competitors to the target mine.

(4) By limiting the market to 50 miles and the truck sales, the complication of Interstate Commerce Commission rate schedules which affect rail shipments primarily is avoided. As a matter of interest a favorable freight rate schedule can determine the market share for the larger firms, especially for long shipments in which frequently the shipping charges are greater than the cost of coal.

The relevant market area for this study includes, therefore, the demand for coal from the ten coal producing counties of southeast Ohio by firms located within a 50 mile radius of the target mine. Counties excluded from this market include those along the Ohio River, which include the two largest coal producing counties in Ohio, Belmont and Harrison counties, and Guernsey County, the last being a very small producing county but nevertheless apparently involved in the eastern Ohio coal market.

AMD Control for a Small Company—Costs and Effects 331

Isolation of the coal market in which the target firm operates is essential in order to understand the nature of economic effects it is hoped to determine. As with any market analysis there are arbitrary limits placed on the area but a reasonable series of assumptions helps define the relevant market.

It is, for instance, impossible to say very much about the demand for coal from the target mine without some delineation of the market and it is similarily impossible to discuss the industry demand without some delineation of the market. To discuss elasticity of demand or shiftability of costs presumes a market area. Consequently, what was done is absolutely critical to the rest of the study. Using the above delineation of the market area, it was concluded that the target mine is selling in what appears to be an oligopolistic market with price leadership by a dominant firm.

Broadening the scope of the market area would alter the market structure somewhat. However, in 1968 firms producing 50,000 tons of coal or less constituted 73% of all mines, although they produced only 10% of the coal. Therefore, a few large firms provide the vast majority of the coal at a price negotiated with electric utility companies who consume 61% of the coal produced in 1969. The large number of small companies then compete for the residual demand.

Besides the difficulty of defining a market area, there is an added difficulty in this study in defining the product. Coal is not a homogeneous product. The quality of coal varies from one seam to another, as well as within a given seam. For any large user of coal, especially electric utilities, each shipment of coal must be tested for chemical characteristics and coal prices reflect these differences. In a given contract situation, other things being equal, the higher the Btu content of coal the higher the price. According to executives of several large coal producers, the rule of thumb is that the price of coal varies $0.05 for each 100 increase in Btu content.

There are other characteristics which influence the coal market. For instance, many users of coal require that it be washed while others do not. The target mine does not wash its coal so it is excluded from selling to those customers who require washing.

Without further examples it should be apparent that not only is a given coal market constrained by overlapping market factors, the very product sold in a given market is subject to considerable variation. All of this tends to limit general statements and conclusions with respect to any specific behavior pattern discernible in a given competitive situation such as the one under study here.

Demand for Coal in the Market Area

As might be expected, coal use in southeast Ohio and coal use by firms buying coal from southeast Ohio resembles national patterns generally. In 1969 electric power utilities were the leading consumers of coal at 52.6%. Oven coke manufacturers used 20.3% and cement mills 2.7%.

The estimate of the demand for coal from southeast Ohio producers theoretically comprises the industry demand curve. It should be apparent that several problems occur in attempting to determine an industry demand for coal on this basis. First, there are many coal users in southeast Ohio who do not report to state agencies, and whose coal usage data are impossible to acquire. These users, according to the target mine executives, include all the manufacturing and other types of firms that exist in southeast Ohio and quite naturally purchase coal from their closest suppliers, assuming quality is acceptable. Related to these users is the fact that many users of coal purchase from suppliers outside or on the periphery of the market area. This problem resembles overlapping market problems. It is largely an insurmountable problem given the availability of coal use data now extant.

A second problem occurs in that there are many municipalities, villages, school districts in southeast Ohio which also use coal but for which data are not available. The State of Ohio keeps detailed records of coal used by Ohio institutions which purchase coal on term contracts through bid arrangements handled by the Department of Finance. There are a number of these institutions located in southeast Ohio which purchase coal from local producers, and have a large amount shipped into southeast Ohio.

Moreover, there are many state institutions located on the periphery of the market area and beyond which purchase coal from southeast Ohio producers. Other Ohio institutions throughout Ohio purchase coal from southeast Ohio producers and from other Ohio producers.

It should be mentioned that coal use has changed over the study period for many of these institutions due to sulfur content restrictions by cities or voluntary change to other energy sources.

In terms of specifications of the industry demand for coal, the public institutions represent a sizeable part of the market for which annual figures are available. The total demand for coal by all public institutions in Ohio in 1966 was 355,000 tons, in 1969 351,000 tons.

The total demand for coal by public institutions represents a part of industry demand that is identifiable and for which the target mine could possibly compete. The other part of our industry demand is that attributable to other industrial users, retail and electric utilities. Electric utilities make up the largest part of coal demand nationally and for the state of Ohio. In recent years electric utilities absorbed 90 to 99% of the target mine's output.

In Ohio the generation of electricity with coal as the energy source accounts for approximately 100% of the total electricity generated. Electricity generating plants change over time in coal usage, generating capacity, quality of coal required and other ways.

As might be suspected the coal supply available to a generating plant has to be reasonably predictable over time. Given the large quantities of coal required for the typical generating plant and the certainty of delivery essential to this plant, it is understandable that long term contractual arrangements would exist between electric utilities and large coal producers. Electric utilities in the market area served by the target firm use long term contractual arrangements for most of their coal purchases.

While the electric utility companies represent a demand for coal slightly less than the total produced in southeast Ohio in 1968, several important factors about

AMD Control for a Small Company—Costs and Effects 333

this demand must be noted. Coal for use by electric generating plants does constitute the largest part of the demand from southeast Ohio producers. According to a private survey by one mine in the market area, 74% of all coal produced in Ohio goes to electric utilities. It is estimated in the same report that 82% of the coal produced in the 10 southeast Ohio counties goes to electric utilities.

Of the 10 producing counties in the market area, 5 ship 99 to 100% of their output to electric utilities based on 1967 data. It may be of interest to note that the same study points out that 50% of Ohio produced coal is exported out of the state and 70% of this goes to electric utilities.

Of the potential users of coal in the electricity generating category which apparently amounts to 75 to 80% of the demand for coal from southeast Ohio producers, only 4 make up potential demand from the target mine. They are the Columbus Division of Electricity, and 3 plants of Columbus and Southern Ohio Electric Company, Picway, Poston and Walnut.

Ohio Power Company is controlled by American Power Systems and most of the coal used by Ohio Power Company plants is produced by captive mines, i.e., those owned by the utility itself. Subtracting Ohio Power usage from the total of 9,595,000 tons leaves a potential electric utility coal demand in 1969 of 2.5 million tons.

Almost half of this is absorbed by the Conesville Plant of C & SO Electric. This plant is located "on the coal" and is jointly owned by the utility company and a coal producer. The rest of the users in the area theoretically represent potential customers for the target mine and its competitors. However, those outside the trucking market are ruled out and this reduces the target mine's potential market even farther.

Further discussion of the relevant demand by electric utilities requires an explanation of how the demand for coal by utilities is satisfied and how the target mine competes for this demand. It must be remembered that C & SO Electric is, in essence, the demand for coal by electric utilities in southeast Ohio from the small southeast Ohio producers.

As with every utility C & SO Electric enters into long term contracts for coal. Usually, the utility contracts for up to 75% of its coal needs from one large producer and obtains the remainder from other producers, typically smaller producers within its area. The remainder of the market is distributed through coal brokers who have most of the small coal producers as clients.

Executives of the target mine stated that all of their small competitors sold exclusively through coal brokers. The coal brokerage business is apparently a highly competitive market in itself. The typical arrangement calls for the client coal producers to be prepared to deliver given quantities of coal on a monthly basis with no chance for negotiation of price.

The price quoted by the broker is always the delivered price and the producers must be able to mine and deliver at the broker's quoted price or forego the contract. The 20 or 30 year contract of the utility with its major producer contrasts sharply with the month to month arrangement of the small coal producers. As payment the broker usually receives $0.20 to $0.30 a ton commission and handles all paper work for the small producer.

Responsibility for delivering the proper quantity and quality of coal on time is up to the mine and failure to do so results in the loss of business. Over the last decade the target mine has delivered between 43 and 99% of its annual production to electric utilities. In those years when electric utility deliveries were down, the coal was trucked primarily to public institutions and a relatively small amount went to industrial users. The firm's retail business is very small as is true for most firms producing in southeast Ohio.

From the above description of how the electric utility demand is translated into coal sales for the target mine, it should be apparent that over the past 10 years the target mine and most of its small competitors have been dependent on the residual demand of C & SO Electric Company.

While it is not known where each of the small competitors sold its coal, many compete for the C & SO business. The relationship with the broker may be explained by the fact that the target mine has no sales force and needs the broker more than the broker needs it. This situation may not prevail under different market conditions such as those now developing. During the study period it prevailed for the target mine and probably for most of its competitors.

Elasticity of Demand for Coal from Southeast Ohio Producers

While the market consists of utility demand, public institution demand, industrial demand and retail demand, it is difficult to specifically state the industry demand in southeast Ohio very accurately. It is known that electric utilities make up 75%, industrial user 20% and retail users 5% of the demand. Of the industrial users data are available for public institutions and this makes up only part of the demand.

Because it was impossible to obtain complete demand data for any period of the study, the options existed of trying to estimate the industry demand (i.e., the southeastern Ohio demand) from very limited data or of assuming a demand curve using available information, economic theory, and general knowledge of the coal market.

Both were attempted but the latter approach was adopted after considerable discussion with several econometricians and after several unsatisfactory attempts at statistical estimation. An inelastic industry demand for coal has been assumed for our study period.

It is not unreasonable to assume an inelastic industry demand for coal in the short run since most coal using firms cannot switch to other energy sources readily and at lower cost. Coal costs would have to rise considerably before reaching the average Btu cost of other energy sources, even with other costs remaining constant.

By the nature of the utility demand-coal supplier contractual arrangement, the industry demand appears to be inelastic since part of every contract requires renegotiation once a year to allow the coal producers to automatically pass forward any incremental costs that have occurred since the previous price was established.

With shipping costs an important determinant of coal sales, geographical limitations lend strength to an inelastic assumption. Moreover, large producers shipping by rail may have a significant freight rate advantage over distant potential suppliers. The dominant firm in the market area appears to have such an advantage. Thus, while C & SO Electric could contract with other large producers, it would need to go outside the market area to find sufficiently large producers. To do so would likely add considerably to its coal costs because of the additional shipping required.

It is likely that the industry demand curve is inelastic throughout the range relevant to the study period and under the conditions existing in the 1960s. Even the dominant firm feels that C & SO Electric could readily contract for coal from other Ohio producers, but this is less likely for other coal users since they are not large enough to affect price very much and using small quantities of coal suggests the likelihood of fewer discounts or economies than electric utilities probably now enjoy.

The important consequence of the assumption about an inelastic industry demand for coal is that coal buyers will be willing to pay higher prices for coal with very little substitution taking place in the short run. Moreover, it implies that coal suppliers would be able to pass forward to coal users almost all incremental costs of production regardless of their nature.

Target Mine's Demand Curve

In the past few years the target firm has supplied all of its output to the electric utility users. Its demand is thus a function of the residual demand of the electric utilities which is not provided by its long term supplier. This represents the best customer for the small producers under 1960-70 market conditions. Under these conditions the target firm has a perfectly elastic demand function since it must accept the broker's offered price and could likely do no better by selling on its own which would necessitate that it provide its own sales force. Finally, the broker's price is not negotiable and remained quite stable over most of the study period.

There is in the southeast Ohio market a dominant producer. The dominant firm and electric utilities negotiate their long term contract and agree on an annual tonnage price, given the Btu content of coal and other quality characteristics that are agreed upon. The dominant producer provides up to 75% of the coal needs of the utility and the small coal producers supposedly provide the other 25% at the same price as that negotiated by the large coal supplier.

This price is adjusted for the broker's commission since most of the residual suppliers sell through brokers. The broker distributes the business on a month-to-month basis to a number of firms.

The small firms in the market do not have to sell through brokers nor do they have to sell to electric utilities. For the stability and certainty it offers, most of these firms choose to operate in the above outlined market. In fact, with the brokers absorbing most of the sales, bookkeeping and other administrative costs, the electric utility may represent a relatively low cost high profit market for many of the small producers.

In this market structure a small firm must know well its costs, particularly hauling costs, and its projected operating expenses in order to profit from the brokerage arrangement. This would be true under any contractual arrangements such as usually occur in the coal industry. Moreover, with some fixed costs from month to month such as debt services, depreciation and administrative costs, the firm may be better off by taking the broker's offer at a small loss than to discontinue operations in the short run.

Target Mine's Supply Function

In order to estimate the target mine's ability to control its acid drainage and continue to operate profitably requires knowledge of its demand function and its supply function. In this section the supply function of the firm is estimated using accounting data provided by the firm. The supply function estimated is in fact the average cost function of the firm for four separate years. Where relevant to the adjustment process, marginal costs were also estimated.

The primary requirement in the decision to statistically estimate the average cost function was to obtain a series of data from the firm for periods in which all factors, except output, remained constant. This is very difficult to do for a strip mine since it typically strips at various locations moving equipment from place to place.

Moreover, the overburden and seam thickness change even within a single cut around the hill. The amount of overburden and coal seam size are very important in the estimation of average cost, although there are many less important individual costs which, when aggregated, are considerable. However, the major variable cost in the stripping operation for the target mine is labor, with repair, gasoline and oil of secondary importance.

Because of difficulties involved in using annual cost and production data, it was decided that monthly figures would provide the best possible source of information for an average cost function. The target mine officials were asked to supply monthly costs for whatever specific operations data were kept. The figures provided were monthly labor, repair, and hauling costs. The firm was asked, furthermore, to provide these data for periods in which the firm was operating in the same relative area of overburden, rock content of overburden and coal seam width. Several such periods existed and these data were used in estimating the average cost functions. No attempt was made to analyze data from an operation in a particular environment that was of less than one year.

Figures 13.1, 13.2, 13.3 and 13.4 present average costs based on monthly labor, repair, and hauling costs and an approximation of all other costs. Total annual labor, repair and hauling costs were subtracted from total annual costs, which leaves total annual costs attributable to all other factors. This residual figure was divided by 12, resulting in a monthly estimate for all other costs. Adding this constant to actual monthly labor, repairs, and hauling costs gave a total monthly cost figure which, when divided by monthly production figures, provided an average cost figure for each month. The years 1965, 1966, 1969 and 1970 are those in which production costs and technology were reasonably constant. This approach gives theoretically appealing downward sloping cost functions. The statistical methods used are given in AD 740 157.

AMD Control for a Small Company—Costs and Effects   337

FIGURE 13.1: ESTIMATED AVERAGE COST CURVE—1965

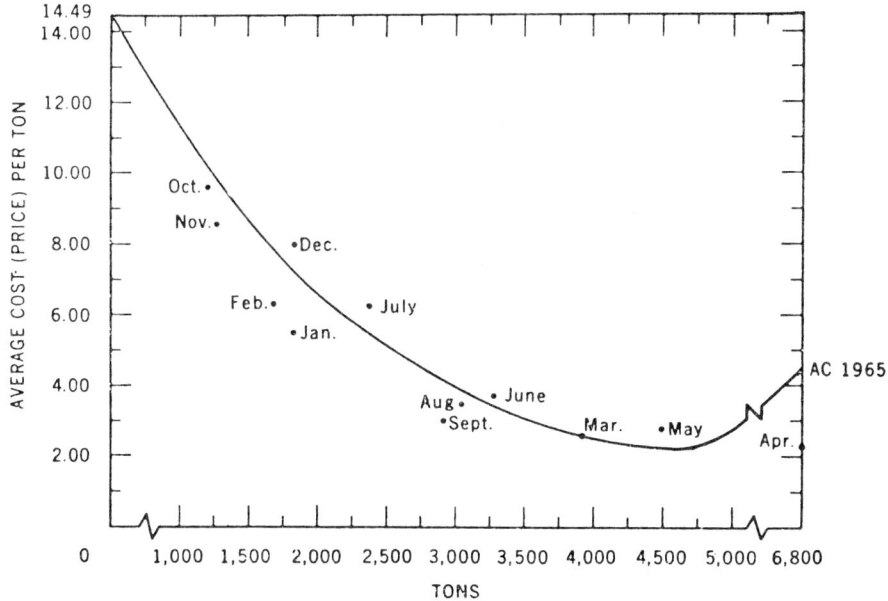

FIGURE 13.2: ESTIMATED AVERAGE COST CURVE—1966

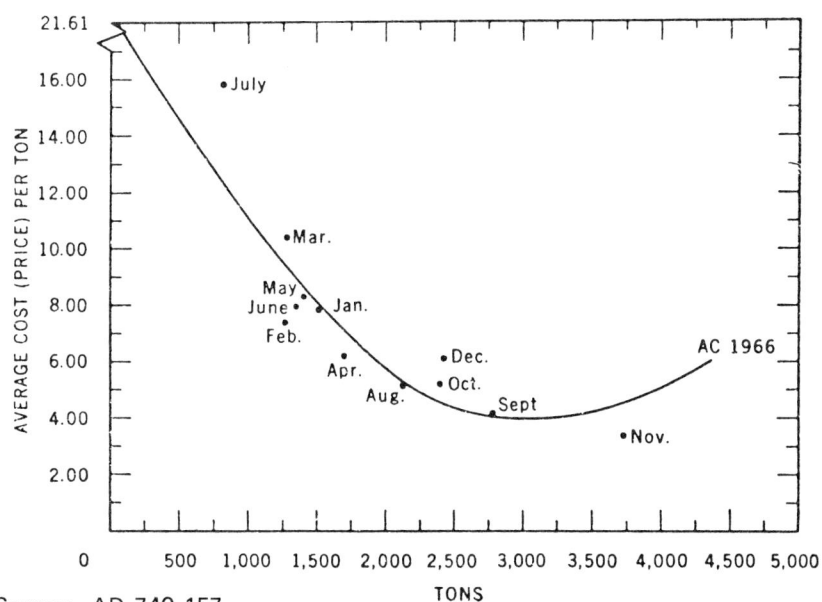

Source: AD 740 157

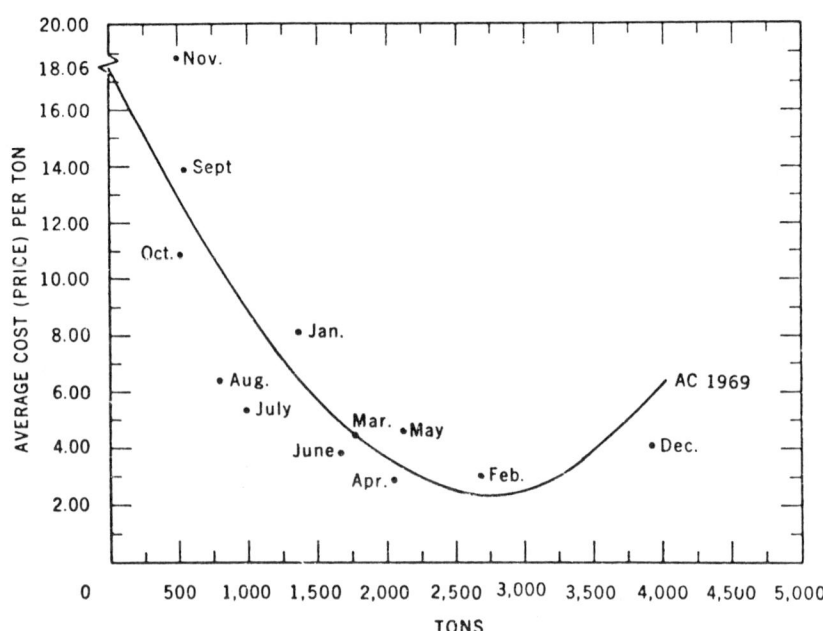

FIGURE 13.3: ESTIMATED AVERAGE COST CURVE—1969

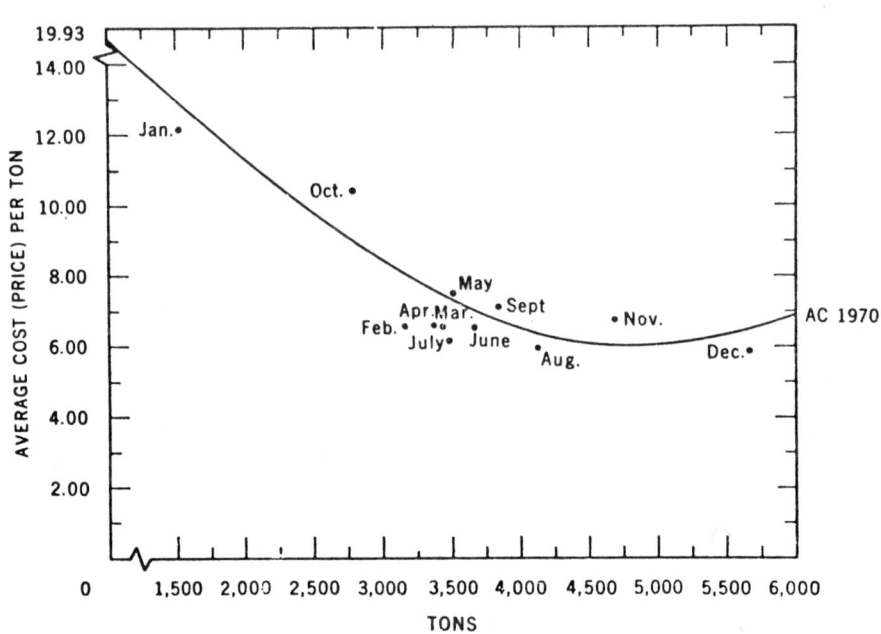

FIGURE 13.4: ESTIMATED AVERAGE COST CURVE—1970

Source: AD 740 157

The Figures 13.1, 13.2, 13.3 and 13.4, which represent the years 1965, 1966, 1969 and 1970 are examined as to average cost curve slope. For each year there is essentially obtained a curvilinear fit to the data. This was done to see if the average cost curve slope would become positive at some level of output rather than have a continuous negative slope, i.e., a declining marginal cost.

The output levels at which the slope of the average cost becomes positive for each year have been calculated and are presented in Table 13.3. For instance, this would occur at an output of approximately 4,971 tons a month in Figure 13.4. Table 13.3 also shows the average monthly and profit maximizing output levels of the target mine.

TABLE 13.3: ACTUAL MINIMUM COST AND MAXIMUM PROFIT OUTPUT LEVELS OF TARGET MINE (TONS)

Year	Actual Average Monthly Output (1)	Minimum Cost Monthly Output (2)	Maximum Profit Monthly Output (MC = AR)* (3)
1965	2,788	4,728	5,113
1966	1,898	3,094	3,179
1969	1,579	2,800	3,005
1970	3,620	4,971	5,031

*MC = marginal cost
AR = average revenue or price per ton

Source: AD 740 157

Several interesting problems are highlighted by the Figures. In 1965, which is one of the most typical experiences of the firm (if a typical year can in fact be identified), the months of November, December, January and February are at the low output high cost part of the curve. In 1970, the months of November and December appear at the other end of the curve.

The firm estimates that five working days a month are missed on average during January through April due to bad weather conditions, while three days a month are lost on average in other months. During winter months efficiency declines even on working days because of the condition of pits and roads.

The 1970 exception to this pattern is easily explained by the fact that the firm fortuitously happened upon a four foot seam of No. 7 coal very close to the top of a hill being stripped. The stripping was being done to uncover No. 6a coal several yards underground. This extremely unusual windfall resulted in the firm's largest monthly output ever produced at relatively low average cost.

The 1970 curve lies above the other years because all cost began rising for the firm in late 1969 and continued rising during 1970. By the firm's account, all costs were relatively stable until 1969, including its hourly labor cost. The only exception was repair costs which the firm estimates doubled during the decade. These cost increases explain the position of the various average cost curves but have little to do in explaining their shapes.

The appropriate way to estimate the average cost function using monthly data would be to have monthly costs for all activities of the mine. Unfortunately, these are not available although the firm provided all of the monthly cost items it could, trying particularly to get accurate labor, repair and hauling costs. Of the four years for which monthly data were provided, the 1965 and 1966 experiences are taken to be most typical for the firm although the firm experienced relatively low repair costs in 1965. In 1969 the firm became involved in non-mining activity and in 1970 unusual cost increases took place, a No. 7 seam was fortuitously discovered, and the market price of coal was rising quite rapidly.

In the next section the economic effects of alternative water quality programs will be presented using the cost curves estimated from the monthly data.

IMPACT OF MINE DRAINAGE ABATEMENT

Ohio water quality legislation had no impact upon the target mine in the 1960s. The long history of mining in the region had so degraded the water quality that the state legally permitted miners to discharge their effluent into the streams without treatment so long as further degradation did not occur. Therefore, any water costs incurred by the mine were associated with removing water interfering with active mining.

The target mine had some reclamation expenses during the sixties, although the amounts are difficult to ascertain because separate records were not kept for reclamation activity. The company separates its toxic materials for future burial and, during the 1960s, reclaimed 35 acres in compliance with state and federal regulations.

The state has been willing to allow firms to postpone reclamation if the company plans to return to the area for further stripping. The target mine deferred much of its reclamation of disturbed land because of its intent to take additional cuts after acquiring larger equipment.

The attitude of the state officials towards deferring reclamation has been altered and the target mine was required to reclaim most of the previously disturbed acreage during the first half of 1971.

The records of the target mine list only annual reclamation costs associated with the purchase of trees acquired for planting on reclaimed land. Company officials estimate that the total cost of grading, backfilling, and reforestation was $150.00 per acre for the 35 reclaimed acres.

Short Run Output Adjustment

While it is difficult to clearly isolate water treatment costs from general reclamation costs in an ongoing mining operation, it is attempted in this section to illustrate the firm's adjustments to these costs and the market effects these additional costs would have. The costs of various water management techniques which the target mine could adopt have been presented and discussed. In this section the four alternative acid control programs and two reclamation techniques will be illustrated with respect to the effect each might have on the firm's output.

The alternative methods are illustrated in Figures 13.5, 13.6, 13.7 and 13.8 by shifting the firm's average cost functions for the years 1965, 1966, 1969 and 1970. In each figure the firm's demand curve is illustrated as perfectly elastic. Finally, the analysis is basically short-run because it is assumed that some costs remain fixed during the adjustment process. Some implications for long-run adjustment are discussed later.

Method 1 would add approximately $0.73 to the firm's average cost of production. Methods 2 and 3 combined would add $0.250 per ton to its average cost and Method 4 would add approximately $0.090 per ton to its average cost. It should be recalled that the target mine had average losses per ton in five of the last seven years of its operation. To avoid larger losses per ton the firm could expand output to absorb the higher cost of the water control program. With declining average costs the firm apparently would have had little difficulty absorbing the higher cost in 1965 and 1969, even for the most expensive treatment method.

The firm could absorb additional water control costs by increasing output and moving down its average cost curve to a lower average cost output level or where its rising marginal cost (MC) = average revenue or price per ton (AR). This output level is presented for the four years in Column 3 of Table 13.3. In 1966, the firm could have increased output to a lesser degree to absorb the acid control costs of Methods 1, 2, 3 and 4. In 1970, it could have absorbed only the additional cost of Methods 1 and 4.

The average cost functions suggest a declining average cost up to some average monthly output which the firm seldom achieved beyond which point it became positively sloped. This output level is presented in Column 2 of Table 13.3. In none of the years was the firm's average monthly output near the minimum cost output level nor the approximate profit maximization output level where MC = AR.

In 1970, the firm approached its most profitable output level more closely than in any other year, i.e., its actual average monthly output level was 72% of the MC = AR level. Again, this is ignoring the rate at which its costs were changing. It is believed and some information indicated that the firm's average cost in the short run begins rising rapidly even before the lowest cost levels shown in the charts. The functions shown may thus not represent the most likely range of adjustment for the firm in the short-run.

The firm estimated that it can produce about 4,500 to 4,800 tons per month under conditions at the time of the study. These capacity conditions existed during several periods of the study but existed consistently in several months of 1970. They include a six-day work week, with time and one-half for employees for all hours over 40 per week. The work day is generally from dawn to dusk and equipment maintenance is quite poor under these circumstances and, in fact, repairs begin to increase during these maximum output periods.

342 Strip Mining of Coal

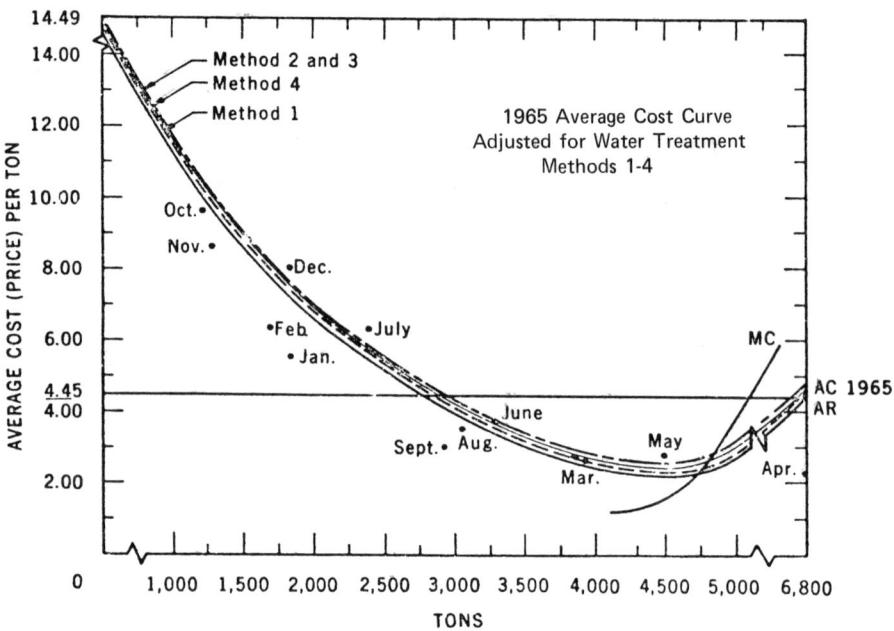

FIGURE 13.5: 1965 ADJUSTED AVERAGE COST CURVE

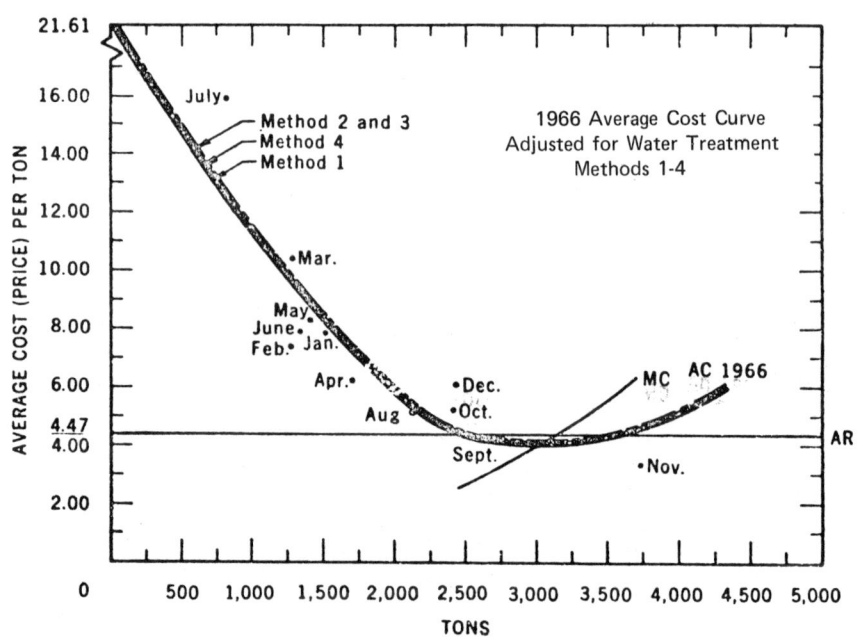

FIGURE 13.6: 1966 ADJUSTED AVERAGE COST CURVE

Source: AD 740 157

FIGURE 13.7: 1969 ADJUSTED AVERAGE COST CURVE

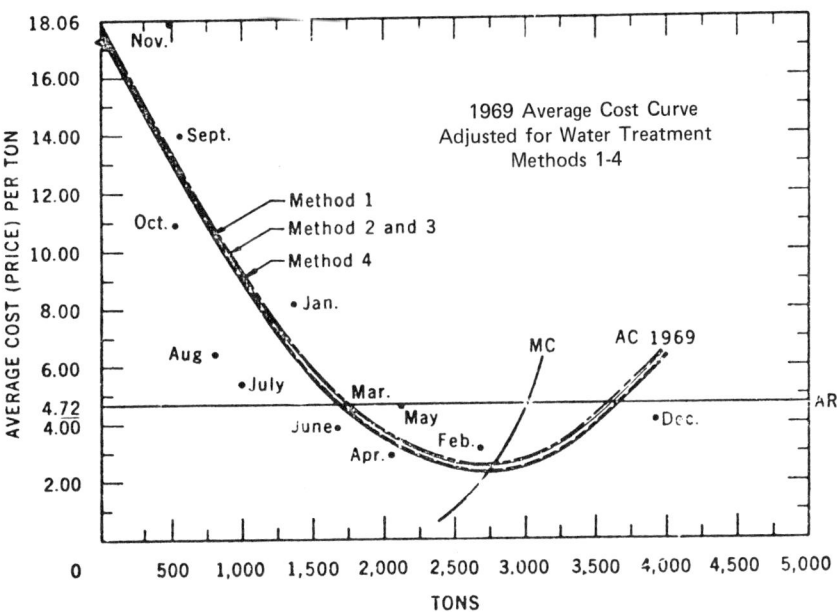

FIGURE 13.8: 1970 ADJUSTED AVERAGE COST CURVE

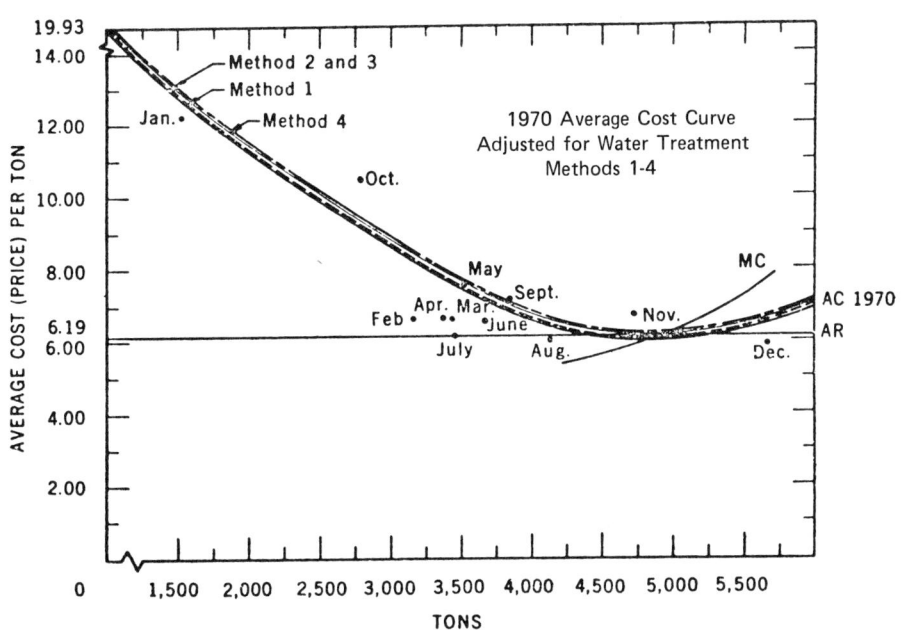

Source: AD 740 157

The average cost functions for the years 1965 - 1970 are discussed further. Thus, while the plotted functions suggest considerable flexibility in expanding output to absorb higher costs especially under 1965 and 1969 conditions, it is concluded that the range within which this expansion could reasonably occur is quite limited in the short run probably resembling the 1966 and 1970 conditions.

If worker productivity and other factors were introduced into the equations, the average cost curves would very likely have different shapes and become steeply sloped below what the firm considers its most efficient capacity level. For instance, this would likely occur at an output below what the firm considers its most efficient capacity level of 4,500 to 4,800 tons per month in 1970.

If the firm's productivity improved, it would result in a shift downward in the average cost function, making the firm's adjustment easier. This would represent a long-run adjustment by usual definitions of long and short-run cost behavior.

Statistics show that the target mine's productivity is considerably below the Ohio or U.S. averages. A similar productivity pattern probably exists for most of its small competitors as well. Moreover, it is thought that the target mine's productivity begins to decline even before it attains what the executives call the maximum capacity output level.

Consequently, the firm might be able to expand sufficiently to absorb the cost associated with acid control Method 1, the additional $0.07 per ton, but if one recalls that the firm had losses in most of the 1960s, and that its only recent profit except for 1970 was $0.01 a ton in 1967, then an additional $0.07 cost is quite meaningful to the firm. To absorb the costs of Methods 2 and 3 would be most unlikely for the firm, assuming its costs rise sharply beyond the 4,500 to 4,800 tons of output per month. Were the firm enjoying profits each year the additional cost of acid control would be easier to absorb, particularly if its average cost functions resembled those for 1965.

Long Run Adjustments

There is another alternative adjustment that the firm could make. This would involve a change in its scale of operations. The firm could increase its capacity by buying additional equipment, hiring more employees and possibly seeking new customers, although the latter seems questionable unless the firm is able to increase production sufficiently to enter into long term utility contracts. There are risks involved in this adjustment.

The firm's owners believe that a larger operation would most certainly require unionization which would add to its costs in several ways. Welfare costs of unions are $0.40 per ton and union wages would need to be paid to its employees, including heavy equipment operators, mechanics, etc.

It is questionable whether the economies of large scale would offset the higher costs, and whether the additional competition for customers would be possible in a market with several large producers with years of experience and large amounts of capital. The ability to get mineral leases would involve considerable expense and competition with larger firms that apparently have leased much of the mineral rights to land in the market area, particularly the land with thick coal seams.

AMD Control for a Small Company—Costs and Effects

There may be advantages in a large scale operation that would offset some of the disadvantages. It is apparent that economies exist in building treatment plants and in other control methods, the costs of which would be considerably easier to absorb for the large firm.

It is reasonable to assume that other operating economies exist at larger output levels, although it is not known at what output levels, or at what rate these economies occur in coal stripping operations. Yet even with its appeal, significant change in size is a most unlikely adjustment for the firm, particularly since the firm is opposed to an enlargement of such a scale.

The recent experience of the firm may provide a solution to its acid control cost difficulties in that market prices of coal have been forced up due to many factors and the rising prices have been partly passed on to the target mine. The rise in price to the dominant firm apparently did not filter down to the small producers in its entirety.

The firm's 1970 average revenue was greater than that for 1969. This was largely due to longer hauls with their attendant higher price per ton. Its average costs in 1970 were also higher than previous years but the price increases were large enough to offset rising costs of production for the firm. Consequently, the firm enjoyed its most profitable year of the last 9 years. Again, this is due to rising market prices which are entirely out of the control of the target mine. Part of this good fortune was the result of finding No. 7 seam coal while stripping for the No. 6a seam late in the year.

SOURCES

The following reports used in the preparation of this book are available from:

>National Technical Information Service
>U.S. Department of Commerce
>5285 Port Royal Road
>Springfield, Virginia 22151

AD 740 157   Dreese, G.R. and Bryant, H.L., *Costs and Effects of a Water Quality Program for a Small Strip Mining Company,* August 1971.

COM-72-10623   Camin, K.Q., Hardy, R.G. and Hambleton, W.W., *Mined-Land Redevelopment: Southeast Kansas Portion of The Ozarks Region,* October 1971.

ORNL-NSFEP-16   Nephew, E.A., *Surface Mining and Land Reclamation in Germany,* May 1972.

PB 187 738   Magnuson, M.O. and Kimball, R.L., *Revegetation Studies at Three Strip-Mine Sites in North-Central Pennsylvania,* February 1969.

PB 191 360   Czapowskyj, M.M., *Experimental Planting of 14 Tree Species on Pennsylvania's Anthracite Strip-Mine Spoils,* 1970.

PB 207 189   Scott, R.B., Hill, R.D. and Wilmoth, R.C., *Cost of Reclamation and Mine Drainage Abatement—Elkins Demonstration Project,* 1970.

PB 208 817   Grube, W.E., Jr., Jencks, E.M., Singh, R.N., Smith, R.M. and Wilson, H.A., *Mine Spoil Potentials for Water Quality and Controlled Erosion,* December 1971.

PB 210 709	Smith, R.W. and Frey, D.G., *Acid Mine Pollution Effects on Lake Biology,* December 1971.
PB 217 872	Vimmerstedt, J.P., Finney, J.H. and Sutton, P., *Effect of Strip-Mining on Water Quality,* January 1973.
PB 219 259	U.S. Department of the Interior, Bureau of Mines, *Surface Mining and Our Environment,* 1967.
PB 219 264	King, D.L. and Simmler, J.J., *Organic Wastes as a Means of Accelerating Recovery of Acid Strip-Mine Lakes,* February 1973.
PB 221 337	Foreman, J.W. and McLean, D.C., *Evaluation of Pollution Abatement Procedures, Moraine State Park,* January 1973.
PB 225 165	McNay, L.M., *Surface Mine Reclamation, Moraine State Park, Pennsylvania,* 1970.
PB 226 905	Carter, R.P., Zimmerman, R.E. and Kennedy, A.S., *Strip Mine Reclamation in Illinois,* December 1973.
PB 230 022	U.S. Department of the Interior, Federal Water Pollution Control Administration, *Stream Pollution by Coal Mine Drainage in Appalachia,* 1969.
PB 231 559	Brenner, F.J., *Ecology and Productivity of Strip-Mine Areas in Mercer County, Pennsylvania,* March 1974.
PB 232 069	Sopper, W.E., Kardos, L.T. and Edgerton, B.R., *Using Sewage Effluent and Liquid Digested Sludge to Establish Grasses and Legumes on Bituminous Strip-Mine Spoils,* March 1974.
PB 233 955	Paone, J., Morning, J.L. and Giorgetti, L., *Land Utilization and Reclamation in the Mining Industry, 1930-71,* June 1974.
PB 238 538	Grim, E.C. and Hill, R.D., *Environmental Protection in Surface Mining of Coal,* October 1974.

STATE AGENCY CONTACTS

The following list was provided by Ronald A. Pense, Program Information Specialist, Bureau of Mines, U.S. Department of the Interior.

STATE	ADMINISTERING AGENCY
Alabama	Mr. Henry T. Williams Chief, Division of Ecology and Inspection Department of Industrial Relations 1816 8th Avenue, North Birmingham, Alabama 35203
Arkansas	Mr. S. Ladd Davies Director, Department of Pollution Control and Ecology 8001 National Drive Little Rock, Arkansas 72209
Colorado	Mr. Norman Blake Deputy Commissioner of Mines Colorado Bureau of Mines 210 Columbia Building 1845 Sherman Street Denver, Colorado 80203
Florida	Mr. Bobby J. Timmons Economic Geologist Bureau of Geology Department of Natural Resources P.O. Box 631 Tallahassee, Florida 32302

State Agency Contacts

Georgia
: Mr. Sanford Darby
Chief, Land Reclamation Section
Environmental Protection Division
Department of Natural Resources
P.O. Box 4845
Macon, Georgia 31208

Idaho
: Mr. Gordon Trombley
Commissioner of Public Lands
State Capitol Building
Boise, Idaho 83701

Illinois
: Mr. Eugene Filer
Supervisor, Surface Mined Land Reclamation
Division of Reclamation
Department of Mines and Minerals
100 East Washington Street
Springfield, Illinois 62701

Indiana
: Mr. Richard McNabb
Director, Indiana Division of Reclamation
Department of Natural Resources
State Office Building
Indianapolis, Indiana 46204

Iowa
: Mr. Marvin Ross
Iowa State Mine Inspector
State of Iowa Mines and Minerals Department
Capitol Building
Des Moines, Iowa 50319

Kansas
: Mr. Michael Johnston
Director of Industrial Safety Division
Kansas Department of Labor
401 Topeka Avenue
Topeka, Kansas 66603

Kentucky
: Mr. John Roberts
Director, Division of Reclamation
Department of Natural Resources
Frankfort, Kentucky 40601

Maine
: Mr. John A. Bader
Director, Maine Mining Commission
State House
Augusta, Maine 04330

Maryland
: Mr. Harry B. Buckley
Director, Bureau of Mines
Maryland Geological Survey
City Building
Westernport, Maryland 21562

Michigan	Mr. Arthur E. Slaughter Chief, Geological Survey Stevens T. Mason Building Lansing, Michigan 48926
Minnesota	Mr. William C. Brice Mineral Resources Environment Coordinator Division of Water Soils and Minerals Department of Natural Resources 345 Centennial Office Building St. Paul, Minnesota 55101
Missouri	Mr. Robert Neuenschwander Director, Land Reclamation Commission State Capital Building Room B-36 Jefferson City, Missouri 65101
Montana	Mr. Ted Schwinden Commissioner, Department of State Lands State Capitol Helena, Montana 59601
New Mexico	Mr. Aaron L. Bond Chairman, State Coal Surface Mining Commission P.O. Box 2348 Santa Fe, New Mexico 87501
New York	Mr. John Dragonetti Chief, Branch of Mineral Resources 50 Wolf Road Albany, New York 12205
North Carolina	Mr. Stephen Conrad Director and State Geologist Office of Earth Sciences Department of Natural and Economic Resources P.O. Box 27687 Raleigh, North Carolina 27611
North Dakota	Mr. Bruce Hagen, Commissioner Public Service Commission State of North Dakota Capitol Building Bismarck, North Dakota 58501
Ohio	Mr. Ken Falk Chief, Division of Reclamation Ohio Department of Natural Resources Building 6, Fountain Square Columbus, Ohio 45202

State Agency Contacts

Oklahoma	Mr. Ward Padgett Chief Mine Inspector Oklahoma Department of Mines Capitol Building Oklahoma City, Oklahoma 73105
Oregon	Mr. R.E. Corcoran State Geologist Department of Geology and Mineral Industries 1069 State Office Building Portland, Oregon 97201
Pennsylvania	Mr. Walter N. Heine Associate Deputy Secretary Mines and Land Protection Department of Environmental Resources Town House Apartments 660 Boas Street Harrisburg, Pennsylvania 17101
South Carolina	Mr. John W. Parris Director, South Carolina Land Resources Conservation Commission 1400 Lady Street Columbia, South Carolina 29201
South Dakota	Mr. Howard Geers Executive Secretary State Conservation Commission Office Building #1 Pierre, South Dakota 57501
Tennessee	Mr. Chase Delony Director, Division of Strip Mining Department of Conservation 2611 West End Avenue Nashville, Tennessee 37203
Virginia	Mr. William O. Roller Commissioner, Division of Mined Land Reclamation Department of Conservation and Economic Development Drawer U Big Stone Gap, Virginia 24219
Washington	Mr. Donald M. Ford Assistant Supervisor, Division of Mines and Geology Department of Natural Resources Olympia, Washington 98504

West Virginia	Mr. Benjamin C. Greene Chief, Division of Reclamation Department of Natural Resources Charleston, West Virginia 25305
Wyoming	Mr. Robert Sundin Director, Environmental Quality State Office Building Cheyenne, Wyoming 82002

OIL FROM COAL 1975

by Francis W. Richardson

Chemical Technology Review No. 53

The extensive coal deposits within the continental U.S. and other countries represent a vast, largely untapped energy reserve. The technology for converting coal into gas and liquid fuels continues to proliferate with increased emphasis being placed on liquefaction. The research efforts of the last decade are now being expanded to include pilot and large scale studies for obtaining liquid hydrocarbons via coal technology.

For liquefaction the energy conversion factor is higher than for gasification, and the products are more easily stored and transported from production site to point of use. Earlier research tried to make a crude oil that could be processed into gasoline by conventional refining techniques. While this remains a primary objective, increasing attention is being given to making a low ash, low sulfur oil, that can be burned as fuel, thus freeing natural crude oil and gas for processing into gasoline.

This book presents the operating details of many coal liquefaction processes which are now undergoing extensive field operations here and abroad. Over 170 processes, as described in the U.S. patent literature, are covered.

A partial and very condensed table of contents follows here. Numbers in parentheses indicate the number of processes per topic or chapter. Chapter headings and some of the more important subtitles are given here.

1. EXTRACTIVE CONVERSION (54)
Conversion to Clean Fuel
  Using Precipitating Solvent
Low Sulfur Liquids
Production of Hydrogen-Rich Liquid Fuels
Combination Extraction & Hydrocracking
Deashing While Liquefying
Deashing Without Hydrogen
Use of Hydrogen Donor Solvents
Use of Hydrogenated Thianaphthenes
Solvation & Depolymerization
Microwave Energy for Liquefaction
Liquefaction in Shell Tube
Colloidal Size Coal in Extraction Zone
Multistage Solvent Systems
Successive Extraction Stages
Filtration vs. Fractional Distillation
Separation of Coal-Oil Suspensions
Thermal Liquefaction
Vaporization for Liquid-Solid Separation

2. HYDROGENATION (92)
Aqueous Medium for
  Ebullated Bed System
Supercritical Water Phase
  for Thermal Cracking
Use of Hydroconversion Catalysts
Deposition of Hydrated Iron Oxide
  on Surface of Coal
Liquefaction and Hydrotreating Zones
  Operated at Moderate Pressures
H-Coal Process Slurry Oil System
Expanded Solid Ash Particulate Bed
Partial Hydrogenation in Fluidized Bed
Promoter Liquids by Hydrogenation
  of Liquefaction Product
Barium-Promoted Cobalt Molybdate
  Desulfurization Catalyst
Noncatalytic Multihydrotorting Process
Carbon Monoxide, Steam
  and Aromatic Solvents
Desulfurization by High Turbulence
  & Fixed Bed Catalysts
Upflow Three-Phase Fluid Bed Reactor
Electrolysis of Water
  to Provide Hydrogen

3. CARBONIZATION & OTHER PROCESSES (22)
Low Temperature Carbonization
Jet Fuel from Blended
  Conversion Products
Carbonization of Coal Tar
Hydrovisbreaking of Coal
Underground Liquefaction of Coal
Coking of Carbonaceous Solids
Pressurized Distillation & Gasification
Carbonization of Lignite
  in Presence of Alkali
Pyrolysis of Coal in a Fuel Cell
Production of Hard Coke
  and Hydrocarbon Oils
Pyrolysis of Hydrocarbonaceous
  Matter in a Fluidized Bed
Depolymerization of Bituminous Coal

ISBN 0-8155-0588-4

387 pages

UTILIZATION OF WASTE HEAT FROM POWER PLANTS 1974

by David Rimberg

Pollution Technology Review No. 14

Energy Technology Review No. 3

Present-day steam-driven turbo-electric power plants in the United States discharge as waste heat an amount of energy roughly equivalent to twice their total electricity-generating capacity. This energy is most difficult to utilize, because it is degraded in temperature. The effluent water is warm, but it is far from being near the boiling point. It constitutes a necessary, but unwanted, by-product of the energy conversion process for generating electricity.

The growing quantities of waste heat discharged, and the increasing ecological anxieties about undesirable growths, energy utilization and thermal discharge problems, have stimulated an examination of methods for productively using energy presently wasted on the environment.

Part I of this book discusses the reasons for present-day ineffective utilization, but It also shows ways and means to promote more efficient energy usage.

Part II assesses the cause, magnitude, and possible effects of heat discharges into water from steam electric power plants and related condenser cooling systems. Also contained in this section is a discussion of the thermodynamics of the electric power generation cycle detailing the reasons for the "inefficiencies" in by-product heat generation.

Ultimately all the accessible waste energy appears as low temperature heat, and the term "waste heat utilization" refers to the performance of useful functions with this heat before it is discharged into the environment. Within the framework of this book, the minimum temperature of the effluent water considered for subsequent use is about 38°C or 100°F. These uses are discussed in **Part III**: Food production in agriculture, hydroponics and aquaculture (fish farming) and the use of heat in wastewater treatment.

In the past the dissipation of waste heat was accomplished by wet and dry cooling towers and cooling ponds, lakes and streams. These methods are now being challenged by some sectors of society, and industry is being forced to consider their environmental impact. In this regard **Part IV** discusses the research needs necessary to equate the complicated interactions of the physical, engineering, biological, and social aspects of this waste heat problem.

This Pollution Technology Review is based on studies conducted by industrial and engineering firms or university research teams under the auspices of various governmental agencies, e.g. The National Water Commission, Oak Ridge National Laboratory Federal Water Pollution Control Administration, Office of Water Resources Research, Environmental Protection Agency, National Science Foundation, and the Department of Commerce.

A partial and condensed table of contents follows here:

PART I
ENERGY CONSERVATION THROUGH EFFECTIVE UTILIZATION
Problems in Energy Supply
Present Uses of Energy in the U.S.
Possible Savings in Use of Energy
Summation of the Problems
Means to Promote Effective
    Energy Utilization

PART II
GENERATION OF WASTE HEAT BY ELECTRIC POWER PLANTS
The Waste Heat Problem
Waste Heat and Cooling
    Water Requirements
Effects of Steam-Electric Power Plant
    Operation on the Water Environment
Steam-Electric Power Plant
    Siting Alternatives
Thermodynamics of the Electric
    Power Generating Cycle
Thermal Pollution Effects
Waste Heat Treatment

PART III
WASTE HEAT UTILIZATION
Concepts and Demonstration Projects
Agricultural
Aquacultural (Fish Farms)
Wastewater Treatment

PART IV
Research Needs
Future Trends

ISBN 0-8155-0555-8

175 pages

OIL FROM SHALE AND TAR SANDS 1975

by Edward M. Perrini

Chemical Technology Review No. 51

This book contains 213 process descriptions related to the retorting and refining of shale oil and the separation of oil from tar sands. Current events continue to point in favor of these crude materials as future sources of petroleum energy.

Oil shale is found in diverse regions of the world, but the richest and most extensive deposit discovered so far is in the Colorado-Wyoming-Utah triangle, where most of the land belongs to the U.S. Government.

Oil recovery from shale rock is admittedly difficult. With the *in situ* methods oil is obtained from shale under ground, by injecting hot gases and collecting the oil through holes in the rock. By alternate methods shale rock is mined and mixed with small preheated balls. Heat from the balls retorts the shale, and the oil vapors are condensed in a condensing chamber. While not in use in the U.S., various techniques using nuclear reactions have been devised.

Vast resources of petroleum oils are also found in tar sands. One of the largest of these reserves is the Athabasca deposit in Canada. Oil is obtained by mining the sands with huge shovels and applying aqueous separation techniques.

This book is a survey of the U.S. patent literature. It contains large, descriptive excerpts. A partial and condensed table of contents follows here. Numbers in parentheses indicate a plurality of processes per topic. Chapter headings and some of the more important subtitles are given.

1. SHALE OIL RETORTING (86)
Gas Combustion
Isolation of Retort Zone
   to Utilize Flue Gases
Preheating Requirements
Mixtures of Rich and Lean Shale
Recovery and Conversion of Oil Mist
Hydrogen as By-Product
Continuous Fluidized Process
Solid Heat Transfer Media
Ball Heating Equilibrator
Special Pellets & Steam Stripping
Horizontal Retort
In Ground Processes
Solution Mining to Form Cavern
Fracturing with Well Boreholes
Nuclear Detonation Systems
Noncombustive Process Using
   Superheated Steam
Electrothermal Pyrolysis
Using a Controlled Nuclear Reactor
Sonic Energy-Induced Separations
Laser Beam Heating

2. SHALE OIL REFINING (25)
Hydrotorting
   Using Hydrogen and Water
   Using Synthesis Gas and Water Injection
Solvent Treating Before Hydrogenation
   to Prevent Plugging
Conservation and Reuse of Hydrogen
Combined Retorting & Cracking Processes
Pyrolysis and Hydrogenation
   in Single Reaction Zone
Heat Transfer Balls
   and Controlled Cracking
Visbreaking in Presence of Hydrogen

3. TAR SANDS SEPARATION PROCESSES (70)
Hot Water Processing
With Alkaline Polyphosphates
   and Surfactants
Hydrocarbons as Flotation Aids
Fluidized Bed Techniques
Micellar Dispersions for *in situ* Treatment
Continuous Centrifugal Separation
Hot Water Clarification Processes
Cold Water Processes
Using Hydraulic Cyclones
Sonic Energy of Oil-Water Separation
Gamma Radiation for *in situ* Processing

4. TAR SANDS RETORTING AND OIL REFINING (19)
Retorting and Coking
Simultaneous Separation & Cracking
Internal Combustion Retorting
Compacted Tar Sands Retorting
Hot Sand Recycling
Fluid Coking Processes
Upflow Liquid Phase
   Hydrogenation System
Hydrogenated Thermal Tar
   as Hydrogen Donor-Diluent
Jet Fuel from Tar Sands

5. RECOVERY OF METAL VALUES (13)
Soda Ash-Caustic Soda Treatment
   for Cell Grade Alumina
Recovery of Sodium Aluminate
Vanadium and Zirconium from Tar Sands
Hydrometallurgical Treatment plus
   Use of Anion Exchange Resins

ISBN 0-8155-0583-3

307 pages

DEPARTMENT OF AGRICULTURAL ECONOMICS
NORTH DAKOTA STATE UNIVERSITY
STATE UNIVERSITY STATION
FARGO, NORTH DAKOTA 58102